# The New Economy in East Asia and the Pacific

Has the new economy permanently changed the way leading economies work? How can its beneficial effects be transmitted to countries in the early stages of economic development? Why is Japan continuing to stagnate in a potentially prosperous age while Australia, with a smaller industrial and information technology base, is doing well?

This book sets out the problems of measuring the effects of technological change on economic progress by using the internet in the Asia Pacific region as a case study. Corporate and industry experience, including changing business organisation and new regulatory issues, are explored as well as policy issues such as the digital divide and the approach to e-commerce in the WTO. Using several industry case studies the contributors compare the IT experience in North America with a number of countries in Asia and the Pacific.

With contributions by a number of distinguished authors, *The New Economy in East Asia and the Pacific* is an essential volume for policy-makers, corporate strategists and scholars concerned with understanding the effects of the new economy and its impact in Asia and the Pacific.

**Peter Drysdale** is Professor of Economics in the Asia Pacific School of Economics and Government at the Australian National University.

**Pacific Trade and Development Conference Series**
Edited by Peter Drysdale, Australia–Japan Research Centre, The Australian National University

Titles published by RoutledgeCurzon in association with the PAFTAD International Secretariat and the Australia–Japan Research Centre, The Australian National University include:

**Business, Markets and Government in the Asia Pacific**
*Edited by Rong-I Wu and Yun-Peng Chu*

**Asia Pacific Financial Deregulation**
*Edited by Gordon de Brouwer and Wisarn Pupphavesa*

**Asia Pacific Economic Cooperation/APEC: Challenges and Tasks for the 21st Century**
*Edited by Ippei Yamazawa*

**Globalization and the Asia Pacific Economy**
*Edited by Kyung Tae Lee*

**The New Economy in East Asia and the Pacific**
*Edited by Peter Drysdale*

# The New Economy in East Asia and the Pacific

Edited by
Peter Drysdale

LONDON AND NEW YORK

First published 2004 by RoutledgeCurzon
11 New Fetter Lane, London EC4P 4EE

Simultaneously published in the USA and Canada
by RoutledgeCurzon
29 West 35th Street, New York, NY 10001

*RoutledgeCurzon is an imprint of the Taylor & Francis Group*

© 2004 PAFTAD International Secretariat for selection and editorial matter;
individual chapters, the contributors

Typeset in Garamond by Australia–Japan Research Centre, Canberra, Australia
Printed and bound in Great Britain by TJ International Ltd, Padstow, Cornwall

All rights reserved. No part of this book may be reprinted or reproduced or utilised in any form or by any electronic, mechanical, or other means, now known or hereafter invented, including photocopying and recording, or in any information storage or retrieval system, without permission in writing from the publishers.

*British Library Cataloguing in Publication Data*
A catalogue record for this book is available from the British Library

*Library of Congress Cataloging in Publication Data*
The new economy in East Asia and the Pacific / edited by Peter Drysdale.
   p.cm. — (Pacific Trade and Development Conference series)
Papers from the Pacific Trade and Development Conference, "The New Economy: Challenges for East Asia and the Pacific", held at the Australian National University, Canberra, 20–22 August 2001.
Includes bibliographical references.
1. Information technology—Economic aspects—Asia—Congresses. 2. Information technology–Economic aspects—Pacific Area—Congresses. 3. Information technology–Government policy—Asia—Congresses. 4. Information technology–Government policy–Pacific Area—Congresses. 5. Technological innovations—Economic aspects—East Asia—Congresses. 6.Technological innovations—Economic aspects—Pacific Area—Congresses. 7. Industrial Policy—East Asia—Congresses. 8. Industrial policy—Pacific Area—Congresses. 9. Digital divide—East Asia—Congresses. 10. Digital divide—Pacific Area—Congresses.
I. Drysdale, Peter. II. Series.
HC460.5.Z9 I556 2004
330.95—dc22

ISBN 0–415–28056–7

# Contents

| | | |
|---|---|---|
| *List of figures* | | vii |
| *List of tables* | | viii |
| *List of contributors* | | x |
| *Preface* | | xii |
| *Abbreviations* | | xiv |
| 1 | The new economy in Asia and the Pacific: an overview<br>*Peter Drysdale* | 1 |
| 2 | What's new in the new economy?<br>*Richard N. Cooper* | 12 |
| 3 | The new economy: theory and measurement<br>*Richard G. Lipsey* | 32 |
| 4 | Telecommunications policy and the evolution of the internet<br>*Robert W. Crandall* | 60 |
| 5 | Factors affecting growth in the region: R&D and productivity<br>*Shandre M. Thangavelu and Toh Mun Heng* | 88 |
| 6.1 | Beyond Silicon Valley: the regional spread of innovation<br>*Juan J. Palacios* | 106 |
| 6.2 | The Indian experience<br>*Mangesh G. Korgaonker* | 108 |
| 6.3 | The Chinese experience<br>*Dong Yuntin* | 130 |
| 6.4 | The Taiwanese experience: impact on production and trade<br>*Sheng-Cheng Hu and Vei-Lin Chan* | 140 |

| 7 | Corporate strategies in information technology firms<br>*Yasunori Baba and F. Ted Tschang* | 172 |
| 8 | Intellectual property protection and capital markets in the new economy<br>*Keith E. Maskus* | 203 |
| 9 | A force for market competition or market power?<br>*Lewis Evans* | 220 |
| 10 | Internet providers: an industry study<br>*Haryo Aswicahyono, Titik Anas and Dionisius Ardiyanto* | 237 |
| 11 | Automobiles: an industry study<br>*Roger Farrell and Christopher Findlay* | 257 |
| 12 | The digital divide in East Asia<br>*Emmanuel C. Lallana* | 273 |
| 13 | E-commerce: the work program of the World Trade Organization<br>*Edsel T. Custodio* | 298 |
| 14 | Implications for APEC<br>*Mari Pangestu and Sung-Hoon Park* | 313 |
|  | *Index* | 319 |

# Figures

| | | |
|---|---|---|
| 3.1 | Models of growth | 40 |
| 4.1 | Internet usage vs internet and telephone usage charges | 66 |
| 4.2 | Broadband penetration, June 2001 | 68 |
| 4.3 | Revenues per line vs average local rate | 82 |
| 5.1 | R&D expenditure to GDP ratios for Korea, Singapore and Taiwan, 1978–97 | 91 |
| 7.1 | Market trend of PlayStation format hardware and software | 178 |
| 7.2 | Evolution of Sony's television game business development in the age of broadband communication | 180 |
| 7.3 | Sony's strategic alliances for introducing networked services | 182 |
| 7.4 | The changing nature of software development activities in India | 188 |
| 7.5 | I-flex's products, product characteristics and markets over time | 191 |
| 10.1 | Proliferation of the internet and access prices | 241 |
| 10.2 | Indonesia: internet subscribers and users, 1996–2000 | 247 |
| 10.3 | Technology as monopoly buster | 252 |
| 11.1 | Automobile industry relationships | 263 |
| 11.2 | Overall assessment: snapshot of the Asia Pacific economies | 266 |
| 11.3 | Cluster analysis of the Asia Pacific economies | 268 |

# Tables

| | | |
|---|---|---|
| 2.1 | Growth indicators for the US economy, 1989–2000 | 14 |
| 2.2 | Sources of US growth | 19 |
| 2.3 | Changes in labour productivity in the business sector | 25 |
| 4.1 | The real price of a three-minute call | 63 |
| 4.2 | Estimated contribution of eventual universal broadband connectivity to US consumer surplus | 74 |
| 5.1 | Multifactor productivity growth for selected Asian countries, 1978–97 | 90 |
| 5.2 | Government and private R&D expenditure in Korea, Singapore and Taiwan, 1978–97 | 92 |
| 5.3 | Technology indicators for selected countries | 93 |
| 5.4 | Wald test on the insignificance of elasticities for R&D capital stock, scale effects and technological change | 98 |
| 5.5 | Elasticity of cost to R&D capital stock $s_{k_r} = -\frac{\partial C}{\partial k_r}\frac{k_r}{C}$ | 98 |
| 5.6 | The indirect, direct and total effects of R&D capital on the productive performance of Korea, Singapore and Taiwan, 1978–97 | 99 |
| 5.7 | The cost elasticity of output and contribution of scale to productive performance of Korea, Taiwan and Singapore, 1978–97 | 100 |
| 5.8 | Multifactor productivity growth for Korea, Singapore and Taiwan, 1978–97 | 100 |
| 5.9 | R&D employment in Japan, Korea, Singapore and Taiwan, 1978–97 | 102 |
| 6.2.1 | R&D expenditure in selected industries in India, 1990–91 | 110 |
| 6.2.2 | Technological services used in India and the role of technology institutions | 111 |
| 6.2.3 | Foreign collaborations in India, 1991–94 | 113 |
| 6.2.4 | Quality ratings for selected industries, 1998 | 115 |
| 6.2.5 | Manufacturing technologies in India: implementation and success | 118 |

| | | |
|---|---|---|
| 6.2.6 | Improvements in manufacturing performance from the introduction of advanced technologies, 1997–2000 | 119 |
| 6.3.1 | Comparison of GDP and value added of the IT industry, 1995–2000 | 131 |
| 6.3.2 | The production of major electronic products in China, 1995–2000 | 131 |
| 6.4.1 | Taiwan's computer hardware production, 1995–2000 | 144 |
| 6.4.2 | Taiwan's computer software production, 1996–2000 | 146 |
| 6.4.3a | Characteristics of the Taiwanese ICT-manufacturing industry | 150 |
| 6.4.3b | Characteristics of the Taiwanese ICT-manufacturing industry | 151 |
| 6.4.4 | Taiwanese GDP growth rates and contributions to GDP by sector, 1982–99 | 154 |
| 6.4.5 | Taiwanese manufacturing industries' growth rate and their contribution by division to GDP growth, 1982–99 | 155 |
| 6.4.6 | Rate of growth of production and contributions by total manufacturing and industrial production subgroups, 1987–99 | 156 |
| 6.4.7 | Taiwanese electronic products: ratios of consumption, investment and exports to final demand | 159 |
| 6.4.8 | ICT expenditure as a proportion of GDP and investment, 1996 | 160 |
| 6.4.9 | Investment in ICT as a proportion of investment by industry division | 161 |
| 6.4.10 | Indicators of ICT infrastructure and access to ICT in Taiwan | 162 |
| 6.4.11 | Growth rate of total factor productivity by industry, 1979–98 | 164 |
| 6.4.12 | Input coefficient A of the Taiwanese electronic products sector, 1986–96 | 166 |
| 6.4.13 | The inversion of the coefficient matrix $(I-A)^{-1}$ or $[I-(I-M)A]^{-1}$ of the Taiwanese electronic products sector | 167 |
| 7.1 | Comparison of combinative strategies of Sony and i-flex | 197 |
| 10.1 | Worldwide e-commerce growth | 239 |
| 10.2 | Indian software labour: cost comparison | 240 |
| 10.3 | Internet access costs, October 2000 | 242 |
| 10.4 | Market share of Indonesian internet service providers | 245 |
| 10.5 | Indonesian internet service provider profiles | 248 |
| 10.6 | Local call costs | 249 |
| 10.7 | Indonesia: price elasticity of telephone services | 254 |
| 11.1 | Introduction of information technology into the Japanese automobile industry | 260 |
| 11.2 | Types of components in a car and a computer | 261 |
| 12.1 | Internet users in East Asia | 279 |
| 12.2 | Adult literacy rates and total enrolments in East Asia, 1999 | 281 |
| 13.1 | Comparative treatment of various disciplines under GATT 1994 and GATS | 312 |
| 14.1 | Selected APEC capacity building initiatives and activities | 315 |

# Contributors

**Titik Anas** is a member of the research staff at the Department of Economics, Centre for Strategic and International Studies, Jakarta.

**Dionisius Ardiyanto** is a member of the research staff at the Department of Economics, Centre for Strategic and International Studies, Jakarta.

**Haryo Aswicahyono** is Senior Research Associate at the Centre for Strategic Studies in Jakarta.

**Yasunori Baba** is a professor at the Research Center for Advanced Economic Engineering at the University of Tokyo.

**Vei-Lin Chan** is an associate research fellow at the Institute of Economics, Academia Sinica, Taiwan.

**Richard N. Cooper** is Maurits C. Boas Professor of International Economics at Harvard University.

**Robert W. Crandall** is a Senior Fellow in the Economic Studies Program at the Brookings Institution.

**Edsel T. Custodio** is Ambassador and Alternate Permanent Representative to the World Trade Organization.

**Dong Yunting** is President of the China Academy of Electronics Industry Development, Vice-director of the Planning Department of the Ministry of Electronics Industry and Chairman of its Department of Policy Research.

**Peter Drysdale** is the Executive Director of the Australia–Japan Research Centre and a professor in the Asia Pacific School of Economics and Management and the Economics Division, Research School of Pacific and Asian Studies at the Australian National University.

**Lewis Evans** is Professor of Economics at Victoria University of Wellington and the Executive Director of the New Zealand Institute for the Study of Competition and Regulation.

**Roger Farrell** is a Centre Associate of the Australia–Japan Research Centre at the Australian National University and a Director of the Centre for Asia Pacific Research.

**Christopher Findlay** is Professor of Economics in the Asia Pacific School of Economics and Management at the Australian National University.

**Sheng-Cheng Hu** is Minister of State, Executive Yuan (the Cabinet), ROC, Taiwan.

**Mangesh G. Korgaonker** is ICICI Chair Professor and Head, Shailesh J. Mehta School of Management, Indian Institute of Technology Bombay, India.

**Emmanuel C. Lallana** is Executive Director of the e-ASEAN Task Force, an advisory body to the Association of Southeast Asian Nations.

**Richard G. Lipsey**, FRSC, OC, is a professor emeritus of economics at Simon Fraser University and Fellow of the Canadian Institute for Advanced Research.

**Keith E. Maskus** is Professor of Economics at the University of Colorado and a visiting research fellow at the Institute for International Economics.

**Juan J. Palacios** is Professor of Regional Development at the Doctoral Program in Social Sciences, University of Guadalajara.

**Mari Pangestu** is on the Board of the Centre for Strategic and International Studies in Jakarta and an adjunct professor at the Australia–Japan Research Centre at the Australian National University.

**Sung-Hoon Park** is Professor of Economics and International Trade at the Graduate School of International Studies, Korea University.

**Shandre M. Thangavelu** works at the National University of Singapore, Department of Economics, Singapore.

**Toh Mun Heng** is an Associate Professor and Deputy Head of the Department of Business Policy, National University of Singapore.

**F. Ted Tschang** is Assistant Professor, Economics and Technology, Singapore Management University.

# Preface

The Pacific Trade and Development (PAFTAD) Conference series has been at the frontier of analysing challenges facing the economies of East Asia and the Pacific since it was first established in January 1968 in Tokyo. The theme of the twenty-seventh conference, held at the Australian National University in Canberra on 20–22 August 2001, was the 'new economy' and its implications for the region. The papers presented at the conference are gathered together in this volume.

The 'new economy' is an illusive concept. But there is general agreement that it is wreaking fundamental changes to the way in which economies operate and the efficiency with which they will operate around the world. There is less agreement on whether the new economy has permanently changed the way in which economies at its forefront, like the United States, work or how its beneficial effects can be transmitted to other economies, such as those at earlier stages of economic development in Asia. And there are questions about why some countries, like Japan, which seem to be well positioned to prosper in the age of the new economy continue to stagnate and other countries, like Australia, with a smaller industrial and information technology base, appear to have done so well. There are questions not only about how national policy regimes best capture the benefits of the new economy but also about how international and regional policy regimes can be directed more effectively towards that end. And there is the question of the 'digital divide' or whether the distribution of its benefits leave some people and some countries in some ways excluded.

A distinguished group of economists from around the region gathered in Canberra at PAFTAD 27 to discuss all these questions, from the perspective of the experience of Asia and the Pacific. The presentation and discussion of papers was followed by their revision for publication in this book in the rigorous PAFTAD tradition.

I am very grateful to all the contributors – paper writers, discussants and referees – who have collaborated so enthusiastically to bring this research to early publication. My debt to the authors of chapters in the volume is obvious. In addition, Hugh Patrick, Ross Garnaut, Yung Chul Park, Peter Petri, Frank

Holmes, Henry Ergas, Wendy Dobson, Fred Hilmer, Zainal Aznam Yusof, Akira Kohsaka, Shu Urata, Mari Pangestu, Hadi Soesastro, Juan Palacios, Raghbendra Jha, Somkiat Tangkitvanich, Allan Fels, Mario Lamberte, John Rimmer, Rohan Pitchford, David Kennedy, Ralph Huenemann, Peter Forsyth, Ed Chen, Wisarn Pupphavesa, Richard Snape, Chia Siow Yue, and Narongchai Akrasanee all made substantial contributions to the clarification of issues and refinement of ideas at the conference and in the process of preparation of the papers for publication. And without the particular help of the PAFTAD International Steering Committee, this work would not have been completed.

I am especially grateful to Jane Drake-Brockman, who assisted me in editing the volume. She helped to revise many of the chapters and worked with the authors of Chapter 14, Mari Pangestu and Sung-Hoon Park, to draw together their presentations to a final panel at the conference into an excellent conclusion to the book.

At all stages of the management of this project, Marilyn Popp and Andrew Deane gave their usual excellent and devoted service.

The PAFTAD program and PAFTAD 27 are supported by a consortium of international donors and serviced by the PAFTAD Secretariat at the Australian National University in the Asia Pacific School of Economics and Management (APSEM).

I record the sincere gratitude of the PAFTAD International Steering Committee and my own personal thanks to our donors, whose support continues to make this important work possible. They include the Ford Foundation, the Canadian International Development Research Center (IDRC), the Japanese Research Institute of Economy, Trade and Industry (RIETI), the Korea Institute of International Economic Policy (KIEP), the Australian Government's Aid Agency AUSAID, the Australian Government's National Office for Information Economy (NOIE), the Asia Foundation, Toronto University, Victoria University (Canada), the National University of Singapore, the Taiwan Institute of Economic Research, Columbia University, Sanaree Holdings and the Australian National University.

Sue Mathews did editorial wonders with the manuscript and Minni Reis did her normal fine job to prepare and pagemake the text for publication. My thanks also go to Heidi Bagtazo and the team at Routledge, with whom it is always a pleasure to work.

This book is a first and important collection of essays on the impact of the new economy on Asia and the Pacific. It opens up a set of questions for further research and, hopefully, makes a useful contribution to thinking about issues that now confront policy-makers concerned with the different effects of the new economy and its impact around the region.

Peter Drysdale
August 2003

# Abbreviations

| | |
|---|---|
| 3G | third generation |
| ADL | Arthur D. Little |
| ANX | Automotive Network eXchange |
| AOL | America Online Inc. |
| APDIP | Asia Pacific Development Information Program |
| APEC | Asia Pacific Economic Cooperation |
| APII | Asia Pacific information infrastructure (APEC initiative) |
| APJII | Asosiasi Penyelenggara Jasa Internet Indonesia |
| APO | Asian Productivity Organization |
| APROC | Asia-Pacific Regional Operations Centre (Taiwan) |
| AR1 | first-order regressive form |
| ARPA | Advanced Research Projects Agency (US) |
| ASEAN | Association of Southeast Asian Nations |
| ASP | application service provider |
| ATM | automatic teller machine |
| B2B | business-to-business |
| B2C | business-to-consumer |
| BIS | Bank for International Settlements |
| BTO | build-to-order |
| CAD | computer-aided design |
| CAGR | compound annual growth rate |
| CAM | computer-aided manufacturing |
| CBO | Congressional Budget Office (US) |
| CCL | Computer and Communications Research Laboratories (Taiwan) |
| CD | compact disc |
| CEA | Council of Economic Advisers (US) |
| CII | Confederation of Indian Industries |
| CLEC | competitive local exchange carrier |
| CMM | capability maturity model |
| CNNIC | China Internet Network Information Centre |
| CPI | consumer price index |
| CPU | central processing unit |
| CRM | customer relationship management |

| | |
|---|---|
| CRTC | Canadian Radio-television and Telecommunications Commission |
| CSFs | critical success factors |
| CTD | Committee on Trade and Development (of the WTO) |
| CTIA | Cellular Communications and Internet Association |
| DDN | Digital Divide Network |
| DGBAS | Directorate-General of Budget, Accounting and Statistics (Taiwan) |
| DOT Force | Digital Opportunity Task Force (of the G-8) |
| DRI | Data Resources Inc. |
| DS–1,2,3 etc | digital signal 1, 2, 3, etc |
| DSL | digital subscriber line |
| DSLAM | digital subscriber line access multiplexer |
| DVD | digital video disk |
| eATF | e-ASEAN Task Force |
| EAPs | Ecotech action plans (of APEC) |
| EDI | electronic data interchange |
| ENX | European Network eXchange |
| ERSO | Electronics Research and Service Organisation (Taiwan) |
| EU | European Union |
| EVA | economic value added |
| EVSL | early voluntary sector liberalisation |
| FCC | Federal Communications Commission (US) |
| FDI | foreign direct investment |
| FFR | federal funds rate (US) |
| FTAA | Free Trade Area of the Americas |
| G-8 | Group of Eight |
| GATS | General Agreement on Trade in Services |
| GATT | General Agreement on Tariffs and Trade |
| GDP | gross domestic product |
| Ghz | gigahertz |
| GICT | Global Information and Communication Technologies Department |
| GNP | gross national product |
| GPT | general-purpose technology |
| GSM | Groupe Speciale Mobile |
| GUI | graphical user interface |
| GUIDEC | General Usage for Digitally Ensured Commerce |
| HDTV | high-definition television |
| HIV–AIDS | human immunodeficiency virus–acquired immune deficiency syndrome |
| IAMAsia | Interactive Audience Measurement Asia Ltd. |
| IBM | International Business Machines Corporation |
| IC | integrated circuit |
| ICANN | Internet Corporation for Assigned Names and Numbers |
| ICE | information, communications and entertainment |
| ICP | internet content provider |
| ICSEAD | International Centre for the Study of East Asian Development |

| | |
|---|---|
| ICT | information and communication technology |
| IDC | International Data Corporation |
| IEEE | Institute of Electrical and Electronics Engineers |
| infoDev | Information for Development Program (a program of the World Bank) |
| III | Institute for Information Industry (Taiwan) |
| IP | internet protocol |
| IPO | initial public offering |
| IPRs | intellectual property rights |
| ISDN | integrated services digital network |
| ISO | International Organization for Standardization |
| ISP | internet service provider |
| IT | information technology |
| ITA | Information Technology Agreement (Ministerial Declaration on Trade in Information Technology of the WTO) |
| ITES | information technology enabled services |
| ITIR | Institute for Technological and Industrial Research (Taiwan) |
| ITRI | Industrial Technology Research Institute (Taiwan) |
| ITSS | information technology software and services |
| ITU | International Telecommunication Union |
| IVCA | Indian Venture Capital Association |
| JAMA | Japan Automobile Manufacturers Association |
| JBIC | Japanese Bank for International Cooperation |
| JIT | just-in-time |
| JNX | Japanese automotive Network eXchange |
| kbps | kilobits per second |
| kwh | kilowatt hours |
| LAM | local area multicomputer |
| LAN | local area network |
| M1 | currency, travellers cheques, checkable deposits |
| M2 | M1 plus savings and small time deposits and retail money market funds |
| M3 | M2 plus large time deposits, institutional money market funds, repurchases and eurodollars |
| Mbps | megabits per second |
| MEPs | manufacturing enhancement programs |
| MFN | most favoured nation |
| MFPG | multifactor productivity growth |
| MNC | multinational corporation |
| MOEA | Ministry of Economic Affairs (Taiwan) |
| MOFTEC | Ministry of Foreign Trade and Economic Cooperation (PRC) |
| NACA | National Advisory Committee on Aeronautics (US) |
| NAFTA | North American Free Trade Agreement |
| NAIRU | non-accelerating inflation rate of unemployment |
| NASSCOM | National Association of Software and Service Companies (India) |

| | |
|---|---|
| NBER | National Bureau of Economic Research (US) |
| NCTS | Northwest China Telehealth Service |
| NGO | non-government organisation |
| NIE | newly industrialised economy |
| NII Plan | National Information Infrastructure Plan (Taiwan) |
| NZEM | New Zealand Electricity Market |
| OECD | Organisation for Economic Co-operation and Development |
| OEM | original equipment manufacturer |
| OES | Opto-Electronics and Systems Laboratories (Taiwan) |
| OTC | Optimization Technology Center |
| PC | personal computer |
| PCB | printed circuit board |
| PCMM | people capability maturity model |
| PCS | personal communications services |
| R&D | research and development |
| RCA | Radio Corporation of America |
| RDBMS | relational database management system |
| REAP | readiness assessment evaluation partnership |
| S&P | Standard and Poor |
| S&T | science and technology |
| SCE | Sony Computer Entertainment |
| S-E model | structuralist-evolutionary model |
| SEBI | Securities and Exchange Board of India |
| SEI | Software Engineering Institute |
| SMEs | small and medium-sized enterprises |
| SUR | seemingly unrelated |
| SVP | Silicon Valley phenomenon |
| TBI | technology business incubator |
| TBT Agrmt. | Agreement on Technical Barriers to Trade |
| TD | technology development |
| TFP | total factor productivity |
| TFT | thin film transistor |
| TI | technology institution |
| TILF | Trade and Investment Liberalisation Framework (of APEC) |
| TMC | Technology Marketing Corporation |
| TNS | Taylor Nelson Sofres Interactive |
| TQM | total quality management |
| TRIPs | trade-related aspects of intellectual property rights |
| TSEC | Taiwan Stock Exchange Corporation |
| TSMC | Taiwan Semiconductor Manufacturing Corporation |
| UMC | United Microelectronics Corporation |
| UN | United Nations |
| UN/CEFACT | United Nations Centre for Trade Facilitation and Electronic Business |
| UNCITRAL | United Nations Commission on International Trade Law |

| | |
|---|---|
| UNCTAD | United Nations Conference on Trade and Development |
| UNDP | United Nations Development Program |
| UNITeS | United Nations Information Technology Service |
| SDNP | Sustainable Development Networking Program |
| URETS | Uniform Rules for Electronic Trade and Settlement |
| VCD | video compact disk |
| VCF | venture capital fund |
| VISC | Visionary Integration Services Corporation |
| VOIP | voice over internet protocol |
| VTR | video tape recorder |
| WAN | wide area network |
| WAP | wide area paging |
| WIPO | World Intellectual Property Organization |
| WPIIS | Working Party on Indicators for the Information Society (OECD) |
| WTO | World Trade Organization |

# 1 The new economy in Asia and the Pacific: an overview

*Peter Drysdale*

The new economy was ostensibly built on the surge in the development and diffusion of a whole raft of new information and communication technologies (ICTs) in the last decade or two of the twentieth century. The epicentre of the new economy revolution was the United States, and its impact on productivity, and economic growth more generally, was pronounced in relatively few developed economies. The policy environment was crucial in achieving the gains for growth from the application of these new technologies. Major centres for the production of ICT goods, such as Japan, seem to have been passed by in terms of the growth-enhancing impact of the new economy, presumably because the macro and micro-economic policy environment did not allow its benefits to be captured. Other countries, like Australia, not significant producers of ICT goods but with policy settings that encouraged the early application and diffusion of the new economy technologies, enjoyed even stronger productivity growth than the United States and an even longer boom. Countries in the developing world without a domestic environment conducive to the new economy, it seemed, would be increasingly marginalised from the globalised production process and the global economy (Mann and Rosen 2001)

The idea of the new economy was not confined solely to the impact of the ICTs on productivity and growth. In the euphoria that accompanied the long boom in the United States, the suggestion was that the traditional business cycle was a thing of the past, long run growth rates had been ratcheted up permanently, and the stock market was on a now continuous high. The American new economy boom lasted for a decade, but the boom has now come and gone, and there remain questions, as Cooper points out in Chapter 2, about what was really new in the new economy. There are also questions about what was new about the digital or productivity divide between developing and developed economies and within the developing economies themselves.

The APEC Leaders' Summit in Shanghai in October 2001 related the power of globalisation to stimulate growth and to improve living standards and social well-being to the impact of the new economy. The new economy, their Declaration proclaimed (pp.6–7), had the potential to raise productivity, to stimulate innovation in economic organisation and entrepreneurship and to create and disseminate knowledge and wealth (APEC Leaders' Declaration 2001, p.2). APEC's e-APEC Task Force noted that the effective application of the new technologies required the strengthening of market structures and institutions, a strong legal and institutional infrastructure for investment and technological development, and capacity building to promote entrepreneurship. Under these conditions, it argued, new ICTs had enormous potential to assist economies in achieving specific economic development goals, not only because of the technology itself but because of its ability to improve communication and the exchange of information. A priority therefore was to promote effective education and training, and technological cooperation and exchange that would foster the application of ICTs and help to close the digital divide.

This book addresses these issues. They are issues of considerable policy interest and importance to the economies of the Asia Pacific region.

The scene is set in the first two chapters, in which Cooper examines the experience of the United States and Lipsey reviews the theory and measurement of the effects of the new economy. Crandall then looks at the evolution of the internet and the role of telecommunications policy in that process. Thangavelu and Toh explore the sources of regional growth for evidence of the impact of the new technologies. The contributions from Palacios, Korgaonker, Dong and Hu and Chan analyse experience in India, China and Taiwan, looking at the nature of the links in the chain of both technological generation and technological diffusion. Baba and Tschang provide an insight into the strategies of IT firms. Maskus considers the intellectual property market issues and Evans the impact of the new economy technologies on competition in markets. Two industry case studies for the region are developed by Aswicahyono, Anas and Ardiyonto and Farrell and Findlay in the next two chapters. And the final chapters by Lallana (on the digital divide), Custodio (on e-commerce in the World Trade Organization (WTO)) and Pangestu and Park (on implications for APEC) address regional policy issues.

The US economy in the 1990s enjoyed a prolonged boom and exceptionally high productivity growth. The boom in equity markets, and in particular high-tech equity markets, was one feature of the boom. Strong growth was accompanied by very low levels of unemployment as well as low rates of inflation. This unusual combination of circumstances encouraged a view that the US economy had changed fundamentally and that the new technologies had indeed created a new economy.

Cooper documents the puncture of the new economy boom and how the high rate of growth was the result of an unexpected jump in total factor productivity. Productivity growth in the 1990s was three times higher than in the previous two decades. As Cooper explains, some have argued that strong productivity growth was largely concentrated in the electronics and related information sectors. But other evidence points to a much more broadly based lift in productivity throughout the economy. A key factor in the widespread impact of the new technologies on productivity performance in the United

States was pervasive deregulation and openness to trade, through which the benefits from rapidly falling costs and technological improvement in the ICT sector were captured for the economy as a whole.

Cooper argues that the sharp rise in productivity was driven by the application of cheap, widely available computational power combined within less and less expensive communication. This was analogous to the introduction of new general-purpose technologies such as steam power, electricity or plastics in the past. The introduction of new technology involves much new investment. It also requires openness to breaking down old ways of doing things, organisational forms and institutional arrangements. The technologies were twenty or thirty years old. What was new was their effective application. The new economy seemed to place an especially high premium on education beyond secondary school levels.

Cooper notes that, surprisingly, very few economies in the industrial world, with the notable exception of Australia, enjoyed a similar above-trend lift in productivity. I shall return to discussion of the Australian case at the end.

Cooper's conclusion is cautionary: five or ten years is a short period on which to base any long-term extrapolations about the future. In the 1980s Japanese production, organisation and management were seen as the way of the future. But assessment of Japanese performance has now changed profoundly.

Chapter 2, by Lipsey, puts US experience into theoretical and historical perspective.

Lipsey highlights the social, economic and political changes brought about by the ICT revolution and the range of products, such as computers, satellites, lasers, fibre-optics, the internet and related technologies, that underpin it. He makes the strong link that APEC leaders make between technology – and its capacity to raise income via productivity improvement and innovation – and human welfare.

Technology is *the* major determinant of long-term growth, according to Lipsey, and technology is embodied in capital and labour, the important vehicles whereby it is delivered. This is why it is so difficult to measure the contribution of technology to economic growth. The interdependence between technological progress, capital accumulation and the growth of the work force makes it impossible to sort out the contribution of technology from other factors of production in the way assumed possible in the neoclassical growth accounting model.

Lipsey elaborates on the distinction between new general-purpose and incremental and specific technologies and the pervasiveness of the effects of the introduction of new waves of general-purpose technologies. And he nominates ICTs and their appearance in the late twentieth century as a substantial new wave of general-purpose technology.

All nations have access to the best practice technology in the world. But there is substantial cost in acquiring these technologies through learning-by-doing or using them. And Lipsey makes the important point that many economies lack the absorptive capacity to be able to introduce them.

In this argument the link between the adoption or the absorption of new technology and what Lipsey calls 'structures' is critical. The introduction of new technologies requires that old structures give way and that new ones are put in place. The structures that Lipsey has in mind here are ways of organising business, institutions that affect the way markets operate and policies that govern the relationships between government and the market. In the process of introducing new technologies, old structures become dysfunctional and have to be discarded.

The problems of catching up with industrial technologies are very different from the problems facing economies at the leading edge of technological advance. Catch-up involves dealing with established technologies, and many of the uncertainties that are associated with the introduction of cutting-edge technologies are absent. Governments and economic agents need to play a very different role in each process. And by implication, the structures and policy strategies that successfully promote catch-up may be positively deleterious to fostering creative innovation at the technological frontier. This is a major challenge for the most advanced East Asian economies as they confront the transition from industrial catch-up to being at the technological frontier.

The internet is one of the key new ICTs, and even in developed economies access to efficient internet services is not wide. A little over 40 per cent of households use the internet in the United States, although a much higher proportion have a personal computer (PC). The telecommunications infrastructure is key to the wider diffusion of the internet, in particular through the spread of broadband access. In Chapter 3, Crandall reviews aspects of telecommunications regulation that inhibit the spread of broadband access and the use of internet services. Distance may be dead, as Crandall observes, but speed is not. Broadband, and the new digital architecture to deliver it, is essential to reliable always-on internet services and to realising the consumer benefits from the large range of services it can provide. Crandall argues that we are only at the beginning of the digital revolution even in economies that are the most advanced users of internet services.

Crandall reviews the deficiency of current telecommunications regulation with its focus on universal service obligations. He concludes that, although 30–40 per cent of people in North America, Japan, Korea, Singapore, Europe and Australasia use the internet, lack of broadband vastly limits its potential for development of e-commerce and a host of other services that can be delivered over it. The principal impediments are not only technical and commercial, but also significantly regulatory. Traditional regulatory policies force providers to earn a large share of their revenues from per-minute charges and charge high rates for high-speed services. This reduces incentives to deploy broadband services. Asia Pacific economies, Crandall suggests, would be unwise to follow North American and European policies in this area.

Chapter 6 examines the role of research and development and productivity growth in the growth performance of economies in East Asia. Thangavelu and Toh apply standard growth accounting techniques to measure multi-factor productivity growth across Asia and growth accounting using the cost function approach to look at the impact of research and development (R&D) expenditure on productivity performance in Korea, Taiwan and Singapore.

Multi-factor productivity contributed significantly to growth in the East Asian economies over the last thirty years. There was a notable decline of productivity growth in Japan, Korea, Taiwan, Thailand, Malaysia, the Philippines and South Asia in the 1990s. The impact of R&D on growth in the three East Asian economies was positive, but scale economies had a much bigger effect. The ratio of R&D expenditure to GDP is very large, especially in Korea, so a question arises about the effectiveness of investment in R&D in the higher-income East Asian economies. That question relates to the challenge for these countries, mentioned earlier, in making the structural transition to industrial maturity.

The links between centres of technology generation in the United States and elsewhere in the region are taken up in the four parts of Chapter 6. Palacios provides the context for these three country case studies, describing the Silicon Valley phenomenon. He observes that the replication of Silicon Valley is harder than it looks, even in the United States. Whether clusters of innovative activity develop successfully elsewhere in Asia and the Pacific, he suggests, is likely to depend critically upon the strategies and involvement of multinational corporations.

Korgaonker provides a comprehensive picture of experience in India. He stresses the importance of economic liberalisation in encouraging innovation, through technology imports being used to improve quality and competitiveness in manufacturing. The earlier government dominance of R&D is being overtaken by vigorous private entrepreneurial activity in partnership with foreign investors. Korgaonker details the growth of the IT sector and the links between Silicon Valley (where over 30 per cent of start-ups have been initiated by entrepreneurs of Indian origin) and the burgeoning industry in India. Again the links with international firms are critical. Global remote-service operations in India represent every element of the value chain: research and design; procurement and logistics; marketing and sales; accounting; IT; and human resource management. The new technologies themselves have allowed the rapid development of an export-oriented IT sector through links with sophisticated customers in the United States and elsewhere.

Dong documents the huge growth in the Chinese IT sector, focused on growing connectivity, PC and mobile phone use and the growth of output and exports of IT goods. IT goods production is led by foreign firms, which account for almost half of the industry's total exports. Entry into the WTO is leading to deep rationalisation of IT goods production and strong improvement in competitiveness. While the largest domestic firms are small by international standards, they are establishing a presence in the international as well as the domestic market.

The Chinese government is now promoting some huge science and research centres in which links between universities and entrepreneurs aim to foster clusters of innovation.

The scale of developments in China overwhelms that in other large regional markets: its mobile phone network already outranks that of Japan. Electronics products have come to dominate labour-intensive exports of manufactured goods as they do in most other East Asian economies. And the IT industry in China is projected to more than double in size over the next five years.

Finally Hu and Chan analyse the impact of the new economy on production and trade in electronics in Taiwan. The ICT sector is the largest single sector in the Taiwanese economy, accounting for more than 30 per cent of value added in manufacturing in recent years. Taiwan is the third largest computer hardware producer in the world. The first-rate telecom network facilitates extremely high market penetration for phones, mobiles, PCs and the internet. The sector accounted for 45 per cent or 2.6 percentage points of GDP growth in 1999. However, there was a marked drop in Taiwan's overall growth between the 1980s and the 1990s, and Hu and Chan are inclined to view over-reliance on the production of ITC goods as damaging to growth prospects. It is difficult to quantify significant spillover effects from the sector for productivity growth.

By far the most important effect on trade and production patterns in the last decade or so has come from the integration of the Taiwanese electronics sector with that in mainland China. There are extremely close production networks between Taiwanese and other firms in China and operations in Taiwan. And China has become Taiwan's leading competitor in computer hardware.

In India, China and Taiwan the common factor in ICT is close interaction with the North American market. The collapse of the high-tech boom in the United States was transmitted very rapidly to all three Asian suppliers of ICT products and software at different points in the supply chain.

In Chapter 7, Baba and Tschang discuss how Asian IT firms cope with rapid changes in the market for IT products, both in taking advantage of emerging market opportunities and in adjusting to competitive pressures from new suppliers and new products. They illustrate the issues at different ends of the market, identifying the elements in the success of Sony in introducing PlayStation 1 and 2 and of the Indian software company i-flex in establishing its international finance product lines. These cases present in microcosm the challenges facing firms, in the case of Sony, at the leading edge of technology, and in the case of i-flex, in the process of upgrading service and product supplied in the course of catching up with international market leaders in a particular field.

Sony, like other IT leaders in Japan, faced the threat of declining margins on its traditional electronics hardware, as it shifted production to lower wage cost production sites. Its response was to embrace a new strategy linking broadband and the development of new market segments. This involved high-risk investment in prototyping products for hypothetical markets, combining hardware and entertainment strengths with managing production and development networks.

In an Indian software industry that is largely based on contracting services, i-flex stands out as a company that successfully moved into the higher value product market. This strategy is one that protected against the eventual erosion of low labour cost advantages by leveraging off initial human resource capabilities and knowledge in banking software. The investment in innovative product and the establishment of brand-name through its affiliation with Citicorp enabled it to enter new markets against the biggest and most deeply entrenched international suppliers.

Intellectual property rights (IPRs) are a central interest in countries wishing to promote innovation, as well as in those wishing to access new ICT and other technologies as rapidly and cheaply as possible. Maskus deals with this issue in Chapter 8.

IPRs protect the returns from investment in ICT and other technologies. But a regime that is too protective slows the diffusion of new technologies, and can distort the allocation of capital as well as slow productivity growth, especially in catching-up economies. Striking a balance between these two conflicting interests, as Maskus argues, makes it likely that one size will not fit all in the approach to the IPR regime across the economies at different stages of development in Asia and the Pacific.

Many of the creations and inventions in the knowledge sector of the new economy, however, depend importantly upon intellectual property protection. Digital products and software can be easily downloaded and copied in the absence of well-enforced copyright law and technologies that prevent copying. This is a sector in which product life-cycles are short and reverse-engineering relatively straightforward.

Maskus identifies reasons why developing economies might want more stringent regimes in this area than in others. The role of alliances with multinational firms in the development of local ICT capacities is one. Another is that weak protection regimes discourage the indigenous development of software capabilities, a circumstance that may be relevant both to Taiwan in the past and to China recently. The commercialisation of research performed in universities and public research institutions is a question in all economies, and Maskus draws attention to the innovative programs in Hong Kong, China and Taiwan to encourage this.

Maskus concludes that the IPR regime is only one of a number of elements that are important in the development of the ITC sector. Others include telecommunications infrastructure and regulatory policies, discussed elsewhere in the book, and the depth and accessibility of capital market and competition structures.

Whether, in turn, the new information economy promotes competition or is likely to reinforce market power is a question that Evans takes up in Chapter 9. He argues that the ICT revolution reduces information costs. A reduction in the costs of information lowers transactions costs, and lower transactions costs curtail monopoly power. Yet information is also central to cooperation between firms, and cooperation between firms may be directed at monopolistic behaviour.

Evans concludes that the increased availability and lower cost of information in the new economy will on balance raise competition in the economy rather than reduce it. The effects of the rapid creation of new markets, the reorganisation of firms because of reduced transactions costs and the introduction of new technologies has increased the scope for competition more than for cooperation. Evans points out that cooperative arrangements are easier when there are low levels of uncertainty about the future path of technological change and the pay-off from investment. But the new economy has increased the level of uncertainty on both fronts. These characteristics of the market in the ICT economy are the same as those identified in Chapters 8 and 3 as important to assessment of IPR and telecommunications policies and in Chapter 7 as important to corporate strategies.

Evans suggests that the technological and related economic changes ushered in by the new economy make dynamic, not static, efficiency the relevant yardstick in competition law.

The next chapters explore the impact of ICTs in two industries – the internet industry in Indonesia and the automobile industry in East Asia.

In Chapter 10, Aswicahyono, Anas and Ardiyanto begin their case study of the internet industry in Indonesia with the important observation that efficient internet services, like efficient ports, are now critical to the inclusion of developing economies in global growth. The internet has revolutionised manufacturing by streamlining product and service delivery, reducing transactions costs. The impact on outsourcing strategies has been profound. This has dramatically increased the scope for developing economies, with their wage and other cost advantages, to be integrated into the operation of multinational firms and global markets. Transformations in trade and industry – which took decades to achieve in years past – can occur very rapidly with effective deployment of new economy technologies. The speed of industrial catch-up can thus be accelerated.

Aswicahyono, Anas and Ardiyanto confirm in the Indonesian case Crandall's point that telecom costs are the major barrier to diffusion of internet use. Telephone charges account for almost 60 per cent of the high cost of internet access in Indonesia. Aswicahyono, Anas and Ardiyanto recommend breaking up Indonesia Telekom's monopoly power by allowing more players into the market. The new Indonesian telecommunications law has set the scene for change, but change is yet to come.

In Chapter 11, Farrell and Findlay trace the impact of new economy technologies in the motor vehicle industry. Information technology has obviously already had a large effect on the design of the product, and the trend is towards more electronically sophisticated automobiles. Toyota's ambition, Farrell and Findlay report, is to transform the automobile into a 'mobile information processing and communications platform'. Applications of ICT in the automobile industry extend well beyond the 'talking' car.

The most important impact in the industry is on the organisation of production and trade. E-commerce complements the lean production and delivery system pioneered in the Japanese industry. But the emergence of auction markets for generic parts and components and the effect of module design and internet communications between parts and components makers and assemblers is breaking down relational dealings and driving a reorganisation of the industry. Lower transactions costs lower prices. These developments introduce more competition and openness into the market for components and parts. Such trends are still more evident in US production and delivery systems than they are in Japan.

ICT is often seen as the enabler of development. Yet the uptake of ITC is uneven and increased access and its exponential growth in some parts of the world economy may also mean that the digital divide is actually growing. Lallana surveys this problem in Chapter 10.

Lallana provides a detailed picture of the patterns of access and the characteristics of internet use in East Asia. Teledensity, PC access and utilisation, and mobile telephone diffusion are closely but not entirely related to income levels. Access is not simply a function of the availability of physical infrastructure but is also a function of literacy, the geographical distribution of population, and demographics, for example; and the application of ITC technologies in business, in the household, for e-government, e-learning and medicine are influenced by government policies and private initiative. Lallana, like Crandall and others in this volume, emphasises the importance of competition in the telecommunications market and the provider market to improving both access and the quality of access. He notes, however, differences in government policy strategies for achieving these objectives: the 'light touch' market-led approach; the cooperative approach to extend reach to target groups; and the comprehensive national approach to extending broadband access with pro-active government involvement in roll-out. He also reviews the regional and global as well as national initiatives in bridging the digital divide.

He concludes that while there are still large gaps, East Asians – especially Japan and China – are well aware of what is at stake and are well poised to be major players in the electronic century.

E-commerce is now profoundly affecting the way international business is being done. In Chapter 13, Custodio reports on progress with work on e-commerce within the WTO. E-commerce has the potential to grow strongly internationally, boosting trade in both goods and services.

The decision to put a moratorium on barriers to e-commerce is intended to contribute to the growth of e-commerce across borders. But, as Custodio points out, there is a worry that the moratorium violates the principle of technological neutrality in respect of mode of delivery, and could divert from trade delivered by other modes. A further suggestion is that barrier-free e-commerce services would be more effectively secured by deepening and widening the presently limited cross-border trade commitments under the General Agreement on Trade in Services (GATS), and by strengthening GATS disciplines, for example on most favoured nation and national treatment (Stephenson et al., 2002).

Because e-commerce straddles the delivery of trade in goods and trade in services, it falls between the General Agreement on Tariffs and Trade (GATT) and GATS law within the WTO. Custodio explains these complications carefully and recommends caution in going forward on any one front.

Many of the hurdles to growth of e-commerce relate to the application of domestic regulations that provide redress against international intellectual property theft and fraud. This is not a set of issues that the WTO can easily mediate. But it is a key field for international cooperation.

The theatre for policy development and initiative on the new economy in East Asia and the Pacific is APEC. Pangestu and Park discuss the role of APEC in Chapter 14.

East Asia and the Pacific include a diverse group of economies displaying the whole range of issues raised by the ICT revolution and the new economy. APEC, with its focus on trade and investment liberalisation and facilitation and on economic and technical cooperation, is an excellent vehicle for international cooperation on the issues that are exposed in this book.

Trade liberalisation and reform are seen as key ingredients to the spread of ICTs. Many economies in the region are strongly competitive in the production of ICT products. But, as the Taiwan study and the Australian case suggest, the production of ICT products is perhaps less important than their use in yielding the macro-economic and micro-economic or efficiency gains from the new economy. A primary APEC role is to build capacity through sharing policy experience, and this is likely to be a powerful force in spreading the benefits of ICTs. New economy issues have become a primary focus of APEC's work program since 2000.

Pangestu and Park note APEC's success in carrying the International Technology Agreement (ITA) through the WTO and initiating the moratorium on barriers to trade in e-commerce.

Many of the main arguments in this book are illustrated in the experience of Australia. There the uptake of ITCs can be seen as a dynamic outcome of comprehensive policy reform which made the economy both more open to trade and more competitive domestically (Parham 2002). The rapid uptake and productive use of ITCs was not the primary strategy. But the reforms gave Australian business strong incentive to exploit the opportunities that ITCs provided for corporate reorganisation and productivity improvement. The exceptional lift in productivity, on which Cooper remarks, was more a consequence of the increased openness, competition and flexibility in the economy and its encouragement of the smart use of ICTs than it was of the production of ICTs themselves. Access to ICT product imports at sharply declining international prices was a bonus terms of trade gain. Especially for relatively small economies, openness is the essential ingredient.

More broadly in East Asia and the Pacific, openness to trade and foreign investment and intense competition in the marketplace are assets in capturing the opportunities of the new economy, just as they were assets in catch-up industrialisation built on integration into the international economy. So is the strong education base. The new ICTs have rapidly accelerated the pace of catch-up. But the challenge for economies at the leading edge of this process, like Japan and Korea, is to effect the fundamental reform of policy and institutions (including education systems and entrepreneurial culture) that is necessary to maintain momentum as these economies reach the frontier of industrial technologies. In part the poor growth and productivity performance of Japan (and Europe) may be a statistical artefact, the product of differences in the way IT expenditure is measured in national accounts (Jorgenson 2003). But the evidence, nonetheless, is that the productivity growth of Japan's and Europe's IT-using industries still lags behind that in America and Australia importantly because the policy and regulatory settings are not sufficiently supportive of change.

For this reason, the patchy productivity performance of some leading economies in East Asian economies over the last decade is an indication that the next phase of development in East Asia is likely to come harder and take a little longer than the last.

## REFERENCES

Jorgenson, Gale (2003) http://post.economics.harvard.edu/faculty/jorgenson/papers/papers.html

Mann, Catherine and Daniel Rosen (2001) 'Background, Overview and Issues Paper: The New Economy and APEC,' *SOM/008, APEC Senior Officials Meeting*, Shanghai.

Parham, Dean (2002) 'Australia's Productivity Surge and Its Determinants', Thirteenth Annual East Asian Seminar on Economics, NBER and Productivity Commission of Australia, Melbourne.

Stephenson, Sherry and Christopher Findlay with Soonhwa Yi (eds) (2002), *Services Trade Liberalisation and Facilitation*, Asia Pacific Press, Canberra.

# 2 What's new in the new economy?

*Richard N. Cooper*

The 'new economy' has become a buzzword to characterise the American economy, with positive connotations but imprecise meaning. Sometimes it is used to refer only to selected high technology sectors, specifically computers, semiconductors, software and telecommunications. But usually the term implies significant changes in the US economy as a whole. At its most dramatic, the term has been used to suggest that the traditional business cycle has been banished, inflation and unemployment have been brought forever under control, US long-term growth rates have increased significantly, and the high-value stock market has not been overvalued and indeed will continue to rise. More modestly, it suggests that the structure of the US economy has changed fundamentally, with the implication, inter alia, that monetary and fiscal measures affect the economy differently from the way they did in the past. Finally, the term has been used to suggest that US productivity growth has returned to, or at least toward, the high levels it enjoyed before the slowdown of the mid-1970s.

This paper will discuss the factual bases for conjecturing that the United States might indeed have a new economy, review the controversies and evidence surrounding that claim, and suggest how the emergence of a new economy, if indeed there is one, might affect economies elsewhere in the world, including in the Asia Pacific region.

## EVIDENCE FOR A NEW ECONOMY

Four factors in particular suggested that the US economy might have experienced fundamental changes during the late 1990s. The first was the long period of uninterrupted growth following the recession of 1990–91. Gross domestic product (GDP) passed $10 trillion during 2000 and in that year showed the longest period of growth since adequate data have been available, surpassing the previous long recovery of 1961–70. In the 'business cycle', a period of economic downturn historically occurred every three to four years, so this long period of growth suggested that perhaps at least the traditional business cycle had been banished. Various reasons, most emphasising better management of inventories by firms, were advanced to explain this.

The second development was the decline in US unemployment from 7.5 per cent in 1992 to 5.6 per cent in 1995 – a normal economic recovery – followed by a continued decline to 4.0 per cent in 2000, a rate that had not been seen since the Vietnam boom years of the late 1960s, when it was associated with a significant acceleration of inflation, to a 6.2 per cent increase in the consumer price index (CPI) in 1969. In the late 1990s, in contrast, inflation remained relatively low and under control, rising to 3.4 per cent in 2000 as measured by the CPI – 2.4 per cent if food and energy are excluded – and 2.3 per cent for the implicit deflator on consumption in the national income accounts. This suggested that an important alleged macroeconomic characteristic of the economy, the so-called non-accelerating inflation rate of unemployment (NAIRU) might have declined significantly.

A third development was an unexpected increase in productivity growth in the mid-1990s. That such an increase occurred is not in serious dispute; but controversy surrounds the magnitude, scope, interpretation and durability of the increase. In particular, to what extent, if at all, did it signal a rise in the long-term potential growth of the US economy?

A fourth development was the rise in equity valuations significantly above historical norms – as related, for instance, to book values or to corporate earnings – and their persistence despite protests by many stock analysts, economists, and even Federal Reserve chairman Alan Greenspan, who complained as early as December 1996 of the 'irrational exuberance' of the stock market. Again, controversies surrounded interpretation of the high stock valuations, and their durability. Table 2.1 shows various economic indicators for the period 1989–2000.

Popularisation of the new economy reached its peak in 1999 and early 2000. Excessive enthusiasm was dampened by a cooling of the stock market in the third quarter of 1999. This was followed by some recovery into early 2000, and then, in April, a dramatic 20 per cent drop in the index of Nasdaq stocks – a drop dominated by high technology, especially information technology (IT), stocks. Growth of the economy slowed dramatically in the second half of 2000 and into 2001. Nasdaq stocks continued to tumble, reaching less than 40 per cent of their peak value by April 2001. The more representative Standard and Poor (S&P) 500 index fell to 80 per cent of its April 2000 peak in April 2001. These declines scotched any notion that stock prices would climb forever, although even in mid-2001 their values exceeded most historical norms. Moreover, by mid-2001 some observers expected economic recession (defined as two successive quarters when GDP declines in real terms), although a downturn had been avoided with growth of 1.0 per cent in the first half of the year, and until the attacks on New York in September most forecasters foresaw a recovery in the second half and into 2002. (Industrial production, however, had declined for nine successive months from the third quarter of 2000.)

Most analysts saw the decline in growth and the decline in stock prices as desirable corrections. Indeed, the Federal Reserve had been trying to engineer

Table 2.1  Growth indicators for the US economy, 1989–2000 (per cent) (1941–43=10)

|      | Real GDP | GDP deflator | Consumer price index[a] | Hourly compensation business sector | Output per hour business sector | Unemployment rate | Federal funds rate | S&P 500 |
|------|---|---|---|---|---|---|---|---|
| 1989 | 3.5  | 3.8 | 4.6 | 2.8 | 1.0 | 5.3 | 9.21 | 323   |
| 1990 | 1.8  | 3.9 | 6.1 | 5.7 | 1.3 | 5.6 | 8.1  | 335   |
| 1991 | −0.5 | 3.6 | 3.1 | 4.9 | 1.2 | 6.8 | 5.69 | 376   |
| 1992 | 3    | 2.4 | 2.9 | 5.3 | 3.7 | 7.5 | 3.52 | 416   |
| 1993 | 2.7  | 2.4 | 2.7 | 2.2 | 0.5 | 6.9 | 3.02 | 451   |
| 1994 | 4    | 2.1 | 2.7 | 2.1 | 1.3 | 6.1 | 4.21 | 460   |
| 1995 | 2.7  | 2.2 | 2.5 | 2.1 | 0.9 | 5.6 | 5.83 | 542   |
| 1996 | 3.6  | 1.9 | 3.3 | 3.2 | 2.5 | 5.4 | 5.3  | 671   |
| 1997 | 4.4  | 1.9 | 1.7 | 3   | 2   | 4.9 | 5.46 | 873   |
| 1998 | 4.3  | 1.2 | 1.6 | 5.5 | 2.7 | 4.5 | 5.35 | 1,086 |
| 1999 | 4.1  | 1.4 | 2.7 | 4.6 | 2.5 | 4.2 | 4.97 | 1,327 |
| 2000 | 4.1  | 2.3 | 3.4 | 6.0 | 3.1 | 4.0 | 6.24 | 1,427 |

Source: Joint Economic Committee, US Congress, *Economic Indicators*, April 2001 and earlier.

Note
a   December to December.

a decline from the vigorous growth of the preceding several years, and especially from mid-1999, since 5 per cent or more was considered well above the growth potential of the US economy, unemployment continued to fall, and inflation was an ever-present danger, although not yet evident. The Fed-targeted federal funds rate (FFR), the rate for bank reserves in the interbank market, was intermittently raised from its recession low of 3 per cent in 1993 to 6.5 per cent in late 2000, with a brief dip in autumn 1998 to help avert a financial crisis. In his July 2000 testimony to Congress, Federal Reserve chairman Alan Greenspan stated:

> For some time now, the growth of aggregate demand has exceeded the expansion of production potential ... It has been clear to us that, with labour markets already quite tight, a continuing disparity between the growth of demand and potential supply would produce disruptive imbalances.

Thus the slowdown per se was not an indication that the US economy was performing badly, contrary to what was assumed by financial journalists and others around the world. Nonetheless, it gainsaid the most exuberant claims for a new economy.

## CHANGES IN MACROECONOMIC STRUCTURE?

That the US economy should have grown so rapidly, and unemployment fallen so far, without triggering a significant rise in inflation was a surprise to many (most?) analysts. On relationships that had obtained since the late 1960s, a sustained drop in unemployment below 5 per cent should have led to a significant rise in wages and, through wages, to an acceleration of price inflation. This in turn would have led an inflation-fighting Federal Reserve to tighten credit conditions enough to thwart the inflation, thus curtailing growth.

As noted above, the Federal Reserve did tighten credit significantly during the late 1990s, with a brief reversal in autumn 1998 to deal with the credit market panic associated with the Russian default on government debt and the near failure of Long-Term Credit Management – at least as measured by its operational instrumental variable, the FFR. Moreover, the growth of M1 (currency, travellers cheques and checkable deposits – a measure of monetary policy stance preferred by some economists) was actually negative in 1995, 1996 and 1997, leading some monetarists to forecast that the Federal Reserve was leading the US economy into recession. However, broader definitions of the money supply, M2 and M3 – which augment M1 by including savings and small time deposits and retail money market funds (M2), plus large time deposits, institutional money market funds, repurchases, and eurodollars (M3) – both grew robustly, M2 by over 5 per cent annually and M3 by over 7 per cent annually in 1997–99, leading some critical observers to argue that the Federal Reserve was unwittingly feeding the stock market boom.

In fact, as Chairman Greenspan emphasised in his semi-annual statements to Congress, the Fed was watching the actual and prospective rate of inflation. Monetary policy was nudged tighter (as measured by the FFR) in order to discourage the acceleration of inflation, after adjusting for special pressures (up or down) arising from food and oil prices. Rapid growth and declining unemployment resulted from this process, rather than being taken in themselves as decisive signals that monetary policy needed to be tightened further.

Fiscal policy was not actively used following the 1993 tax increases that were designed to reduce the large budget deficit that continued from the early 1980s. It has been argued that fiscal tightening, combined with continued restraint on spending increases, permitted the Federal Reserve to be more relaxed about monetary policy than it would otherwise have been, and also facilitated a decline in long-term interest rates, from over 8 per cent on the ten-year government bond in 1990 to under 6 per cent a decade later, despite rapid economic growth. Indeed, with rapid growth in the economy, the federal budget moved into surplus in 1998, for the first time since 1969. Growing budget surpluses were significantly blunted by a slowdown of the economy as well as by a tax reduction in spring 2001, providing some fiscal stimulus in the second half of that year.

Did the drop in unemployment to 4 per cent without noticeable inflation signify a major structural change in the US economy? In particular, did it signify a decline in NAIRU or a durable shift in the Phillips curve relating inflation to unemployment?

The US Congressional Budget Office (CBO) has calculated a NAIRU for the US economy over the past four decades. It shows very little change, rising slowly from 5.5 per cent in 1962 to just over 6 per cent in the late 1970s, and slowly falling to 5.5 per cent in the late 1990s (reported by Brainard and Perry 2001, p.62). These movements reflect mainly demographic changes, with many more young people entering the labour force in the late 1970s than before or since. The estimations by Robert Gordon (1998, p.321) show a somewhat greater decline in NAIRU, from 6.4 per cent in 1988 to 5.7 in 1998 for the GDP deflator.

How can these calculations be reconciled with experience in the late 1990s? After late 1995, unemployment was under 5.5 per cent – mostly substantially under. With a NAIRU of 5.5 per cent, this should have generated accelerating inflation. The CPI rose from a 2.5 per cent increase in 1995 to 3.6 per cent in 1996, but then registered increases of only 1.7, 1.6 and 2.7 per cent in the following three years. It rose by 3.4 per cent in 2000, suggesting some acceleration after 1998. But if food and energy are excluded, on the grounds that their prices are exceptionally volatile and determined largely in world markets, inflation was a full percentage point lower in the last years of the decade, with unemployment around 4 per cent, than it was in 1991–93, when unemployment exceeded 6 per cent.

Wage settlements conform more closely with the expectation based on NAIRU, allowing for lags: compensation per hour in the business sector grew by around 2 per cent in 1994–95, 3 per cent in 1996–97 and 5 per cent in 1998–2000, a clear acceleration. However, much of this increase came from the movement of labour into higher-paying jobs; when we control for shifts among occupations and industries, wage and salary increases rose only one percentage point between 1994–95 and 1998–2000, from 2.8 per cent annually to 3.8 per cent (another 0.8 percentage point came from increases in fringe benefit costs). Rich and Rissmiller (2001) could find no significant shift in their equations for estimating aggregate wages over the period 1967–2000. In other words, the structure of the labour market with respect to wage determination does not seem to have changed appreciably – at least on one specification for wage determination, which includes growth in labour productivity. Rich and Rissmiller concluded that any shift in the inflation–unemployment tradeoff occurred outside the labour market.

In discussing the reasons for the lower than expected inflation, Robert Gordon (1998) has emphasised the importance of positive supply shocks, of which he identifies five: improved food and energy prices, a fall in import prices (due in part to appreciation of the dollar), an acceleration in the decline of computer and related prices, a slowdown in the rise of health care costs, and improvements in the measurement of price increases. Allowing for these five factors, however, goes only a little more than halfway toward explaining the 'shortfall' in inflation in the late 1990s.

One possible reconciliation of the apparent breakdown of inflationary expectations based on conventional NAIRU under low unemployment is that productivity growth jumped unexpectedly in the late 1990s, allowing somewhat higher wage settlements with no increase in unit labour costs, and thus no inflationary pressures on product prices. Indeed, as Table 2.1 suggests, productivity growth was notably higher during 1996–2000 than it had been earlier. According to this thesis, advanced by CEA (2001, pp.73–74), the increase in productivity growth was unexpected, and hence not taken into account in wage bargaining, either by labour or by employers. If the higher productivity increases continue, they will cease to be unexpected, and wage settlements can be expected gradually to incorporate the higher productivity growth. Thus NAIRU was only temporarily reduced by the unexpected growth in productivity; it can be expected to return to its normal, higher level as the new data are incorporated into wage bargaining.

An alternative explanation is that NAIRU does not exist, or rather is not stable over time, and hence does not provide a useful parameter either for policy-making or for understanding the performance of the American economy. Brainard and Perry (2001) have examined the determination of US wages and prices over the period 1960–98, using a variety of statistical techniques (recursive regression, and contemporary and backward Kalman filters).[1] They concluded that NAIRU is not only not useful but, on the slowly changing CBO version, would have provided extremely poor policy guidance during most of the period they examined, when policy-makers are assumed to be interested in both keeping inflation low and maintaining high employment and output.

Brainard and Perry found (p.54) that the unemployment rate consistent with maintaining low inflation rose substantially from 1965 to a peak in 1980, and then receded by 1998 to levels slightly above those in the late 1960s. This is of course a descriptive statistic, and can be interpreted – as John Taylor, in his comments to Brainard and Perry (2001), points out – as shifts in the Phillips curve over time, first to the right, then to the left. A key question is to what extent those shifts are endogenous to inflation itself, and to perceptions of policy toward inflation.

A third possibility, suggested by James Stock in his comments on Gordon (1998), is that the Phillips curve collapsed altogether in the 1990s – that is, became horizontal. His tests suggest a sharp break in the unemployment–inflation relationship in early 1993, consistent with a flat curve thereafter. In other words, unemployment has ceased to be a good measure of tightness of aggregate demand in the economy, due possibly to changes in the labour market or to other changes in the structure of the economy.[2] After all, the Phillips curve is an empirical relationship between two endogenous variables;

it is not well grounded in theory, and such empirical relationships can be expected to change over time, or even disappear altogether.

Is one of the attributes of the new economy that monetary policy works differently? Boivin and Giannoni (2001) reported that output seems to have become less sensitive to FFR, but found no evidence that firms and households had become less sensitive to changes in interest rates (their analysis, however, extended only through 1995). The Fed has responded more quickly to changing economic conditions in recent years, which arguably has reduced the variability of output and inflation, as observed by John Taylor in his comments on Brainard and Perry (2001). But the shocks were also notably lower during the 1990s than they were in earlier decades (Mankiw 2002 forthcoming).

Monetary policy (in the form of changes in the FFR) is usually assumed to influence the US economy through three channels. First, lower FFR reduces borrowing rates, thereby stimulating investment and consumption. Second, it depreciates the dollar, thereby stimulating net exports. Third, it raises asset values, thereby stimulating consumption and investment (via Tobin's 'q', the market value of corporations divided by the replacement cost of their assets). Bruce Kasman of Chase Bank has analysed the Federal Reserve macro model, according to which a one percentage point reduction in FFR will raise GDP by 0.6 per cent after one year and 1.7 per cent after two years.[3] The January–May 2001 cut by 2.5 percentage points in FFR should have lifted share prices by 22 per cent within a year, depreciated the trade-weighted dollar by 5 per cent, and reduced long-term bond rates by 0.75 percentage points. By late July, however, despite an additional 0.25 point FFR reduction in late June, the S&P 500 had fallen by 10 per cent, the trade-weighted dollar had appreciated by 7 per cent, and bond and mortgage yields had changed little since the beginning of the year.

With respect to the exchange rate, perhaps foreigners now expect stock prices to rise following a decline in the FFR, and therefore buy rather than sell dollars, despite lower short-term yields. That would represent an important change in behaviour, particularly if it were symmetric, and would weaken the impact of a given change in the FFR on the economy.

## A WIDE AND DURABLE RISE IN PRODUCTIVITY GROWTH?

US growth slumped to below 1 per cent in the first half of 2001, but it had accelerated in the late 1990s, to over 4 per cent a year in 1995–2000. These figures were significantly above the previously assumed potential growth of the US economy of 2–2.5 per cent a year. Of course, unemployment fell by 1.4 percentage points during this period, permitting growth higher than potential. Somewhat more than half the increase in GDP from the early 1990s can be explained by increased inputs of labour and conventional capital. But total factor productivity growth jumped considerably, to three times its average over the period 1973–95 (see Table 2.2).

*Table 2.2* Sources of US growth (per cent per annum)

|  | 1948–73 | 1973–90 | 1990–95 | 1995–99 |
|---|---|---|---|---|
| Gross domestic product | 3.99 | 2.86 | 2.36 | 4.08 |
| Hours worked | 1.16 | 1.59 | 1.17 | 1.98 |
| Average labour productivity | 2.82 | 1.26 | 1.19 | 2.11 |
| Contribution of capital deepening | 1.45 | 0.79 | 0.64 | 1.24 |
| Information technology | 0.15 | 0.35 | 0.43 | 0.89 |
| Non-information technology | 1.3 | 0.44 | 0.21 | 0.35 |
| Contribution of labour quality | 0.46 | 0.22 | 0.32 | 0.12 |
| Total factor productivity | 0.92 | 0.25 | 0.24 | 0.75 |
| Information technology | 0.06 | 0.19 | 0.25 | 0.5 |
| Non-information technology | 0.86 | 0.06 | -0.01 | 0.25 |
| *Addendum* | | | | |
| Labour input | 1.95 | 1.97 | 1.7 | 2.18 |
| Labour quality | 0.79 | 0.38 | 0.53 | 0.2 |
| Capital input | 4.64 | 3.57 | 2.75 | 4.96 |
| Capital stock | 4.21 | 2.74 | 1.82 | 2.73 |
| Capital quality | 0.43 | 0.83 | 0.93 | 2.23 |

*Source*: Jorgenson (2001).

These developments raise several questions. First, are the measurements accurate? Second, if the measurements are broadly correct, how can the acceleration in growth be interpreted, and in particular how durable is it likely to be? Third, particularly if it is judged to be durable, what is the explanation for the acceleration? In this section we take up each of these questions in turn.

### Measurement

It is well known that measuring productivity growth in a modern complex economy is a difficult assignment; with the gradual switch of the labour force from production of goods to production of services, the task has become increasingly difficult, because it is hard to measure real output in many service sectors (for example, education and health).[4] Government statisticians do not even try to measure real output in the education sector; rather, output is measured by inputs, and productivity growth is *assumed* to be zero. The situation is nearly as bad in several other sectors, including customised software.

Sectors where output is measured primarily or exclusively by input account for 23 per cent of US GDP (Landefeld and Fraumeni 2001, p.29) This measurement problem is directly related to the measurement of price increases in a modern economy: to calculate real output, increases in the value of nominal output are deflated by measured price changes. In 1996, the Boskin Advisory Committee to Study the Consumer Price Index reckoned that the US CPI exaggerated average price increases by over one percentage point a year – implying, if correct, that US growth was significantly understated. Since then, the Bureau of Labor Statistics (which is responsible for calculating the CPI) has made a series of adjustments that have lowered the US inflation rate by about 0.45 percentage points a year. Tables 2.1 and 2.2 show the new, revised figures, so these measurement issues do not resolve the question; in the view of Jorgenson (2001), however, true (constant quality) price reductions are understated both for some software and for some communications equipment.

Nordhaus (2001) has produced figures on what he considers 'well-measured' GDP, that is, GDP less those sectors where measurement of real output is especially problematic. His figures cover the goods producing sectors, transportation and utilities, and wholesale and retail trade, in 1999 together accounting for 43 per cent of GDP; they exclude construction, finance, insurance, real estate, other services and government. Growth in labour productivity in well-measured GDP was both higher and accelerated even more rapidly than that in total GDP, from 2.24 per cent annual growth in 1990–95 to 4.65 per cent in 1996–98 (Nordhaus, Table 2.7).

Nordhaus identified another, more subtle, measurement problem. The measurement of sectoral growth requires consistent deduction of inputs into each sector. This is not typically done in the official US figures, which measure sectoral outputs from production (or sales) and sectoral inputs from income earned. The error would be negligible if both output and income were accurately measured, since in a consistent set of accounts total income must equal total output, after appropriate allowance for the necessary adjustments such as business taxes. Errors of measurement lead to the situation where total income does not equal total output. In recent years, measures of income have exceeded corresponding measures of output.[5] When business sector labour productivity growth is calculated consistently using income data, the figures indicate an even greater acceleration in labour productivity than do the official figures, from 1.26 per cent in 1990–95 to 3.16 per cent in 1996–98 (Nordhaus, Table 2.6).

Thus the official measures that provide the factual basis for most quantitative discussions of the new economy if anything understate the acceleration of US growth in recent years.[6]

## Scope and durability

If the change is real, how widespread is it and how durable is it? These questions arise especially because of the claims by Northwestern University economist Robert Gordon (1998, 2000) that productivity growth has shown extraordinary concentration in just a few sectors, notably semiconductors, computers, and computer-associated equipment. US prices of these items are measured on a 'hedonic' basis – that is, on the basis of characteristics useful to users, such as computational speed and memory capacity – and have shown extraordinary declines over the past two decades. The price declines accelerated in the mid-1990s: computers went from declines of 16 per cent a year in 1990–95 to 32 per cent a year in 1995–99 (Jorgenson 2001, p.10). Since total demand has continued to grow, 'real' output (defined as the index of total demand deflated by the index of price changes) has shown extraordinary growth. While these sectors comprise only a small portion of expenditure in the US economy, the growth has been so great as to affect total growth.

Productivity growth showed some acceleration in other sectors as well, but Gordon deemed such acceleration to be barely more than what could be accounted for by the impact of a boom in demand, with output rising more than employment when demand is high.

Several analysts have explored Gordon's claim that exceptional productivity growth was concentrated in relatively few sectors. Using just his well-measured output, Nordhaus (2001, Table 2.11) found a near doubling of labour productivity between 1990–95 and 1996–98, from 1.60 per cent to 3.09 per cent, even when the IT sectors (computers, software, and telecommunications) were excluded. Within manufacturing, the acceleration was heavily concentrated in machinery, both electrical and non-electrical, but wholesale and retail trade also experienced large increases (Nordhaus 2001, pp.43–47).

Kevin Stiroh (2001) found that after 1994 productivity growth by sector was highly correlated with earlier IT investment in the 61 sectors he examined. High productivity growth was experienced not only by the IT-producing sectors but also by the IT-using sectors. Indeed, IT-using sectors – which make up nearly two-thirds of the economy – accounted for the bulk of the growth in average labour productivity between 1987–95 and 1995–99, whereas non-IT-intensive sectors actually showed some slowdown in productivity growth in 1995–99 (Stiroh 2001, Table 2.8). On this evidence, the productivity acceleration is intimately linked to IT, partly in IT production, but mainly because of earlier investment in IT by other sectors. Oliner and Sichel (2000) also found productivity increases in many IT-using sectors.

Of course, durability can be tested only after the passage of time. Empirical work cited here ended in 1998 or 1999. Overall productivity growth in 2000 was even greater than that in earlier years; but this slowed markedly (to 2 per cent) in the second half of 2000, and still further (to 1.4 per cent) in the first half of 2001. That is normal following a decline in demand, since production falls before labour is shed, resulting in a slowdown in recorded labour productivity growth, or even a decline. If demand resumes its earlier growth, this slowdown will be transitory; if weak demand continues, firms in the United States will gradually shed labour and productivity will be restored after several quarters. Thus a test of the durability of the increase in productivity growth will come only when growth in aggregate demand recovers, presumably later in 2003.

We now know that some of the extensive investment in IT in the late 1990s, especially in numerous so-called dot.coms, was quite foolish, made possible by a ready availability of venture capital. Not all IT investment has a high payoff; indeed, in mid-2001 there was something of a glut of 'used' IT equipment on the market, as failed firms liquidated their recently acquired assets.

However, a number of analysts have raised their estimate of the long-term growth rate for potential output of the US economy. The Council of Economic Advisers (CEA 2001, p.78) has suggested that average labour productivity in the non-farm business sector will increase by 2.3 per cent a year over the period 2000–08, up from 1.4 per cent in the period 1973–90 and 2.2 per cent over the decade of the 1990s. When augmented by the anticipated growth in the labour force, potential GDP is expected to grow by 3.4 per cent a year over the next decade (actual GDP may grow by somewhat less because of a rise in unemployment to 5 per cent). Data Resources Inc. (DRI), a highly reputable forecasting firm, has also raised its estimate of potential growth to 3.4 per cent a year – roughly a percentage point higher than it was considered to be in the mid-1990s.

## Explanations

Firm explanations for the rise in productivity must await information on how durable the rise proves to be. But it is possible to speculate on why such a rise might have occurred when it did.

It is widely recognised that cheap, widely available computational power, particularly when combined with inexpensive communication, represents a new general-purpose technology – analogous to the introduction of steam power, steel, electricity, plastics and other human-made chemicals. Such new technologies change radically the way economic activity is carried out. But it takes time, perhaps a generation or two, for such new technologies to be fully absorbed by economic and social structures. Partly this is because the introduction of a new technology involves much new investment, as yet not fully tested. Partly it is because mature human thinking, and especially human organisations, have a high resistance to radically new ways of doing things. Partly it is because new technologies involve taking financial and career risks that are not fully understood, inducing a cautious approach to their introduction.

Significant improvements in computation and communication date from the early 1970s, followed by a constant stream of innovative products, including the personal computer (PC), which today has computational power in excess of mainframe computers twenty-five years ago, at a tiny fraction of the cost, and the internet. Robert Solow famously lamented in 1987 that 'we see computers everywhere but in the productivity statistics'. Computers were already in widespread (but not universal) use, but their potentialities were not fully used. Realising the full potential of a new technology involves not merely introducing the new technology physically, but reorganising the work flow and even the output of the enterprise, to take full advantage of it.

Paul David (1990) compared the introduction of IT to the introduction of electricity in the late nineteenth century. Initially electricity was viewed simply as a substitute source of power, replacing steam engines or water wheels, and of illumination, safer than the gas that was widely used in American and European cities. It was not until the 1920s that electricity was well integrated into the workplace, making it much cleaner and quieter as the motive power was distributed through wires rather than through the belts, pulleys and gears that had attended single-source steam or water power.

Perhaps an analogous process is taking place with IT. Investment in equipment is a necessary but by no means sufficient condition for full integration of new technology. At first, PCs were often used simply as substitutes for typewriters; secretarial efficiency decreased while the secretary mastered the word-processing program, but improved thereafter (for example, because making text corrections was easier). Full integration requires supplemental improvement through associated innovation (for example, better software, modems and printers), new investment, training and re-training, reorganisation of the workplace, and even reorganisation of the firm.

Risk-averse management is unlikely to take all the steps required unless compelled to do so through competitive pressure. In time the new technology will be fully integrated, largely because older management is gradually replaced by people who have grown up with the new technology and are both aware of its potential and accustomed to it. Hence there is a generation-long process of integration.

Full integration can be accelerated by high competitive pressure, leading firms to constantly search for ways to reduce costs, improve products, and otherwise appeal to customers. Thus the competitive environment is important, as is the availability of capital, particularly risk capital. In these respects the United States perhaps has an edge on other countries.

Deregulation has been widespread during the past decade or two, but has particularly occurred in the United States: compared with other countries, there is less concern to protect profits and employment in the industries being deregulated, so firms are compelled to adapt or go out of business. For example, America's international flagship air carrier for nearly five decades, Pan American Airways, no longer exists; it went bankrupt, as did several railroads. Few countries have been willing to see that occur. America's once near-monopoly telephone company, AT&T, now operates in a highly competitive environment, where even its survival can be questioned. Neither of these firms was publicly owned, but both were subject to heavy regulation.

An important source of competitive pressure has been openness to international trade. In the late 1990s, over 60 per cent of total US domestic computer purchases was imported, and over half of US computer production was exported (CEA 2001, p.46). Thus American firms must compete with the best and least expensive products elsewhere in the world. There are no import restrictions on most computers and related products.[7] However, extremely advanced computers are subject to export control, stimulating development of such products by non-US firms and governments.

In a period of new, general-purpose technology, the United States has been fortunate to have an abundance of venture capital – in times of enthusiasm, perhaps even too much. This is partly due to the willingness by many Americans to risk funds for the sake of substantial gain (and the confidence, subject to income taxation,[8] that they will receive the gain if it occurs). It is also partly due to an institutional framework, largely in the form of investment banks, for evaluating new ventures and investing risk capital in the most promising of them. American firms in the IT sector increased from under 70,000 in 1990 to 150,000 in 1997 (CEA 2001, p.36); by 2000, many had still not made a profit. The typical pattern is for one or several people with a bright idea to start a firm by drawing on their own time and savings and those of friends willing to invest. When the idea has been developed sufficiently, and inserted into a plausible business plan, they approach a source of venture capital. This source may provide funds and managerial advice for the several years of effort needed to develop the idea further, convert it into a marketable product, and generate enough buying customers to run a profit. Once the product develops a good enough reputation (which may occur even before the new firm is profitable), the firm is 'taken public' through an initial public offering (IPO), whereby the shares are sold to the general public, under strong Securities and Exchange Commission rules regarding disclosure, accounting standards, etc. The venture capitalists typically get their return by selling their equity in the firm to the buying public, thus replenishing their venture capital. During late 1999 and early 2000 in particular, there were an extraordinary number of IPOs, raising over $100 billion in the four quarters from July 1999, equivalent to nearly 10 per cent of total non-residential fixed investment in the United States. An impressive $40 billion was raised through IPOs in Japan, Germany and the United Kingdom during the same period (BIS 2001, p.107).

## IMPLICATIONS FOR THE REST OF THE WORLD

To appraise the implications of the new economy for the rest of the world it is first necessary to discover whether the rest of the world has experienced the improvement in economic performance that has blessed the United States. The answer, with the exception of Australia, seems to be negative. Table 2.3 (from BIS 2001, p.21) shows growth in productivity in the business sector of 14 rich countries for three periods. Among them, only Australia experienced an increase in productivity growth in the second half of the 1990s; many

*Table 2.3* Changes in labour productivity in the business sector (annual rates in percentages and percentage points)

|  | Output per hour 1996–99 | of which: Capital[a] | MFP[b] | Output per hour 1990–95 | of which: Capital[a] | MFP[b] | Output per hour 1981–89 | of which: Capital[a] | MFP[b] |
|---|---|---|---|---|---|---|---|---|---|
| Australia | 3.1 | 1.0 | 2.1 | 1.8 | 0.6 | 1.2 | 1.5 | 0.5 | 1.0 |
| United States | 2.3 | 0.5 | 1.8 | 1.0 | 0.2 | 0.8 | 1.3 | 0.2 | 1.1 |
| Germany | 2.1 | 1.0 | 1.1 | 2.2 | 1.2 | 1.0 | – | – | – |
| Japan | 2.1 | 1.2 | 0.9 | 2.9 | 1.6 | 1.3 | 3.1 | 1.1 | 2.0 |
| Switzerland | 1.9 | 1.0 | 0.9 | 0.7 | 1.2 | –0.5 | – | – | – |
| Sweden | 1.7 | 0.6 | 1.1 | 2.1 | 0.9 | 1.2 | 1.5 | 0.6 | 0.9 |
| France | 1.6 | 0.5 | 1.1 | 2.3 | 1.4 | 0.9 | 3.4 | 1.1 | 2.3 |
| United Kingdom | 1.5 | 0.5 | 1.0 | 1.8 | 0.6 | 1.2 | 3.4 | 0.5 | 2.9 |
| Norway | 1.4 | 0.3 | 1.1 | 3.2 | 0.7 | 2.5 | 1.4 | 0.9 | 0.5 |
| Canada | 0.9 | 0.6 | 0.3 | 1.4 | 1.1 | 0.3 | 1.4 | 1.3 | 0.1 |
| Denmark | 0.9 | 0.6 | 0.3 | 3.7 | 1.3 | 2.4 | 2.5 | – | – |
| Italy | 0.7 | 0.8 | –0.1 | 2.7 | 1.4 | 1.3 | 2.3 | 0.9 | 1.4 |
| Netherlands | 0.4 | –0.2 | 0.6 | 2.9 | 0.9 | 2.0 | 3.4 | – | – |
| Spain | 0.4 | 0.3 | 0.1 | 2.6 | 2.0 | 0.5 | 3.9 | – | – |

*Source*: US *Federal Reserve Bulletin*, October 2000 (based on OECD data).

*Notes*
a  Capital deepening.
b  Multifactor productivity.

other countries experienced a marked reduction, both from the early 1990s and especially from the 1980s.

Two other points in Table 2.3 should be noted. The first is that in the late 1990s recorded productivity growth in several countries, including Germany and Japan, was nearly as high as it was in the United States. The second is that many other countries showed much greater productivity growth in the 1980s and even in the early 1990s than did the United States. These observations should, at a minimum, warn against making sweeping generalisations based on data for just a few years. Furthermore, the details of price measurement differ from country to country; in particular many countries have not adopted the hedonic measures used in the United States. Data on price increases, and hence on productivity increases, are therefore not strictly comparable across countries. In particular, productivity growth in several countries would be somewhat higher than that recorded in Table 2.3 if they had used hedonic price indices, insofar as they produce in abundance the products – especially computers and semiconductors – whose hedonic prices have fallen so rapidly.[9]

During the late 1990s, growth in emerging markets was blunted by various financial crises – in Mexico, Korea, Southeast Asia, Russia and Brazil – that also affected their competitors and trading partners. Poland and Vietnam experienced sharp increases in labour productivity between the 1980s and the 1990s, but this was due mainly to a switch from central planning to market pricing. Ireland also experienced a renaissance in economic growth, in which 'new economy' factors undoubtedly played some part but not the major role. Among larger countries, the United States stands out, with Australia, for its acceleration of growth.

Suppose the higher growth in the United States is not a fluke, and can be expected to endure for a decade or longer. What then are the implications for the rest of the world?

## Higher US growth

First, of course, the incomes of Americans will grow more rapidly – on CEA projections, a full percentage point more rapidly. Americans have shown a marked willingness to increase consumption as their incomes rise. Thus US demand for goods and services, including imports, will continue to rise rapidly. In this respect the United States will continue to be a locomotive for the world economy. Of course, output will be rising by the same amount, so Americans will produce more goods and services, some of which will be desired and competitive in the rest of the world, possibly displacing some more traditional products and creating new demands.

Second, the structure of employment and output in the United States will change even more rapidly than it has. Output and employment have grown rapidly in the IT sectors, along with research and development (R&D) and patent awards. Yet the number of people engaged in the production of all goods (agriculture, mining, manufacturing) has already declined to only 17 per cent of the labour force in the United States (another 5 per cent is engaged in construction), leaving 78 per cent of the labour force engaged in the production of all kinds of 'services'. The relative decline in manufacturing employment has occurred with no decline in manufacturing output; indeed, manufacturing production rose by 50 per cent from 1989 to 1999. These trends will continue, with machinery continuing to be substituted for labour in goods production. It will not be long before all goods production occurs with less than 15 per cent of the labour force.

At the same time, the returns to new capital do not seem to be declining. Technical change is buoying these returns, even as the capital:labour ratio rises. To put it another way, the 'quality' of new capital is continually improving. Thus the capital:output ratio has been declining in the United States: less new investment is required for a given increase in output. As shown in Table 2.2, the US capital stock has grown increasingly more slowly than GDP since 1973. This rise in productivity of capital explains in part how the United States can continue to grow with low savings rates. Gross investment, embodying new, high-yield technology, is much more important for growth than the net additions to the capital stock that have been emphasised by many economists.

Good returns to capital in the United States will continue to draw investment from around the world, where returns are generally lower or (in emerging markets) higher but less reliable. Thus the dollar is likely to remain strong and the large current account deficit is likely to continue for some time – though not necessarily to remain as strong and as large as in 2001 (see Cooper 2001).

The new economy also seems to place a premium on education beyond secondary school levels. The gap between compensation to college-educated employees and high-school-educated employees has been growing (although some narrowing occurred in the late 1990s as unemployment dropped to 4 per cent); this seems to be related mainly to technical change, which leads increasingly to new capital being a substitute for unskilled labour but complementary to educated labour. Expertise and reliability are also important, as manifest in increased wage dispersion even within educational or skill or professional cohorts. These developments will encourage labour to upgrade its skills.

## International diffusion of the technology

If indeed the new economy in the United States is due to the arrival and gradual absorption of a new general-purpose technology, IT, use of the new technology will gradually spread to the rest of the world. One channel will be foreign direct investment, especially where the full use of the new technology requires organisational changes. But the speed of the diffusion and its degree of dislocation will depend on local circumstances, particularly on the characteristics of labour markets and the business environment.

If firms are protected, whether by import restrictions or through regulatory protection, they are unlikely to feel the pressure to make the efficiency-enhancing changes permitted by the new technology. Export-oriented firms, of course, must do so sooner or later to maintain their international competitiveness.

Workplaces must be organised in new ways to achieve optimal use of the new technology. This should not be a great problem for poor or middle-income countries that are developing rapidly, partly by pulling labour in from the countryside. Such labour is relatively flexible. There the problem is likely to be too little educational background for at least some jobs.

In richer countries, with highly structured and better educated urban labour forces, the problem is more likely to be both the entrenched attitudes of organised labour and the framework of regulations and accepted practices that have accumulated over the years to prevent labour from being treated as a 'commodity' and being moved around at will by management. Ironically, one characteristic of much new technology is that it is knowledge-oriented, so the protections of labour accumulated through collective bargaining or political action are outdated. High morale and dedication to work, increasingly reinforced by incentive compensation, are often necessary to achieve the high productivity made possible by the new technology.

Finally, for some activities a high respect for intellectual property is necessary to foster the continual advancement of the new technology, particularly software. Without such respect, reinforced if necessary by legal protection, people will not make the effort to generate the many applications made possible by ever-advancing computational and communications capacity.

Americans have no monopoly on new ideas or on willingness to translate them into lucrative applications. Silicon Valley is properly famous for being a melting pot, bringing together not only Americans but also British, Chinese, French, Germans, Indians, Vietnamese and people of many other nationalities to translate new ideas into viable applications and successful business ventures.

US patents awarded for IT applications rose sharply during the 1990s. Half those awards were to residents of countries other than the United States, up from about a quarter before 1980. In other words, foreign patents in the US have risen more rapidly than awards to US residents. Japan accounts for more than any other country.

The framework for business in the United States is especially conducive to innovation. And the US labour market is relatively flexible, at least compared with that in many other rich countries. Individuals expect to have many different employers during their lifetimes, and will leave jobs that are considered unsatisfactory. The average period of employment with the same employer is four years for males, less for females. Firms are willing to hire older workers when they have the appropriate qualifications; and it is relatively easy to enter self-employment of many kinds (although not necessarily easy to become highly successful doing so!). Unemployment in the United States is relatively brief, except during periods of recession, averaging only six weeks. Finally, temporary and part-time employment have become well established, for men as well as women. In short, the US labour market exhibits great institutional and attitudinal flexibility for both employees and employers. These characteristics should be kept in mind when assessing the official 'safety net' in the United States, which is seen as weak by many Europeans.

As noted above, the new economy may not function optimally without reorganisation of the workplace and of the organisations that produce goods and services. Under competitive pressure, Americans have been reorganising in this way; in the late 1990s firms undertook an unparalleled number of mergers and acquisitions, both horizontally and vertically. During the same period, firms did an extraordinary amount of 'outsourcing' – buying goods and (especially) services from outside firms that had once been provided within the firm – since new methods could often be used more efficiently by specialised independent firms.

These possibilities are likely to arise across national borders as well as within the United States. A reasonable forecast is that international mergers and acquisitions will continue at a high pace while the new technologies are being diffused; indeed, foreign direct investment will be one of the principal vehicles for such diffusion, through what might be called organisational arbitrage. Countries with a high resistance to foreign ownership will also create resistance to the new technologies.

Similarly, there will be more outsourcing across national boundaries, made possible by quick, reliable and inexpensive long-distance communication, as can be seen in the increasing internationalisation of telephone call services. There will be new opportunities for those quick-footed and flexible enough to take advantage of them.

In summary, continuation of the new economy in the United States will affect the rest of the world through many channels: foreign trade (both imports to satisfy growing US demand and exports of new, technology-related products); foreign investment into the United States, attracted by the new possibilities and relatively high returns to capital there; US investment abroad, a form of organisational arbitrage through mergers, acquisitions, joint ventures, and outsourcing; and more general diffusion of the new ideas through international conferences, professional journals, journalistic reporting, and extensive education abroad, especially in the United States, where over half a million non-residents are studying. Increased education will command a wage premium world-wide, not just in rich countries; and there will be a premium on flexibility and adaptability in the regulatory environment, especially with respect to labour and the formation of new enterprises.

At the same time, five years is a short period on which to base sweeping generalisations and lengthy extrapolations into the future. Caution is indicated by recalling that it was barely more than a decade ago that Japanese approaches to production, organisation and management were seen as superior to those elsewhere, to be the wave of the future.[10] The economic environment can change rapidly and unexpectedly.

## NOTES

1  The Brainard–Perry analysis has been criticised by Ray Fair in his comments on Brainard and Perry (2001) for using the CPI as a measure of domestic price inflation, since it includes imported goods; for failing to include productivity in their wage equation, since it is included in the price equation; and for failing to include cost shocks in their analysis. Fair conjectured that a more complete specification would reveal greater stability in the estimated coefficients. Gordon (1998) found no influence of productivity on wages; Rich and Rissmiller (2001), in contrast, found that real wages are strongly influenced by changes in productivity.
2  Stock, in his comments on Gordon (1998) (p.340) suggested that capacity utilisation rates, new building permits, manufacturing production, employment growth and trade sales all did a better job at predicting inflation in the mid-1990s than did unemployment rates.
3  Reported in *The Economist*, 30 June 2001, p.70.
4  In the United States, total production of goods – agriculture, forestry, mining, manufacturing – now takes only 17 per cent of the labour force, with 78 per cent

devoted to provision of 'services' (including government services), a collective expression too broad to be very useful, and 5 per cent in construction.
5   The Council of Economic Advisers (CEA 2001, p.78), using the income-side measurement, 0.3 per cent above the official product-based measure, found that non-farm business output grew by 4.2 per cent annually from 1990 to 2000.
6   For a discussion of measurement issues by two of the officials responsible for compiling US GDP, see Landefeld and Fraumeni (2001).
7   The absence of import restrictions on information technology products was 'multilateralised' in the Information Technology Agreement of 1997, under which the rich countries agreed to eliminate import duties by January 2000 and other signatories agreed to eliminate them in the subsequent five years. See Wilson (1997).
8   The maximum federal income tax rate from 1993 to 2001 was 39.6 per cent; the addition of state income taxes brings this into the mid-40s, after allowance for deduction from federal taxable income, in those states with income taxes, which include California, New York, and Massachusetts. The top federal tax rate on capital gains on investments held for more than one year was 20 per cent in 2001, down from 28 per cent in 1997.
9   According to Maddison (2001, p.138), hedonic indices are not used by Belgium, Finland, Germany, Italy, Japan, Spain, or the United Kingdom.
10  For documentation and a reasoned assessment of Japan's economic prowess, see NRC (1992).

## REFERENCES

BIS (Bank for International Settlements) (2001) *71st Annual Report*, Basel, Switzerland, June.
Boivin, J. and Giannoni, M. (2001) 'The monetary transmission mechanism: has it changed?' Federal Reserve Bank of New York, April.
Brainard, W.C. and Perry, G.L. (2001) 'Making policy in a changing world,' in G. Perry and J. Tobin (eds), *Economic Events, Ideas, and Policies: the 1960s and after*, Washington DC: Brookings Institution. See also comments by Ray C. Fair and John B. Taylor.
CEA (Council of Economic Advisers) (2001) *Economic Report of the President*, January 2001.
Cooper, R.N. (2001) 'Is the US current account deficit sustainable? Will it be sustained?' *Brookings Papers on Economic Activity*, 2001, 1.
David, P.A. (1990) 'The dynamo and the computer: an historical perspective on the productivity paradox,' *American Economic Review*, 80, 2, May: 355–361.
Gordon, R.J. (1998) 'Foundations of the goldilocks economy: supply shocks and the time-varying NAIRU,' *Brookings Papers on Economic Activity*, 1998, 2. See also comments by James Stock.
Gordon, R.J. (2000) 'Does the "new economy" measure up to the great inventions of the past?', *Journal of Economic Perspectives*, 14, 4, Fall: 49–74.
Jorgenson, D.W. (2001) 'Information technology in the U.S. economy', *American Economic Review*, 91, March: 1–32.
Landefeld, J.S. and Fraumeni, B.M. (2001) 'Measuring the new economy', *Survey of Current Business*, March: 23–40.
Maddison, A. (2001) *The World Economy: A Millennial Perspective*, Paris: OECD Development Centre.
Mankiw, N.G. (2002 forthcoming) 'U.S. monetary policy during the 1990s', in J. Frankel and P. Orszag (eds), *American Economic Policy in the 1990s*, Cambridge MA: MIT Press, spring 2002.

Nordhaus, W.D. (2001) (January) 'Productivity growth and the new economy,' Cambridge MA: National Bureau of Economic Research, NBER Working Paper 8096.

NRC (National Research Council) (1992) *Japan's Growing Technological Capability: Implications for the U.S. Economy*, Washington: National Academy Press.

Oliner, S.D. and Sichel, D.E. (2000) 'The resurgence of growth in the late 1990s: is information technology the story?' *Journal of Economic Perspectives*, 14, 4, Fall: 3–22.

Rich, R.W. and Rissmiller, D. (2001) 'Structural Change in U.S. Wage Determination,' New York: Federal Reserve Bank of New York Staff Report No. 117 (March).

Stiroh, K.J. (2001) 'Information technology and the U.S. productivity revival: what do the industry data show?' New York: Federal Reserve Bank of New York Staff Report No. 115.

Wilson, J.S. (1997) 'Telecommunications liberalization: the goods and services connection', in G.C. Hufbauer and E. Wada, *Unfinished Business: Telecommunications after the Uruguay Round*, Washington, DC: Institute for International Economics.

# 3  The new economy: theory and measurement

*Richard G. Lipsey*

**WHAT IS THE NEW ECONOMY?**

The term 'new economy' means different things to different writers, which is a cause of some confusion when discussing it. Sometimes the term refers to an economy in which the laws of supply and demand no longer hold and there are neither business cycles nor inflations. Not many academic economists were gullible enough to take this view seriously. Nonetheless, new technologies do alter many economic relations, such as when 'natural monopolies' are turned into highly competitive industries, and vice versa.

Dale Jorgenson (2001) defines the new economy as the sector producing computing power and related things. He produces some valuable material on this sector, which is one of the driving forces of the information and communication technology (ICT) revolution. His approach shows the new economy as only a small fraction of the whole economy. It also leads him to argue that if progress stopped in the IT sector, the growth and change attributable to the new economy would slow or even stop.

Robert Gordon (2000) defines a new economy as occurring when the rate of improvement in new products and services is greater than in the past (p.39) and there is thus an acceleration in the rate of productivity growth. In doing this, he is following most growth economists who use models based on an aggregate production function in which technological change is visible only by its effects on productivity. Such models equate changes in technology with changes in productivity.

My brief is to present my own views on these issues. This I have done with a large number of references to my own published work – while not, of course, ignoring all other relevant material.

As I use the term, the new economy refers to the social, economic and political changes brought about by the current revolution in ICTs.[1] That revolution is being driven by the computer, lasers, satellites, fibre-optics, the internet and a few other related communication technologies, many of which were developed with the assistance of computers. It is an economy-wide *process*; it is not located in just one high technology *sector*, any more than the new economy initiated by electricity was confined to the electricity-generating sector.[2]

## TECHNOLOGICAL CHANGE DRIVES LONG-TERM GROWTH[3]

Long-term economic growth is driven by technological change – that is, by changes in product, process and organisational technologies. Calculations of per capita gross domestic product (GDP) radically understate the impact of economic growth on the average person. Although each person living in the United States and Western Europe has about ten times as much 'real purchasing power' as did their forebears 100 years ago, they consume it largely in the form of new commodities made with new techniques. Those who lived at the beginning of the nineteenth century could not have imagined modern dental and medical equipment, penicillin, painkillers, bypass operations, safe births, control of genetically transmitted diseases, personal computers, compact discs, television sets, efficient automobiles, opportunities for cheap, fast worldwide travel, affordable universities, safe food of great variety free from ptomaine and botulism, or the elimination of endless kitchen drudgery through the use of detergents, washing machines, electric stoves, vacuum cleaners, refrigerators, dish washers, and a host of other labour-saving household products that we take for granted today. Nor could they have imagined the clean, robot-operated, computer-controlled modern factories that have largely replaced their noisy, dangerous factories that spewed coal smoke and other pollutants over the surrounding countryside.

The point is important. Technological advance not only raises our incomes; it transforms our lives through the invention of new, hitherto undreamed of things that are made in new, hitherto undreamed of ways.[4] In the long term, these new technologies transform our standards of living, our economic, social and political ways of life, and even our value systems.

### Capital accumulation versus technological change

Economists have sometimes debated whether investment or technological change is the more important cause of long-term growth. This debate is important for our purposes because it concerns the sources of long-term economic dynamism and how to interpret, among other things, the so-called Asian miracle.

My colleagues and I use a simple thought experiment to argue for the primacy of technological change. Imagine holding technological knowledge constant at the level of, say, 1900, while continuing to accumulate more 1900-vintage machines and factories and use them to produce more 1900-vintage goods and services, as well as training more people longer and more thoroughly in the technological knowledge that was available in 1900. Today's living standards would then be vastly lower than those we now enjoy (and pollution would be a massive problem). This thought experiment illustrates what most economic historians and students of technology are agreed on: technological change is *the* major determinant of long-term economic growth.

This conclusion does not imply that saving, investment and capital accumulation are unimportant. Because most new technology is embodied in new capital equipment, technological change and investment are complementary, the latter being the vehicle by which the former enters the production process.

In spite of the long-term importance of technological change, short-run growth is affected more by variations in investment than by variations in the accumulation of technological knowledge. This is because at any moment in time there is a large pool of existing technologies that have not been fully exploited. Variations in the rate of investment cause immediate variations in the rate of exploitation of these technologies. The strong short-run relation between growth and investment has led some observers to conclude erroneously that investment is the major determinant of long-term growth and that technological change is relatively unimportant. But without further technological change, there would be no new opportunities to exploit, and existing ones would be gradually used up.

## Capital accumulation and technological change in developing countries

If technological change is the prime source of the long-term economic growth of the developed industrialised nations over the last 250 years, can we say the same about the growth of the newly industrialising countries over the last half century? In particular, was the 'Asian miracle' led by the four 'tigers', Taiwan, South Korea, Singapore and Hong Kong, due to anything more than simple investment in physical and human capital?

I accept the interpretation of the success of the four Asian tigers and their followers given by such economists as Pack and Westphal (1986) and Westphal (1990) and some of the qualifications found in Stiglitz and Yusuf (2001). Lipsey and Wills (1996) and Lipsey and Carlaw (1996) have also discussed this issue. The governments of the 'Asian tiger' countries followed policies of market orientation, combined with substantial intervention of various sorts. Although the amounts of intervention differed greatly among countries, their policies had some common features. Many active policies sought to transfer and adapt technologies that existed elsewhere and sometimes to create new ones.

For example, the Taiwanese electronics industry was virtually created by the Taiwanese government; when it had got beyond the infant industry stage, it was transferred to private agents. Other local industries were initially sheltered by tariffs and provided with substantial government assistance. Once these industries were well established, policies for industrial growth switched from import substitution to export promotion. The new industries were required to export, contending in the fierce world of global competition. This forced the industries to compete dynamically in a way that was not necessary for import-substituting industries serving protected home markets.

They had to learn about, and then copy, international standards of design, product quality and marketing.

Export orientation also provided a cut-off for unsuccessful firms and industries. With import substitution, inefficient firms with static technology could serve a protected home market more or less indefinitely; with export orientation, those that did not make it in the fierce world of international competition had their support cut off. This experience is a long way from the position stated by many growth theorists that all nations share the same production function. All nations have access to international best practices, but only at a substantial cost of learning by doing and by using – and many may lack the capacity to absorb this knowledge.

A neoclassical counter-attack against this interpretation has been led by economists such as Alwyn Young (1992 and 1995). They were able to fit a neoclassical aggregate production function to the data for the newly industrialised economies (NIEs) with only a small unexplained residual, and therefore argued that NIEs merely accumulated physical and human capital. The small remaining residual, or total factor productivity (TFP), was assumed to measure technological change – although Lipsey and Carlaw (2002) argue that it does not. In this view, the NIEs' technology transfer and development policies did little or nothing to promote their growth. Here are two representative quotations.

> A growth-accounting exercise [conducted by Alwyn Young] produces the startling result that Singapore showed no technical progress at all.
> (Krugman 1996, p.55)
>
> Singapore will only be able to sustain further growth by reorienting its policies from factor accumulation toward the considerably more subtle issue of technological change.
> (Young 1992, p.50)

One of the biggest problems with the measurements that underpin this interpretation is the difficulty of separating embodied technological change from the pure accumulation of physical and human capital. Let me consider human capital first. In the past century, the increase in years of schooling that occurred in most countries was partly used to train people in the rapidly increasing stock of technological knowledge. To isolate the concept of the pure accumulation of human capital, with no contamination from embodied technological knowledge, one would need to estimate the effects of educating people longer and better in only the knowledge that was available in some base period, such as 1900. This is, in practice, impossible to do, so measures of human capital accumulation are always to some extent measures of the accumulation of new technological knowledge (possibly totally new, or possibly just new to the country in question).

Analogous problems exist for physical capital. For example, suppose a new machine is invented and the present value of the additional output is just enough to recover the research and development (R&D) cost and yield a normal return on the developer's investment. The new machine will then be sold at a price that just earns a normal return to the purchaser. In this case, there will be technological change but no change in TFP, because the appropriately weighted percentage change in the value of capital will just equal the percentage change in output.[5] Similarly, let a developing country import new plant and equipment embodying techniques hitherto unused in that country. Assume that the price of the machine is equal to its marginal product. Now output will rise and new technology will be put to use, but there will be no change in TFP. For these, and many other reasons studied by Lipsey and Carlaw (2002), we can conclude that: '...whatever TFP does measure – and there is cause for concern as to how to answer that question – it emphatically does not measure all of technological change (Carlaw and Lipsey 2001, p.49)'. An excellent initial survey of the interpretation and misinterpretation of TFP measurements, particularly in the context of developing nations, can be found in Chen (1997).

The NIEs saved and invested heavily in capital that embodied technologies that were typically new to them. Adopting, and often adapting, these new technologies was no simple matter; nor was learning to use them efficiently to produce commodities at a standard demanded by advanced countries and to market them. One can debate whether or not they got the mix of market and interventionist policies right, but there can be no doubt that the policies followed were a long way from laissez-faire and that changes in the locally used technologies played a big part in what happened. Whether or not the 'Asian miracle' was a miracle in TFP, whatever that would mean, it was a miracle in rapidly rising living standards and in rapid changes in the kinds of technology used by workers.

In a related criticism of the neoclassical counterattack, Nelson and Pack (1999) showed that separation using an aggregate production function cannot distinguish clearly between movements along a constant returns aggregate production function and shifts in a diminishing returns function. In relatively low-income countries, given the importance of such fixed factors as land, diminishing returns to the accumulating factors are a distinct possibility. Also, as Pack (2001) observes:

> ...the huge accumulation of factors [in the NIEs] was successfully absorbed, with no decline in TFP *levels* ...Compared with this achievement, the division of the total growth rate between accumulation and TFP is a second-order question. In contemporary terms, very few economists looking forward from 2000 would argue that Bangladesh, Bolivia, or Tanzania could avoid a decrease in their TFP levels if an inflow of aid increased their accumulation rates to those...for Korea and Taiwan.

Given enormous rates of capital accumulation, achieving zero TFP growth was an amazing accomplishment. The 'Asian miracle' was one of massive investment-driven growth with constant or slowly rising TFPs due to the importing of technologies new to these countries. This is a long way from the view of capital accumulation without technological change put forward by Krugman (1996) and Young (1992).

### General-purpose technologies

Next, we need to enquire in a bit more depth into the kinds of technological changes that drive long-term economic growth. They range from small incremental improvements in existing technologies to the introduction of new general-purpose technologies (GPTs), which are sometimes called transforming or enabling technologies. GPTs share some important common characteristics. First, they begin as fairly crude technologies with a limited number of uses. Second, as they diffuse throughout the economy, they evolve into much more complex technologies, with dramatic increases in their efficiency, in the range of their use, in the range of economic outputs that they help to produce, and in the range of new product and process technologies that incorporate or otherwise depend on them.[6]

The most important GPTs, which I will call 'transforming GPTs', have major impacts on the economic, social and political structures. These initiate 'new economies' on scales equal to, or exceeding, the present ICT-induced transformations. Lipsey and Bekar (1995) and Lipsey, Bekar and Carlaw (1998a) have identified and studied about 20 of these truly transforming GPTs, starting with the Neolithic agricultural revolution. I have space here only to list their names, grouping them into the five classes used by these authors.[7]

- *Information and communication technologies.* This category includes *writing, printing* with moveable type, the *computer* – which, along with several related technologies, is driving the current ICT revolution – and the *internet*, which was enabled by the computer, but can be regarded as a GPT in its own right.
- *Materials.* This category includes *bronze, iron* and the current ability to create *made-to-order materials* invented specifically for use in newly developed products and processes.
- *Power delivery systems.* This category includes *domesticated animals, the water wheel, the steam engine, electricity,* and *the internal combustion engine.*
- *Transportation.* This category includes the *wheel*, the *three-masted sailing ship, railways, the iron steam ship,* the *motor vehicle,* and *commercial aircraft* (the latter two of which were enabled by the internal combustion engine).
- *Organisational technologies.* This category includes the *factory system, mass production* and *flexible manufacturing* (also known as lean production or Toyotaism).

At any one time, each of the above categories may contain GPTs that are still affecting the economy.[8] As each new GPT diffuses through the economy, it creates a research program for entrepreneurs to apply its principles to create new processes and new products and to improve old ones. These, in turn, create other new opportunities, and so on in a chain reaction that stretches over decades, even centuries. Note, for example, the myriad ways that innovators have found to use electronic chips; how these ways have multiplied as the power and reliability of chips have increased; how some of these ways have in turn enabled other developments; and so on in a complex concatenation of related innovations.

It is important to note that many of the responses to a new GPT cannot be modelled (for measurement or any other purpose) as the consequence of changes in the prices of flows of factor services produced by the previous GPT. This is because most of the action is taking place in the structure of capital. The new possibilities depend on how one technology is related to another, not on how a given technology can respond to a change in price.

For example, the most profound effects of electricity came not from a fall in the price of power, but in making possible new products and new processes that were not technically available with steam. There was a major increase in productivity from the revolution in the layout of factories in which machine tools, each with its own independent power source (the unit drive), were rearranged on the shop floor according to the logic of production rather than their power demands.[9] This new layout in turn made possible the assembly line, with its extensive restructuring of all manufacturing production and further large productivity gains. Electricity also made possible the household machines that revolutionised household work and freed women, or their servants, from millennia of household drudgery. No steam engine could have been attached to the carpet sweeper to turn it into a vacuum cleaner, to the ice box to turn it into a refrigerator, or to a washing tub to turn it into a clothes washing machine. Indeed, none of the above changes would have occurred if the steam engine had remained the main source of power, even if the price of its power had fallen to zero.

Nonetheless, equivalent price change measures are often used. For example, we might think of comparing the steam engine and the electric motor with a hedonic index that relied on horsepower or British thermal units produced by each motor for equivalent amounts of inputs. But, as noted above, the major economic gains from the introduction of the electric motor came from its ability to permit a reorganisation of production in ways that were technically impossible with steam power. Similarly, most of the gains from a practical quantum computer, which will arise from its ability to do things that could not be done on any conceivable conventional computer, would not be measured if a hedonic index was used to compare their performance characteristics.

## MODELS OF GROWTH

Not content with merely describing economic growth, growth economists attempt to model it in order to better understand it. Without claiming complete coverage, we can identify two distinct approaches to this work.

## The aggregate production function

The first approach is based on an aggregate production function such as

$$GDP = AK^{\alpha}H^{\beta}L^{\gamma} \quad \alpha + \beta + \gamma = 1 \qquad (1)$$

where $K$ is physical capital, $H$ is human capital, $L$ is labour and $A$ is a productivity parameter.

Approaches that use some variant of equation (1) treat technological changes as featureless shifts in the productivity parameter $A$ or in the efficiency of $L$ (which are here empirically indistinguishable). This neoclassical approach is illustrated in the first part of Figure 3.1. Inputs are passed through the production function to produce the nation's GDP. Institutions, and all other structural components, are hidden in the 'black box' of the aggregate production function. Presumably they help to determine its form.

In this model, technological change is observable only by its effects on productivity. For example, the model does not allow us to observe the coexistence of rapid technological change and slow productivity growth. It is the non-separation of those two phenomena that gives rise to productivity puzzles in periods when independent evidence suggests that technology is changing rapidly but productivity is changing only slowly or not at all. To discover the circumstances in which rapid technological change will or will not be accompanied by rapid productivity growth, we cannot employ a model that equates technological change with productivity growth.

## Microeconomic approach: the 'structuralist-evolutionary' model

The second approach seeks to understand growth by studying the microeconomics of technological change (see, for example, Rosenberg 1976 and David 1991), the changes in economic structures and institutions that co-evolved with technology (see, for example, Freeman and Perez 1988; Perez 1983; Lipsey, Bekar and Carlaw 1998b; and Nelson 1999). Many names are associated with this work and I have mentioned only a few for purposes of illustration. Long before macro-growth economists discovered endogenous technological change, these microeconomists had established its existence and were studying its consequences (see in particular Smookler 1966 and Rosenberg 1982.)

Cliff Bekar, Kenneth Carlaw and I have developed a formulation designed to separate changes in technology from changes in productivity and to reveal some of the elements of the neoclassical black box that research shows to be important for economic growth. This model makes the economy's structure explicit, and is also in line with much microeconomic research on the evolution of technology, so we call it a 'structuralist-evolutionary' (S-E) model. Since it is disaggregated from the aggregate production function, we refer to it as an 'S-E decomposition'.

40   *The New Economy in East Asia and the Pacific*

*Figure 3.1* Models of growth

**(a) Neoclassical approach**

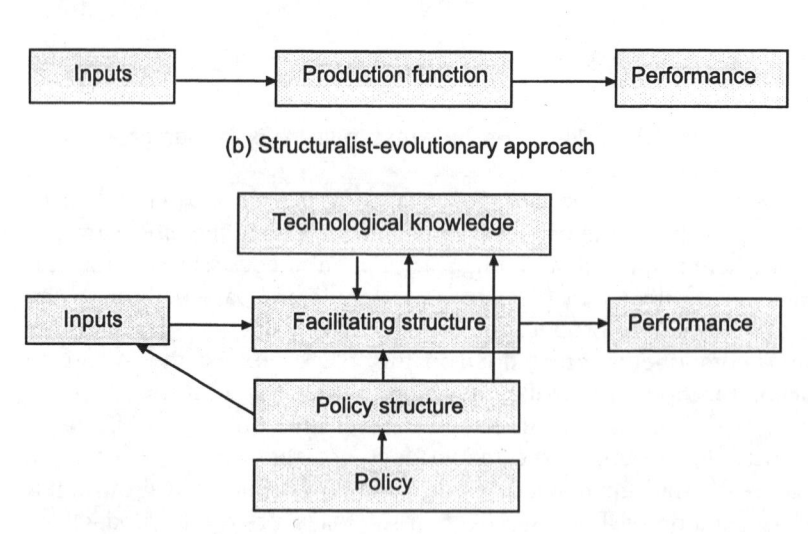

Inputs → Production function → Performance

**(b) Structuralist-evolutionary approach**

```
                    Technological knowledge
                            ↕
Inputs  →      Facilitating structure      →  Performance
     ↘                      ↑
            Policy structure
                    ↑
                 Policy
```

*Notes:* In the neoclassical approach (a), inputs of labour, materials and the services of physical and human capital flow through the economy's aggregate production function to produce economic performance, as measured by total national product. Everything that influences the form of the production function is hidden in the black box of that function. In our structuralist-evolutionary approach (b), technological knowledge is embodied in the facilitating structure; inputs pass through that structure to produce economic performance; and policy acts through the policy structure to influence some inputs, some elements of the facilitating structure and some technologies directly. The arrows indicate major directions of causation.

In ordinary parlance, the term 'technology' is used rather loosely to refer to specifications, designs and blueprints on the one hand and their embodiment in specific items of capital equipment on the other. In contrast, we make a sharp distinction between technological knowledge on the one hand and its embodiment in specific items and organisations on the other. Thus, our scheme separates technological knowledge from the capital goods that embody much of it, making the latter a part of what we call the economy's 'facilitating structure'. The model also separately specifies public policies and the policy structures designed to give them effect.

*Definitions*

Figure 3.1b shows the six main categories in the S-E model; the following discussion provides more detail about them.

*Technological knowledge*

We define technological knowledge as the idea set of everything that can create economic value. The classes of technological knowledge are product technologies, which are the specifications of the goods and services that can be produced; process technologies, which are the specifications of the processes employed to produce them; and organisational technologies, which are the specifications of the organisation of value-creating activities such as R&D, production, management, distribution and marketing. Our concept of technology is wider than usual; it covers all codifiable knowledge of how to create all forms of economic value. Tacit knowledge is part of human capital and hence part of the facilitating structure.

*Facilitating structure*

We define the facilitating structure as the realisation set of technological knowledge; it embodies that knowledge. To be useful, the great majority of technologies must be embodied in one way or another. The facilitating structure is made up of all physical capital; people and all human capital that resides in them, including tacit knowledge of how to operate existing value-creating facilities; the organisation of production facilities, including labour practices; the managerial and financial organisation of firms; the geographical location of firms and industries; industrial concentration; all infrastructure; and all private sector financial institutions, and financial instruments.

*Policy and policy structure*

We separate public policy, which is the idea set covering the specification of the objectives of public policy as well as the specification of the means of achieving them, and policy structure, which is the realisation set that provides the means of achieving public policies and is embodied in public sector institutions and the people who run them.

*Economic performance*

We refer to the system's economic performance rather than just its output because, in addition to GDP, we wish to consider such variables as employment, unemployment, the distribution of income, and the production of such 'bads' as pollution.

*Behaviour*

At any particular time, the facilitating structure, in combination with primary inputs, produces economic performance. That structure is in turn influenced by technological knowledge and public policy. The introduction of any important new technology, or a radical improvement in an old technology, induces complex changes in the whole of the facilitating and policy structures. As the reverse arrows in Figure 3.1(b) indicate, changes in the facilitating structure and the policy structure can induce changes in technology.

I want to discuss four important characteristics of the link between technology and structure. First, new technologies are typically first operated in a structure designed for their predecessors. This can be seen at all levels, from major GPTs (for example, the initial introduction of electronic computers into organisations designed for verbal and hard copy communication) to relatively trivial applications (for example, the use of radio techniques in early television productions such as early episodes of *I Love Lucy*).

Second, if elements of technology change, various elements of the facilitating structure (and policy and the policy structure) need to change adaptively before the new technology becomes fully effective. For example, the technology of electricity generation had to be embodied in new electric motors, new generating stations (first using steam, then water power), new distribution networks and a host of other new types of capital. When electricity replaced steam as a source of power in factories, the optimal layout of the plant changed and its size increased, as did concentration in manufacturing industries. Geographic location was also altered because power could now be consumed far from the place where it was generated. Human capital changed because machine operators required far less skill to handle the reliable, electrically driven machine tools that replaced the less reliable belt-driven machines of the steam-powered factory.

Third, most elements of the facilitating and policy structures show substantial resistance to change. These inertias tend to be particularly strong in public policy, which often takes decades to adapt to new technologies because policy structures are difficult to alter. For example, most public bureaucracies still operate on the organisational structure used by firms in the pre-computer age. They have been slow to change to the looser, more laterally linked, forms of organisation that now characterise most firms, in spite of the fact that more and more of the problems they deal with cut across traditional departmental lines.

Fourth, this period of adjustment is often 'conflict ridden' (the term used by Freeman and Perez 1988). Old methods and organisations that worked well, often for decades, are often dysfunctional in the new situation. The uncertainty accompanying radical new innovations implies that there will be many different but defensible judgments of what adaptations are actually needed. Users may mistrust new technologies and/or take a long time to accept and adjust to them. For example, consider the resistance to many of the applications of biotechnology: some concerns are well founded but others are purely emotional.

*Applications*

One important insight from our S-E decomposition is that adopting a new technology is not usually sufficient for a large change in economic performance. Imported technologies often fail to have the expected results on economic performance because the required changes in the facilitating and policy structures have not been made – possibly because required public policy adjustments were not made or because public policies inhibited the adjustments that private agents would have made.

A second important insight is that there are no necessary relations among the magnitudes of changes in technology, in the facilitating and policy structures, and in performance. In neoclassical models, changes in technology are observable only by their effects on productivity, so economists typically assume that big changes in technology must be associated with big changes in productivity and are puzzled when they are not. But there is no reason to be puzzled. New technologies will replace older ones as long as they promise some gain. Sometimes the difference between the productivity of the old and the new is large; at other times it is small. Neoclassical observers often doubt the reality of observed major technological changes when productivity is not changing rapidly. But, as the S-E decomposition emphasises, there is no necessary relation between these variables. Current changes in productivity may or may not be well measured (and they probably are not), but there is no paradox in there being major changes in technology with no accompanying rapid changes in measured productivity.

Third, there is a natural history of GPTs. In the first phase, the GPT is introduced into a facilitating structure designed for the GPT it is challenging. In the second phase, the facilitating structure must be redesigned to fit the new GPT – a stage that is often long-drawn-out and full of uncertainty and conflict. In the third phase, the principles of the new GPT are applied to produce many new products, processes and organisational forms within the confines of facilitating and policy structures well adapted to it. This third phase is the time when the new technology tends to yield the largest payoffs in terms of productivity, real wages and investment booms. In the fourth phase, the opportunities to apply the GPT to new product, process and organisational technologies begin to peter out; if new GPTs are not introduced, the growth process will slow. Consider, for example, what the range of new innovation possibilities and the rate of return on investment would now be if the last GPTs to be invented were steam for power, the iron steamship for transport, steel for materials (no human-made materials) and the telegraph for communication (the voltaic cell but not the dynamo). The arrival of new GPTs rejuvenates the growth process and the phases repeat themselves – in general form although, of course, not in every particular. Carlaw and Lipsey (2001) present a formal model of this process.

What happens to productivity growth during this process depends on the lengths of these phases, which differ from one GPT to another, and on the scope for further improvements in the existing technologies, which depends on the GPT that is being challenged. After all, a new GPT does not typically overwhelm the one it challenges overnight. For example, hand-loom weaving survived for half a century after being challenged by automatic equipment, the water wheel lasted well over a century after being challenged by the steam engine, and sail survived for more than a century after being challenged by the steamship. During this period of competition, the old technology often achieves many efficiency-increasing improvements. For example, the golden age of the clipper ships came after sail was challenged by steam, and many of the most important improvements in the water wheel came during the nineteenth century.

## THE ICT REVOLUTION

### ICT as a general-purpose technology

Like all GPTs, the electronic computer started as a crude, specific-purpose technology (in the early 1940s) and took decades to be improved and diffused through the whole economy. Its effects began to become really visible in the late 1970s when the string of technological changes enabled by the ICT revolution – in products, processes and organisational forms – began to create another 'new economy'. Since the 1980s, we have been living through a profound transformation of economic, social and political structures. It is arguable that in scale and scope, if not in specific details, this transformation is on a par with the first industrial revolution. In Lipsey and Bekar (1995) and Lipsey, Bekar and Carlaw (1998b), we have discussed a GPT's evolutionary path in detail and noted that the major GPTs have tended to create 'new economies' many times in the past.

Since some economists are still sceptical about this statement (and very many more were when I first started to talk about the ICT revolution early in the 1990s), I will support it by listing just a few of the many changes that have occurred since 1970, organised around the categories of the S-E model. Many of the things in the list affect more than one category; I place them in the category where I judge they will have most effect and in some cases indicate other categories in which I judge they will also have major effects. We are dealing with general interrelations among the elements of the economic, political and social structures where almost everything affects almost everything else, so there is an element of arbitrary judgment in the classifications adopted, although they remain useful as organising devices. Notice that many of these changes were completed well before the 1990s, when the new economy was first noticed by many economists.

### *Technology*

- Many of the technologies of the ICT revolution, such as lasers, satellites, fibre-optics and the internet, were developed with the assistance of computers and/or could not operate without computer chips.
- Driven by the internet, English is becoming a lingua franca for the world; unlike Latin in the Middle Ages, its use is not limited to the intelligentsia. I chose to classify language as a technology, although it could be classified as an element of the facilitating structure; in any case, these new developments will affect performance in many ways.

- Looking into the future, the computer is enabling most of what is happening in the biological revolution, and will do so in the forthcoming revolution in nanotechnology and nanoelectronics. Here we see technological complementarities at their strongest, when one GPT makes possible others that follow it.
- Computer-assisted design of new products allows 'virtual' testing, which greatly reduces the need for learning by using such as was studied by Rosenberg for the aircraft industry (Rosenberg 1982, Chapter 7).

*Facilitating structure*

The following examples cover all of the items of the facilitating structure listed in the earlier section on 'Definitions'.

- The factory has been reorganised along the lines of lean production, using computerised robots that have eliminated many of the high-paying, low-skilled jobs of the old Fordist assembly line factories. Many labour practices, such as rigid job demarcation, have had to be altered.
- The music industry has changed in many deep ways, including the introduction of the virtual band in which the output of several singers and instrumentalists are all produced by one singer and one instrumentalist whose varied performances are amalgamated digitally. These changes also have effects in the 'Performance' category below.
- The management of firms has been reorganised as direct lines of communication opened up by computers eliminated the need for the old pyramidal structure in which middle managers processed and transmitted information. Today's horizontally organised, loose structures bear little resemblance to typical management structures up to the end of the 1960s.
- New skills have been demanded of a significant part of the labour force and many old skills, such as those of middle managers, are in much diminished demand.
- Just as the first industrial revolution took work out of the home, the ICT revolution is putting much of it back, as more and more people find it increasingly convenient to do all sorts of jobs at home rather than 'in the office'.
- The 'e-lance economy' is growing as more and more groups of independent contractors come together to do a single job and then disperse. At the same time, traditional firms are continuing to disintegrate. For example, very few firms in Silicon Valley produce physical products now. In other industries, the role of the main firms is increasingly to coordinate subcontractors who do everything from designing products, through manufacturing them, to distributing them.
- The ease of long-distance communication, combined with the fall in transport costs associated with such non-ICT innovations as containerisation as well as the reduction in trade barriers, has globalised many markets

and increased the number of transnational firms. Globalised trade is not new, but what is new is the two-way flow of manufactured goods between developed and developing nations. The globalisation of trade in manufactured goods would have been impossible without modern ICTs. It has caused massive changes in both the opportunities and challenges facing nations that wish to develop, as well as those that are already developed. These changes also have effects in the 'Performance' and 'Policy and policy structure' categories below.
- The new technologies have made some industries, such as telecommunications, much more competitive than they were 30 years ago and increased the degree of concentration in others, such as automobile assembly.
- We have had to construct whole new infrastructures of such things as satellites, fibre-optic cables, and internet servers.

*Policy and policy structure*

- The computer-enabled internet is revolutionising political activity. For example, non-government organisations (NGOs) are able to organise activities to protest against World Trade Organization (WTO) efforts to reduce trade barriers, and the push for the Free Trade Area of the Americas (FTAA). Never again will trade negotiations take place in the relative obscurity that they enjoyed from 1945 to 1990. Also, consumer boycotts of firms that are thought (sometimes correctly and sometimes incorrectly) to be violating environmental codes are organised with an ease and effectiveness unknown in the past.
- Dictators find it much harder to cut their subjects off from knowledge of what is going on in the outside world, and hence find it harder to sustain their control.
- Changes have had to be made in a whole host of existing laws, rules and regulations governing such diverse activities as telephone communication, air travel, banking, insurance and other financial operations.
- New regulatory activities have had to be set up, such as those overseeing the internet and other new forms of communication.
- Globalisation has led to the need for new international organisations. Significant amounts of national power have been transferred to such international authorities as the WTO, the European Union (EU), the North American Free Trade Agreement (NAFTA), MERCOSUR, and some branches of the United Nations (UN). This process has been resisted, sometimes violently, by many individuals and NGOs, and hence has been fraught with conflict.

*Performance*

The items listed below primarily affect performance. Because new products often require new processes and new forms of organisation, many, if not most, items also cause changes in the facilitating structure. For example, distant surgery may require significant alterations in the organisation and regional specialisation of hospitals, and distant learning will probably eventually do the same to universities.

- Automatic teller machines (ATMs) make it easy to access one's bank account and obtain funds in any currency in almost any part of the world – in sharp contrast to the major difficulties experienced in the past when one was caught short of cash on a weekend at home or in another city at any time. The convenience of this wonderful, computer-driven innovation is hard to measure, but those who have travelled in earlier times know just how great it is. This change also has effects in the 'facilitating structure' category above.
- Subscriber trunk dialling has replaced operator-assisted calls that were expensive, slow to complete, and all too often interrupted. This change also has effects in the 'Facilitating structure' category above.
- The computer-enabled internet is changing interpersonal relations. Chat rooms make relations possible on a scale never seen before. Children and adults have email pals in all parts of the world.
- Children do school work by consulting the internet. Instead of hearing only the received wisdom from their teacher and the prescribed texts, they are now exposed to a battery of diverse knowledge and opinion, and at a very early age have to learn how to cope with more than one view on any subject.
- Computerised translation is now a reality and will go from its present crude form to high degrees of sophistication within the lifetimes of most of us. We are witnessing the arrival of Douglas Adam's vision in *The Hitch Hiker's Guide to the Galaxy*: the ability to hear in one's own language words spoken in any other, and to be understood in any other language while speaking one's own. The only difference is that instead of inserting a fish into one's ear, as did the hitch hiker, a small computer will be attached to one's body.
- Research in everything from economics to astronomy has been changed dramatically by the ability to do complex calculations that were either impossible or prohibitively time-consuming without electronic computers. This change also has effects in the 'Facilitating structure' category above.
- Crime detection in the computer age is much more sophisticated than it was in the past. This change also has effects in the 'Facilitating structure' category above.
- Traffic control in the air and on the ground is being, or has been, revolutionised in many ways; cars will soon receive real time information on traffic conditions at all points in their projected journey. This change also has effects in the 'Facilitating structure' category above.

- Smart buildings and factories already exist and will grow rapidly in number. Among many other advantages, their power consumption is adjusted continually in response to real time price signals sent out by the electricity supply company and calculated in response to current loads. This change also has effects in the 'Facilitating structure' category above.
- The electronic book looks like it might do an end run around consumer resistance to reading books on screen. (But then again, in an illustration of the uncertainty associated with all these new developments, it may fail to gain consumer approval.) The book's blank pages fill up on demand, with any one of a hundred or more books stored in a chip that is housed in its cover. A touch of a button, and one is reading a Physics 101 text in what looks like a conventional book; with only another touch, a Chemistry 202 text replaces the other on the book's leaves.
- Distant education is growing by leaps and bounds; today many North Americans never (or only rarely) set foot inside the institution that they are 'attending'. This change also has effects in the 'Facilitating structure' category above.
- Surgery on hips, knees and other intricate parts of the body is increasingly done by computers, which are more accurate than even the most skilled surgeons. Distant surgery is only a short time away: patients in remote parts of the world will routinely be operated on by specialists located in major urban hospitals. This change also has effects in the 'Facilitating structure' category above.

The above list is only a selection of the many social, economic and political transformations brought about by the ICT revolution. As the list suggests, the new economy is having a greater impact on services than on durable consumer goods. Robert Gordon (2000) correctly points out that the ICT revolution has brought with it nothing like the series of radical new consumer durables that accompanied the electricity revolution. Indeed, each transforming GPT has different effects, and this one has been felt mainly on process and organisational technologies, in the political structure, and in the output of services. This does not make it any less of a revolution than that accompanying electricity and mass production, but it makes it different.

I marvel at how many economists can assert, first, that all of these rich events can be adequately summarised in one series for productivity (usually TFP) and, second, that the existence or non-existence of the entire ICT revolution depends on how that number is now behaving in comparison with how it behaved over the past couple of decades!

Some worry that, if computers cease their hitherto relentless improvements, much of the force might go out of the new economy. We have observed, however, that new GPT creates research programs to improve the GPT itself and to apply it across the whole economy in new processes, new products, new organisational forms, and new political and social structures. Even if the price of computing power stopped falling tomorrow (a very unlikely event), decades would pass before the developments implicit in what has already happened in the ICT sector (including the internet) have fully worked through the economy.

## The ICT revolution and productivity[10]

Concern has been expressed over the slow increase in measured productivity in most industrialised economies over the last quarter of the twentieth century. I argue that the expectations of a productivity bonus *necessarily* accompanying the introduction of a new GPT in advanced industrialised economies are ill founded. Growth economists typically have these expectations because they work with a model that cannot have technological change without productivity growth. I reject these expectations for several reasons.

First, although a new technology will be instituted whenever it promises to be profitable, there is no guarantee that each new GPT will have larger, or even the same, effects on profits and productivity than the ones that preceded it (however these are measured). As already observed, a new GPT presents a research program to apply it to everything in the economy – a program that itself evolves as the power and efficiency of the GPT itself is steadily improved. The resulting innovations spread over decades in a process of linked inventions and innovations. There is, however, no reason to expect that the total impact of successive GPTs should stand in any temporal relation to each other. One may introduce a rich program that brings large changes in products, processes and organisational arrangements, and perhaps productivity; another may introduce a program that is less rich. Furthermore, one that has a larger impact than another may take longer to work through the economy and thus show smaller gains in each year. For example, many of the effects of the ICT revolution on new design and production methods occurred in the period 1975–2000, so the gain was spread over decades.

Second, the extent to which the new technology pervades the economy and the extent of the induced changes in the facilitating and policy structures, right up to deep social transformations, bear no necessary relation to the induced changes in productivity or real wages. There is only a paradox when neoclassical growth theory, which cannot distinguish between changes in technology, the facilitating structure and productivity, is used to interpret what is going on.

Third, if no further GPTs were invented to provide new research programs, the number of derivative technological developments would eventually diminish. There would be further innovations using existing GPTs, but their number and their productivity would be much less than if further GPTs were to become available. New GPTs, such as computers, electricity and steam, prevent the number of efficiency-increasing innovations from petering out. This prevents what would otherwise be a steady decline in the return on investment and in productivity-increasing innovations. But there is no reason to believe that each GPT will increase the average rate of productivity growth over all previous GPTs. If each did, we would see a secular trend for productivity to rise as each GPT succeeded its predecessor. Carlaw and Lipsey (2001, 2002 forthcoming) discuss GPTs as research programs, and discuss and model formally relation to productivity growth.

Finally, when a new GPT is introduced, there are reasons why it may slow the growth of productivity below the average over its lifetime. It takes much time for the range of a GPT's use and applications to evolve, for ancillary technologies to be developed, and for changes to be made in all of the elements of the facilitating and policy structures that support it. Typically, several decades are required for a GPT to make a major impact – and that impact may then stretch over more than a century as new technologies that are enabled by the GPT are developed. Electricity is a prime example, as argued by Paul David (1991). If there is a long lag, there may be a slowdown in productivity growth in the early stages of any one GPT, followed by an acceleration to its average rate after the facilitating structure has been fully adopted but its full potential has not yet been worked out. But this is not necessary. The possibility of a slowdown is even more problematic because at any one time there are likely to be several GPTs – at least one in each of the categories listed above, and each at various stages of its development. Moreover, the existing GPT in any one category has not usually been fully exploited when another challenges it.

I conclude that there may be a productivity bonus around the corner but there is no guarantee of this. Furthermore, the existence or non-existence of such a bonus tells us nothing about how profound are the transformations that have already been brought about by the current ICT revolution, or that are now being brought about. I hasten to add that the ICT revolution does seem to be improving correctly measured productivity, so the productivity slowdown is being at least partially reversed. Nonetheless, there is no reason in technology or economic theory why productivity growth over the next couple of decades should equal or exceed growth during the last period, 1945–75, when GPTs were being exploited within a stable structure that was well adapted to them.

This discussion of the 'productivity paradox' is mainly directed at developed countries that are operating at or near the technological frontier. Catch-up economies whose problem is to learn how to adopt, adapt and effectively utilise technologies already in existence can, if they succeed, achieve several decades of rates of growth of per capita income that advanced countries could not sustain for more than a very few years.

## What is next?

Because innovation takes place under conditions of genuine uncertainty, predicting the course of future technological developments is hazardous. For example, for some decades in the second half of the twentieth century, many thought that atomic power was to be the next power-generating GPT. Those early hopes were dashed, although they may be revived in coming decades. But some GPTs can be spotted, if not when they first emerge, at least when they are still in the early stages, long before they begin to transform the economic and policy structures. Biotechnology is clearly a coming GPT with myriad applications across almost the entire economy, as well as having the power to transform human life itself. Nanotechnology, the ability to make materials and machines upwards one atom at a time rather than downwards by shaping existing materials, has enormous potential. It is almost sure to be one of the great transforming GPTs of the second half of the twenty-first century. I have written about these emerging technologies in more detail (Lipsey 1999). Finally, quantum computing and quantum energy may provide powerful GPTs in the more distant future, but this really is speculative.

## SOME POLICY IMPLICATIONS[11]

Neoclassical general equilibrium theory is structureless. Its equations apply to all markets everywhere, and it produces a single set of policy prescriptions applicable to all places and all times: remove market imperfections. These are defined as anything that prevents the achievement of an optimal allocation of resources. In the case of invention and innovation, a positive externality is recognised and, therefore, a generalised 'non-distorting' subsidy to R&D is recommended.

Neoclassical microeconomic theory is invaluable for dealing with a wide range of problems, but it runs into trouble in dealing with endogenously determined technological change. As Ken Carlaw and I have pointed out in detail in a series of publications, referenced in the footnote to this section's heading, this neoclassical advice does not take into account what is known through both empirical studies and S-E theory concerning endogenous technological change. Like Paul Romer's macro endogenous growth theory,[12] S-E theory recognises that unique optimal policies cannot be derived for knowledge creation. Romer emphasises the non-rivalrous but partially appropriable nature of knowledge, which upsets the standard requirements for an optimal allocation of resources – perfect property rights and competitive markets. S-E theories emphasise the (Knightian) uncertainty associated with technological advance. The key point about uncertainty in this context is that two equally well-informed agents seeking similar technological goals can rationally adopt differing policies and neither can be assessed as right or wrong ex ante. Competing firms do this all the time. In effect, the agents are backing different horses in a race in which the odds are unknown.

Because there is no optimum allocation of resources when technology is changing endogenously under conditions of uncertainty, there is no set of scientifically determined, optimum public policies for technological change in general and R&D in particular; nor is there such a thing as a neutral or non-distorting set of policies. In the world described by S-E models, dynamic efficiency is as inapplicable a concept as is static efficiency. If there are no unique optimum rates of R&D, innovation or technological change, policy with respect to these matters must be based on a mixture of theory, measurement and subjective judgement.[13]

As Ken Carlaw and I have observed elsewhere (Lipsey and Carlaw 1998a: p.47):

> The need for judgment does not arise simply because we have imperfect measurements of the variables that our theory shows to be important, but because of the very nature of the uncertain world in which we live. Although a radical idea with respect to micro-economic policy, the point that policy requires an unavoidable component of subjective judgment is commonly accepted with respect to monetary policy. For two decades from the mid-1950s to the mid-1970s, Milton Friedman tried to remove all judgment from the practice of central banking by making it completely rules-based. When his advice was followed by several of the world's central banks, the monetary rule proved ineffective in mechanically determining policy, as many of his critics had predicted it would. Today, the practice of central banking is no different from the practice of most economic policy: it is guided by theoretical concepts; it is enlightened by many types of empirical evidence (including those emphasised by Friedman) that are studied for the information that they provide; and, in the end, all of these are inputs into the judgment calls that central bankers cannot avoid making.

Thus, both uncertainty and the non-rivalrous nature of technological knowledge create a context-specific, path-dependent world in which there are better and worse policies but no unique optimal set for all times and all places.

## Policies for technological dynamism

The thrust of the above analysis is to undermine the neoclassical position that there is a unique set of policies applicable to all times and places. This makes science and technology policies context-specific in many ways.[14]

### *Catch-up versus leading edge*

As one example, the technological problems facing catch-up economies are different from those facing economies that are at or near the leading edge of technological advance. Catch-up economies, especially in their early stages, have the advantage of dealing with established technologies. Although there are uncertainties associated with local adaptations of generic technologies and the acquisition of tacit knowledge, many of the uncertainties that are associated with cutting-edge advances are absent.

The belief that civil servants knew better than private sector agents, and could efficiently dictate R&D decisions to them, seldom if ever achieved good results, either in catch-up or in leading-edge situations. But many market-oriented Asian countries in the catch-up stage have championed consultative processes whereby the government agency and the main private sector agents pooled their knowledge, came to a consensus on where the next technology push should be and then jointly financed it. This policy often worked well in catch-up situations, and it still can work well when all private agents are pushing for a fairly well-defined, smallish advance in technology. But the inevitable uncertainties associated with more major technological advances are usually more efficiently accomplished through many investigations, each pursued with the minimum funds required. In these cases, concentrating effort, even after a national consensus has been reached, is usually worse than the apparent 'wastefulness' of uncoordinated experimentation in the free market.

*Government institutional competence*

Another illustration of context specificity lies in the government's institutional competence to frame and carry out various policies. A policy that meets textbook requirements may be a great success or a total failure, depending on the nature of the governmental structure that devises it and the civil servants who administer it.

One important context is the basic constitution. For example, the central governments of large and diverse federations with strong regional governments must broker regional interests in ways that are not required in smaller, more homogeneous countries. Various constitutions make it easy or difficult for special interest groups to capture specific policies and turn them to their own uses, which seldom coincide with the national interest. Also, various public sector institutions have different capabilities, based partly on constitutional differences, partly on the power relations between various special interest groups, partly on the nature of their civil services, and partly on the accumulated 'learning by doing' in operating their country's typical set of policy instruments in the past.

Thus, it is not enough to have policies that are well designed in the abstract. Policies must work through an institutional structure, and their success depends to a great extent on the institutional competencies of those administering them. One implication of this is worth mentioning. As long as they are understood to be talking about the institutional capabilities of their own public sectors, US economists may be right in asserting that minimal government is preferable to activist government, while many development economists may also be right in asserting the opposite about *some* developing countries and *some* activist policies. Once again, whether a policy is good or bad depends on the context.

*Support for infant industries*

Only a few countries have developed a modern industrial structure without protection of one kind or another for emerging local industries in the early stages of their development. Many countries, including the United States, Canada and the countries of continental Europe, went through the early stages of industrialisation with substantial tariff protection for their infant industries.

Economists have usually sought to rationalise the infant industry argument by an appeal to static economies of scale. However, the real significance of the infant industry argument turns on two aspects emphasised in S-E theory.

The first is endogenous technological change. When technological change is endogenous, the encouragement of a successful infant industry, such as electronics in Taiwan, can shift the whole set of cost curves downwards as new technologies are imported, and/or developed locally, to produce at lower cost than anything previously in existence. This is not an easy thing to do, as is attested to by countless failures to develop such industries through public policy. But neither is it impossible, as is attested to by some major successes, such as the Japanese car industry after World War II and Taiwanese electronics. The main point is that standard textbook treatments misstate the infant industry problem as one of moving downwards along a pre-existing, negatively sloped, long-run cost curve based on constant technology. The real problem is to develop an industry whose rate of technological change compares favourably with those of its foreign competitors.

The second aspect of S-E theory that is relevant to infant industries is the dynamic path that an industry must follow as it develops through the creation and acquisition of both technologies and the needed elements of the facilitating structure, including human capital. Technologies that are wholly new to a country typically require new facilitating structures. Creating many new elements of this structure can impose net costs on the firms that begin a cluster of related firms and give net benefits to those that enter the cluster later. Thus, many of the benefits from creating new elements of the structure cannot be reaped by those who do the creating, because the externalities take the form of scale effects.[15] Infant industry protection, provided it is judiciously applied and 'sunsetted', can help to develop a structure that would not arise solely from profit-motivated actions of private firms. Examples of clusters that were created, or at least strongly assisted by, public policy include the highly successful one at Austin Texas.

### *Technologies can be singled out*

Neoclassical theory calls for a generalised subsidy to all R&D as *the* optimal for encouraging new technologies. It is opposed to policies that focus on any subset of the economy. But, as I have already observed, the intellectual basis of this advice – the existence of a unique optimal allocation of resources – is undermined by both Romer's emphasis on the non-rivalrous nature of technological knowledge and the S-E emphasis on uncertainty. The pros and cons of specific assistance must, therefore, be decided on context-specific empirical experience.

In practice, many important technologies have been encouraged in their early stages by public sector assistance. Publicly funded US land grant colleges have done important agricultural research from their inception in the nineteenth century. The 'green revolution' was to a great extent researched by public funds. In its early stages, the US commercial aircraft industry received substantial assistance from the National Advisory Committee on Aeronautics (NACA). The airframe for the Boeing 707 and the engines for the 747 were both developed in publicly funded military versions before being transferred to successful civilian aircraft. Electronic computers and atomic energy were largely created in response to military needs and military funding. For many years, support for the US semiconductor industry came mainly from the military, whose high standards and quality controls helped to standardise practices and diffuse technical knowledge. The US government's heavy involvement in the early stages of the US software industry produced two major spin-offs to the commercial sector. One was an infrastructure of academic experts, built largely with government funding; the other was the establishment of high and uniform industry standards. The US biotechnology sector was heavily assisted by public funds during the early stages of its development. Indeed it can be argued that the superiority of US public funding helps to account for the current US lead in private sector biotechnology.[16]

These events did not occur in a laissez-faire economy. Although the list could be greatly extended, the examples seem sufficient to illustrate that knowing when and how to use public funds to encourage really important new technologies in their early stages is an important condition for remaining technologically dynamic, at least in some areas. I hasten to add that this is no easy task. Picking winners is a policy bestrewn with many failures and a few spectacular successes. In most countries, decision-taking by bureaucrats and capture of policies by special interests are sufficient to create failure. The special conditions needed for success are discussed in Lipsey and Carlaw (1998b). I also hasten to point out that focused encouragement merely means more focused than is provided by such economy-wide policies as R&D subsidies or R&D tax relief. Attempting to advance a new technology by encouraging one national champion and/or dictating specific research directions is usually a route to failure. But picking some broad new technology and assisting many firms to push towards developing it in competition with each other, relying to a significant extent on their own funds, is an intermediate policy that cannot be rejected on current evidence.

### The market-oriented consensus: necessary or sufficient?

As a result of experience over a wide range of countries – including the successful and less successful developing nations, the former Union of Soviet Socialist Republics, and the countries of Eastern Europe – a new consensus on economic policy for developing nations emerged in the late 1980s and early 1990s.[17] According to this consensus, policies that ignore, or work against, market forces are usually counterproductive. The new view calls for a more outward-looking, trade-oriented, market-based route to development.

It calls for accepting the value of the price system as a device for coordinating the decentralised decisions of agents who are reacting to prices that correctly signal the relative scarcities of resources and products.

The acceptance of the need for market orientation raises the key question for developing countries: is market orientation enough? Many economists argue that creating a market-oriented environment is a sufficient goal for the public policy of developing nations. They say that, if that environment is created, the activities of domestic agents and foreign multinationals will bring growth and development without any need for a more proactive policy.

Others argue that, although necessary, instituting the measures of the market-oriented consensus is not sufficient. They argue that newer theories in the S-E tradition show the need for more focused policies – always understanding that these are in addition to, not substitutes for, the market-orienting measures just outlined.

These ideas are both powerful and dangerous. They are powerful because they suggest ways to go beyond neoclassical generic policy advice to more context-specific advice. They are dangerous because they can easily be used to justify ignoring the market-oriented consensus and accepting only the interventionist part of the S-E policy advice (forgetting that this is meant to supplement the advice of the consensus, not to replace it).

Those who believe that market liberalisation is only a necessary condition argue that, just as the original NIEs followed many proactive industrial and science and technology policies, so must those countries that follow do the same. This group points out that, even in developing countries that are growing rapidly, incomes are often polarising as increasing inequalities create social pressures that may undo growth programs[18] (see, for example, Peters 2000). They also point out that many South American countries that have liberalised their economies but done little more by way of proactive growth policies are doing less well than they were under the old import substitution programs. Katz (2001) is one author who discusses this issue. No government wants to return to the full set of old policies, raising prohibitive tariffs and stepping aside from the global economy. But some feel that for countries at an early stage of development there is a need for more active policies to encourage industry and innovation, because such countries have strong and frequent impediments to purely private sector growth. Others dissent for a host of reasons, both theoretical and practical. The outcome of this debate will influence developing nations for decades to come.

## NOTES

1 Plus the early stages of the revolution in biotechnology.
2 Carlotta Perez first wrote about this phenomenon in 1983; she and Christopher Freeman took it up in 1988 (Freeman and Perez 1988). Both framed their analysis in terms of what they called a 'technoeconomic paradigm', defined as a systemic relationship among the economy's technology, structure and economic

performance, typified by a few key products with wide application, a few key materials whose costs are falling over time, a characteristic way of organising materials whose costs are falling over time, a characteristic way of organising economic activity, and a characteristic supporting infrastructure. Their work has been influential in Europe but largely ignored by growth economists in the United States. I first wrote about the 'new economy' in 1991 (although I did not use that term, referring instead to the ICT revolution). Originally, I used the technoeconomic paradigm as an organising principle, but later my co-authors and I shifted, for reasons explained in Lipsey and Bekar (1995), to the closely related concept of a structuralist-evolutionary (S-E) model, which is described later in this paper.

3   This section relies heavily on Lipsey and Bekar (1995).
4   Accepting the overwhelming importance of technological change in determining our long-term economic situation does not imply economic determinism (that technology is sufficient to determine all social and economic outcomes). The same technology introduced into different social and institutional structures typically produces different results, which would not be the case if technology determined everything. A classic example is the very different effects that the technology of television has had on the political processes of the United Kingdom and the United States.
5   Such cases are studied in detail in Lipsey and Carlaw (2002 forthcoming) and lie behind the arguments advanced by Jorgensen and Griliches (1967) that total factor productivity measures only the externalities (free gifts) created by technological change.
6   For a detailed consideration of these characteristics see Lipsey, Bekar and Carlaw (1998a).
7   See, for example, Lipsey and Bekar (1995) and Lipsey, Bekar and Carlaw (1998a) for past technologies, Lipsey (1999) for some future ones, and Lipsey, Bekar and Carlaw (1998b) for some generalisations about the impacts of new GPTs.
8   In contrast, virtually all theoretical models of GPTs admit them only one at a time and assume that they all have identical effects. The models in Carlaw and Lipsey (2001) have removed the latter, but not the former, restriction.
9   In a steam-driven plant a large engine drove a central drive shaft off which ran belts to each machine. To reduce the large power loss caused by belt slippage, the machines that used most power were placed closest to the shaft. When electrical drives eventually took the form of a separate engine on each machine, it was possible to rearrange the machines in the order of their use.
10  This section relies heavily on Lipsey (2002b).
11  A full development of the theoretical analysis in this section requires much more space than can be allocated here. For elaboration, see a number of publications by my colleagues and I: Lipsey (1997, 2000 and 2002a forthcoming) and Lipsey and Carlaw (1996, 1998a and 1998b).
12  The relation between S-E theory and Romer's approach is discussed in detail in Lipsey (2000).
13  Abandoning the concept of an optimum allocation of resources upsets what I call the formal defence of the price system but leaves intact the much more powerful and persuasive informal defence. See Lipsey (2002b).
14  For discussion of how this applies to developing countries see Lipsey (2002a).
15  Bekar and Lipsey (2002) discuss these agglomeration effects in more detail.
16  These examples and several others are studied in more detail, and lessons are drawn from them, in Lipsey and Carlaw (1996).
17  At least the consensus is accepted among governments and economists. Many members of NGOs, such as those who protested at the opening of the WTO

talks in Washington in December 1999 and the FTAA talks in Quebec City in May 2001, profoundly mistrust large firms and free markets.

18 The growing inequities that sometimes accompany the early stages of growth are not typically the result of adverse effects of growth being localised in some areas or groups. Instead, they more typically result from an uneven distribution of the gains, with some groups and areas benefiting more than others.

## REFERENCES

Bekar, C. and Lipsey, R. (2002 forthcoming) 'Clusters and economic policy', Isuma, Special Issue on Policies for the New Economy.

Carlaw, K. and Lipsey, R.G. (2001) 'Externalities versus technological complementarities: a model of GPT-driven, sustained growth', Paper presented to the conference in honour of the 20th anniversary of Nelson and Winter's book *An Evolutionary Theory Of Economic Change*, Aalborg, Denmark, 12–15 June 2001. To be published; now available at http://www.sfu.ca/~rlipsey

—— (2002) 'Externalities, technological complementarities and sustained economic growth', *Research Policy*, 31: 1305–1315.

Chen, E.K.Y. (1997) 'The total factor productivity debate', *Asian-Pacific Economic Literature*, 11, 1: 18–38.

David, P. (1991) 'Computer and dynamo: the modern productivity paradox in a not-too-distant-mirror', in *Technology and Productivity*, Paris: OECD: 315–347.

Freeman C. and Perez, C. (1988) 'Structural crisis of adjustment: business cycles and investment behaviour', in Dosi, G., Freeman, C., Nelson, R.R., Silverberg, G. and Soete, L. (eds) *Technical Change and Economic Theory*, London: Pinter.

Gordon, R.E. (2000) 'Does the 'New Economy' measure up to the great inventions of the past?', National Bureau of Economic Research Working Papers: 7833.

Jorgensen, D. (2001) 'Information technology in the U.S. economy', *American Economic Review*, 91: 1–31.

—— and Griliches, Z. (1967) 'The explanation of productivity change', *Review of Economic Studies*, 34, 3: 249–83.

Katz, G. (2001) *Structural Reforms, Productivity and Technology Change in Latin America*, Santiago: Economic Commission for Latin America and the Caribbean.

Krugman, P. (1996) '*The Myth of Asia's Miracle*' in *Pop Internationalism*, Cambridge MA: MIT Press.

Lipsey, R.G. (1991) *Economic Growth: Science and Technology and Institutional Change in a Global Economy*, Toronto: Canadian Institute for Advanced Research, Publication 4.

—— (1997) 'Globalization and national government policies: an economist's view', in Dunning, J. (ed.), *Governments, Globalization, and International Business*, Oxford: Oxford University Press.

—— (1999) 'Sources of continued long-run economic dynamism in the 21st century', in Michalski, W. (ed.), *The Future of the Global Economy: Towards a Long Boom*, Paris: OECD.

—— (2000) 'New growth theories and economic policy for the knowledge economy', in Rubenson, K. and Schuetze, H.G. (eds), *Transition to the Knowledge Society: Policies and Strategies for Individual Participation and Learning*, Vancouver BC: University of British Columbia Press.

—— (2002a) 'Some implications of endogenous technological change for technology policies in developing countries', in *The Economics of Innovation and New Technology*, (EINT), 11, 4–5: 321–351.

—— (2002b) 'The productivity paradox: A case of the emperor's new clothes', *ISUMA: Canadian Journal of Policy Research*, 3, 1: 120–126.
Lipsey R.G. and Bekar, C. (1995) 'A structuralist view of technical change and economic growth', in Courchene, T.J. (ed.), *Technology, Information and Public Policy*, Kingston ONT: John Deutsch Institute.
——Bekar, C. and Carlaw, K. (1998a) 'What requires explanation', Chapter 2 in Helpman, E. (ed.), *General Purpose Technologies and Economic Growth*, Cambridge MA: MIT Press: 5–54.
—— (1998b) 'The consequences of changes in GPTs', Chapter 8 in Helpman, E. (ed.), *General Purpose Technologies and Economic Growth*, Cambridge MA: MIT Press: 194–218.
Lipsey, R.G. and Carlaw, K. (1996) 'A structuralist view of innovation policy', in Howitt, P. (ed.), *The Implications of Knowledge Based Growth*, Edmonton: University of Calgary Press.
—— (1998a) 'Technology policies in neoclassical and structuralist-evolutionary models', in *Science Technology Industry Review*, No. 22, Special Issue.
—— (1998b) 'A structuralist assessment of technology policies: taking Schumpeter seriously on policy', Industry Canada Working Paper Number 25, Ottawa: Industry Canada.
—— (2002) 'What does total factor productivity measure?', to be published as a Statistics Canada Technical Paper. Now available at http://www.sfu.ca/~rlipsey
Lipsey, R.G. and Wills, R. (1996) 'Science and technology policies in Asia Pacific countries: challenges and opportunities for Canada', in Harris, R. (ed.), *Growing Importance of the Asia–Pacific Region in the World Economy: Implications for Canada*, Edmonton: University of Calgary Press.
Nelson, R. (1999) 'The Co-Evolution of Technology and Institutions', manuscript.
Nelson, R.R. and Pack, H. (1999) 'The Asian growth miracle and modern growth theory', *Economic Journal*, 109: 1–21.
Pack, H. (2001) 'Technological change and growth in East Asia', Chapter 3 in Stiglitz and Yusuf (2001).
Pack, H. and Westphal, L.E. (1986) 'Industrial strategy and technological change: theory or reality?' *Journal of Development Economics*, 22: 87–128.
Perez, C. (1983) 'Structural change and the assimilation of new technologies in the economic and social systems', *Futures*, 15, 5: 357–375
Peters, E.D. (2000) *Polarizing Mexico: the Impact of Liberalization Strategy*, Boulder: Lynne Rienner.
Rosenberg, N. (1976) *Perspectives on Technology*, Cambridge: Cambridge University Press.
—— (1982) *Inside the Black Box: Technology and Economics*, Cambridge: Cambridge University Press.
Smookler, J. (1966) *Invention and Economic Growth*, Cambridge MA: Harvard University Press.
Stiglitz, J. and Yusuf, S. (2001) *Rethinking the Asian Miracle*, Oxford: Oxford University Press.
Westphal, L.E. (1990) 'Industrial policy in an export-propelled economy: lessons from South Korea's experience', *Journal of Economic Perspectives*, 4, 3: 41–60.
Young, A. (1992) 'A tale of two cities: factor accumulation and technical change in Hong Kong and Singapore', National Bureau of Economic Research Macroeconomic Annual, Cambridge MA: MIT Press.
—— (1995) 'The tyranny of the numbers', *Quarterly Journal of Economics*, 110, 3: 641–80.

# 4 Telecommunications policy and the evolution of the internet

*Robert W. Crandall*

The rise of the internet has been dramatic. Ten years ago there was no internet; the World Wide Web had just been born in a laboratory in Switzerland. The networking of computers was limited to universities, research laboratories, and a few large businesses. As late as 1994, many individuals in the developed economies were still only vaguely aware of the existence of a ubiquitous network through which they could access certain information and send messages. Even by the dawn of the new millennium, only 40 per cent of United States households had internet connections,[1] and the percentage was substantially lower in other Organisation for Economic Co-operation and Development (OECD) countries.

Other countries are now catching up with the United States. Soon, most households in OECD countries will be connected to the internet. But will the internet continue to evolve to reach its full potential, particularly in the developing countries? What are the barriers to the full exploitation of its potential? Clearly, no one knows for sure, but I believe that an economist who is familiar with network industries in general and telecommunications in particular can offer some advice even before the evidence is in on the recent 'deregulatory' movement in telecommunications. The growth of the internet is impossible without telecommunications policies that encourage the development of the required infrastructure and use of that infrastructure.

In this paper, I provide a brief description of the evolution of telecommunications in the past two decades, focusing on the characteristics of modern telecom services that are relevant to the development of the internet: the 'death of distance' and the importance of high-speed communications. I then describe how the continued growth of the internet and the development of e-commerce requires the widespread deployment of high-speed, 'broadband' circuits to households and small businesses. Unfortunately, providing such high-speed internet access with the current infrastructure is neither easy nor inexpensive.

Many countries are considering some form of subsidy to accelerate the deployment of high-speed lines and the exploitation of the network effects from such deployment. In so doing, however, they risk repeating earlier errors in telecommunications policy that have discouraged the use of the telephone network for both voice and internet applications. Traditional telecommunications policies that recover fixed network costs from high local and long-distance calling charges are often defended as necessary to allow all households to connect to basic telephone services, that is, to ensure 'universal service'. I will show that such policies are unnecessary and are a major impediment to the development of the new broadband architecture that is crucial to the further development of e-commerce. Asian countries – both developed and developing – would be well advised to learn from these errors and avoid pursuing regulatory policies that are antithetical to the development of modern broadband internet services.

## THE DEATH OF DISTANCE – THE RISE OF SPEED

For decades, 'telecommunications' generally meant voice telephony, usually over limited distances. Long-distance calls – national and international – were a luxury because of their prices. At first, these prices reflected the economics of transmission through copper-wire facilities, but later they were the result of a deliberate regulatory policy to generate revenues for cross-subsidies. Over time, however, changes in technology were so dramatic that the cost of sending a voice message through fibre-optic cables halfway around the world fell to less than the price of local calls in many OECD countries.[2] We are now at the threshold of distance-insensitive pricing, because it costs very little more to call New York from Canberra than to call across town. In fact, the cost of measuring the call and billing the customer may soon exceed the cost of letting the subscriber make unlimited calls at a zero price. This technological imperative has been dubbed 'the death of distance' by Frances Cairncross of *The Economist* (Cairncross 2001).

If distance is dead, speed is not. At the dawn of the internet, we simply connected to the network through a converter, or modem, that allowed us to receive data through the telephone network's copper wires at speeds that were as slow as 9.6 kilobits per second (kbps), but that have since risen to a maximum of 56 kbps through ordinary dial-up connections, a rate that is still much too slow. For new applications, such as video streaming, even ordinary household connections will require speeds of 1 megabit per second (Mbps) or even more. The old telephone network is quite simply incapable of delivering the necessary speed, and many regulators have failed to notice the problem. Modern training techniques have made today's athletes much faster than their ancestors; today's telephone network is in urgent need of similar 'conditioning'.

## MODERN TELECOMMUNICATIONS POLICY IS BASED ON FAULTY NOTIONS OF 'UNIVERSAL SERVICE'

Telephone service is still highly regulated in most countries despite the recent wave of 'deregulation' policies. Incumbent carriers' rates and services are carefully controlled by regulators, even after the market has been opened to competition, under the rationale of assuring 'universal service' to all households. Because the value of telecommunications to any user depends on the number of others connected to the network, policy-makers are understandably concerned about assuring that a large share of households choose to subscribe to ordinary telephone services. Therefore, there is at least a theoretical reason, rooted in these 'network externalities', for keeping residential rates for network connection quite low. Unfortunately, this regulatory strategy creates a revenue deficit for telecommunications carriers that must be recovered somehow. Regulators typically recover this deficit by allowing rates for local and long-distance calls to be far above their costs and by establishing similarly high rates for most business services. As I shall show, these 'universal service' pricing policies are a serious impediment to the development of the internet.

In developed economies, the universality of telephone services is not affected by such policies because virtually every household would subscribe to telephone services at virtually any market-determined rate.[3] Rather, these policies are driven by principles of income redistribution from urban residents to rural ones, from intensive network users to those that make few calls, and from businesses to residences. As with most such cross-subsidy schemes, a large share of the rents created by the high calling prices is dissipated in inefficient incumbent-firm operations. The share that reaches its target population has a value that is far less than the value forgone in reduced calling, and the losses in consumer surplus due to this over-pricing of calls are enormous.[4]

Once in place, universal service policies develop a political constituency because of the skewed distribution of using the old voice network (Crandall and Waverman 2000). A large share of households simply do not make many calls; therefore, they like the low monthly rates and very high call charges. Rural residents are subsidised by urban residents, not only through artificially high prices for agricultural commodities, but through low rates for telephone (and electricity) services. Wealthy French farmers or US ranchers enjoy low rates for connecting to the telephone network, even if they are using four or five lines each to engage in modern telecommunications. Unfortunately, this subsidy can be sustained only if the incumbent telephone company continues to obtain supra-competitive prices for calls and urban connections.[5]

Competition is beginning to reduce the price of long-distance or 'national' calls in most countries. The real price of a three-minute or ten-minute national call in the European Union (EU) has declined by nearly 50 per cent in the last three years and is now about equal to national calling rates in the United States and Canada (Commission of the European Communities 2000). Unfortunately, during the same period, the real price of a three-minute local call in the EU has actually risen. Indeed, there has been very little change in local calling charges in the EU since 1990 (see Table 4.1). During the 1990s,

*Table 4.1* The real price of a three-minute call (1995 local currency units)

| Country | 1990 | 1999 |
| --- | --- | --- |
| *European Union* | | |
| Belgium | 6.72 | 5.67 |
| France | 0.81 | 0.71 |
| Germany | 0.27 | 0.20 |
| Italy | 166[a] | 207 |
| Luxembourg | 5.7 | 4.8 |
| Netherlands | 0.32 | 0.26 |
| Portugal | 12.0 | 16.4 |
| Spain | 5.6 | 13.5 |
| United Kingdom | 0.016[b] | 0.011 |
| *North America* | | |
| United States | 0 | 0 |
| Canada | 0 | 0 |
| *Asia and Oceania* | | |
| Australia | 0.22[c] | 0.28 |
| New Zealand | 0 | 0 |
| Hong Kong | 0 | 0 |
| Japan | 10.7 | 9.8 |
| Korea | 37.1 | 37.9 |
| Malaysia | 0.16 | 0.08 |
| Philippines | 0 | 0 |
| Singapore | 0 | 0.19 |
| Thailand | 3.79 | 3.13 |

Source: ITU (2001).

Notes
a  1991 figures.
b  1992 figures.
c  1991 figures.

the cost of switching a local call has fallen dramatically due to innovation in circuit-switching technology. Today, the incremental cost of a three-minute local voice call is no more than 0.04 French francs, 0.014 German marks, or 0.012 Australian dollars,[6] but local rates in most countries remain far above this cost to defray the costs of unnecessary 'universal service' policies. Fortunately, local calling rates have remained at zero in North America and in some Pacific Basin countries, such as New Zealand, Hong Kong, and the Philippines.

Modern telecommunications networks must be redesigned to accommodate much higher speeds and the switching of digital packets transmitted over the internet.[7] But the pricing of modern telecommunications services that are delivered in packets cannot be based on elapsed time, because the packets, once dispatched, do not tie up any given circuit and do not arrive at their destination at predictable intervals. Rather, they course through the network, sharing space with other packets, arriving at their destinations in a period of time that may only be a few milliseconds. The internet is not 'free'; it simply operates on a very different principle from ordinary, circuit-switched analogue telephony.

As more and more traffic is digitised and converted into packets, the old circuit-switched telephone network becomes obsolete. Were telephone companies not regulated, they would now be converting their networks much more rapidly to fully digital, packet-switched architectures. Voice calls would simply become one type of communications that are converted into packets, albeit one that accounts for a small share of the total packets distributed through the network. These voice calls would no longer be charged on the basis of elapsed time. But this world cannot exist if regulators insist on using excess charges for voice calls to subsidise network connections. The carriers will not invest in the new architecture if such investment deprives them of the call-based revenues that support their entire operations. Universal service policies are, in reality, 'suppression of service innovation' policies.

These problems of recovering non-traffic-sensitive costs from per-minute charges are quite separate from the problem that per-minute charges pose for internet use of any kind. In many countries with local calling charges, special flat-rate plans have been developed for internet service providers (ISPs) or large rebates have been given to them. But, as Table 4.1 shows, regulators are not abandoning local per-minute charges altogether; they are simply bending a little. Soon they will have to bend until they break.

## EXTENSION OF UNIVERSAL SERVICE TO THE INTERNET WORLD

As visionaries dream of exploiting the telecommunications network at the speed of light, regulators and politicians in many developed countries struggle with the problem of extending 'universal service' policies to the new digital world. The United States, for example, has encumbered its telecommunications landscape with a new set of subsidies for extending advanced services to schools and libraries that is likely to become permanent even though it is unnecessary (Crandall and Waverman 2000).

### The 'digital divide'

More to the point, one now hears complaints of the opening of a 'digital divide' that regulators or politicians must close. At first, this divide was the one between home-computer haves and have-nots.[8] However, as the prices of entry-level home computers fall to levels that are no higher than the price of a colour television set, political support for subsidising home computer purchases has begun to abate. Besides, more than 60 per cent of US households now have a personal computer (PC), and the proportion is growing rapidly in the United States and throughout the developed world (see ITU 2001).

The 'digital divide' has now shifted to new chasms in the United States and Europe. With virtually everyone having access to a computer, a new concern has arisen: access to high-speed internet services. Because these high-speed or 'broadband' services[9] are new and require major investments by the carriers that offer them, they are most likely to be introduced first in areas of high density and, presumably, high incomes. This is a sensible strategy, but its political vulnerability is obvious. Because poor and rural households are not among the first to be offered broadband services, they are allegedly being deprived of the opportunity to join the digital age. Something must be done, according to politicians taking up the cause, to assure that these households are not left behind. A new 'universal service' policy is one answer, albeit an unsatisfactory one.

There is no denying that there are large consumption externalities in most telecommunications services. You and I benefit from having another person join the network, particularly for new services, such as email or 'instant messenger'. For traditional telephone services in developed countries, however, there is no need to price access to the telephone network below cost in order to attract additional subscribers to exploit these externalities. The demand for access is so price insensitive that 95 per cent of all households will subscribe at a competitive, cost-based price (Crandall and Waverman 2000, Chapter 5). For the newer digital services delivered over the internet, particularly those requiring broadband connections, this conclusion does not hold.

## Telecommunications policy and the internet

The adaptation of the telecommunications network that was built largely for voice communications to the new demands of the internet has not been easy or very smooth. Modems had to be designed to convert digital signals into analogue communications. At first, these modems were extremely slow; now they are simply slow. However, in countries with high local calling rates, more was required. As long as users faced prices of 1 to 5 cents (US) per minute for time spent on the internet, there would be very little general interest in using the internet for activities that involved long holding times. Just one hour per day could translate into local charges of US$18.00 to US$90.00 per month for the privilege of net surfing, and these charges would not include any ISP charges.

At first, telephone companies in countries with high local call rates would offer to pay ISPs for the time that their subscribers spent using the internet, eagerly exploiting the very high price-cost margins realised from expanding local telephone network use at regulated rates. However, this was an insufficient incentive because users continued to pay incremental charges for use of the telephone network. As a result, pressure has mounted for flat-rate telephone tariffs solely for internet use, but such flat rates inevitably invite arbitrage of voice traffic through internet protocol (IP) telephony.[10] Therefore, most countries do not allow zero-priced internet use during peak traffic periods; the zero calling rates apply only to off-peak periods. The result is that those countries with zero local calling charges – the United States, Canada and New Zealand, for example – have forged ahead in internet use. Figure 4.1 demonstrates the inverse relationship between internet use and the price per minute of logging on to the network. Korea is clearly an exception.

## The evolution to broadband

Getting the prices of traditional telephone services right will surely help stimulate internet development, but it will not deliver the speed that is required to provide the next generation of internet applications. In developed economies, the focus is now on high-speed (broadband) services.

There are essentially five alternative sources of broadband connections to households and small businesses: digital subscriber line (DSL) service over copper wires, cable modems, satellite circuits, mobile wireless services, and fixed wireless services.

DSL services may be delivered by existing telephone companies or by new entrants, but only after substantial investments in plant upgrades and electronics. Cable modem services may be delivered by existing cable television companies or by new entrants, but also require substantial investment in new electronics. At present, satellite and fixed wireless services are probably uneconomic for widespread deployment to dispersed residential and small business subscribers except in rural areas.[11] Mobile wireless connections are under development as the 'third generation' (3G) of cellular wireless services, but they are not likely to be a serious alternative to the wire-based technologies in the foreseeable future.[12]

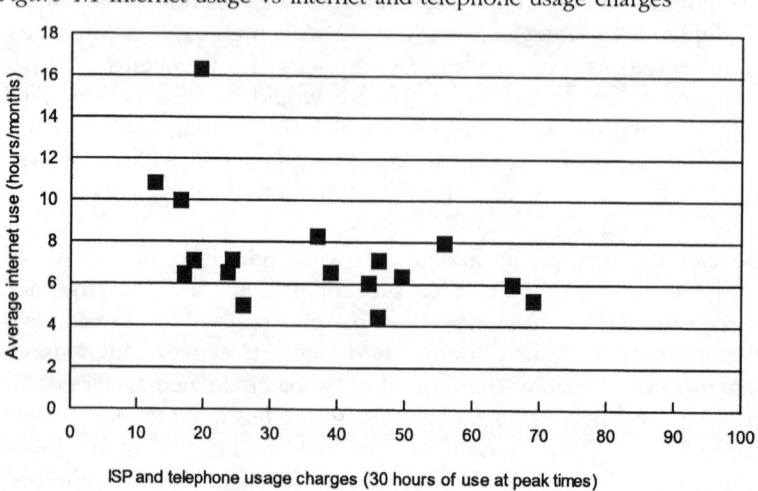

*Figure 4.1* Internet usage vs internet and telephone usage charges

*Source*: ITU (2001) and Neilsen at www.Neilsen-netratings, 8 March 2001.

For the present, therefore, the choice of broadband technologies for small, dispersed subscribers in most countries reduces to the two terrestrial wire-based technologies: DSL and cable modems. In many developed countries and most developing countries, cable television services are rather sparse, but virtually all countries have telephone networks whose copper wires pass a very large share of households and businesses. However, even in countries like the United States, Canada and Germany – where cable and telephone wires are virtually ubiquitous – an investment of $600 to $1,000 per subscriber is required to deliver a broadband internet service to most subscribers. I shall not review in this paper the technical requirements for converting one-way coaxial cable systems or traditional telephone systems to allow the delivery of broadband internet connections.[13] It is sufficient to point out that such conversion can increase the required investment per subscriber by 50 per cent or more (Crandall and Jackson 2001a), and that this investment is much more risky than investments in traditional telephone plant have been.

At present, there is very little residential broadband in most countries.[14] Figure 4.2, drawn from an unpublished OECD report, shows that Korea has much greater broadband penetration than any other country in the world. Canada is a distant second, but all other countries have fewer than three broadband connections per 100 persons. Only Korea has a large share of households connected to DSL, the service provided by regulated telephone companies over copper wires. Canada, Austria, the United States and the Netherlands have a fairly substantial number of cable modem subscribers but few DSL subscribers. Outside Korea and Canada, two countries with relatively non-intrusive regulatory regimes for incumbent telephone carriers,[15] there is very little residential broadband in even the world's most developed economies. Three of the largest EU countries – France, Germany, and Britain – are essentially without residential broadband services, despite increasingly intrusive regulation of their incumbent carriers that is designed to stimulate broadband competition. It is likely that another approach to encouraging broadband deployment will be required.

### Is a new subsidy program needed?

The few empirical estimates of the demand for broadband service suggest that it is still quite price sensitive.[16] Therefore, reductions in subscription prices, through direct subsidies or other policies, could have important effects on the rate at which broadband spreads throughout the residential sector. As broadband becomes more widespread, new, attractive services will develop that require broadband connections, further spurring demand for broadband access. This is the 'network externality' in action.

An obvious problem with any policy designed to spur broadband deployment is that it is likely to outlive its usefulness, much like agricultural price supports. Therefore, such a policy – if any – should be designed to help cover the substantial initial costs of enrolling new subscribers – the marketing costs, the

68   *The New Economy in East Asia and the Pacific*

*Figure 4.2* Broadband penetration, June 2001

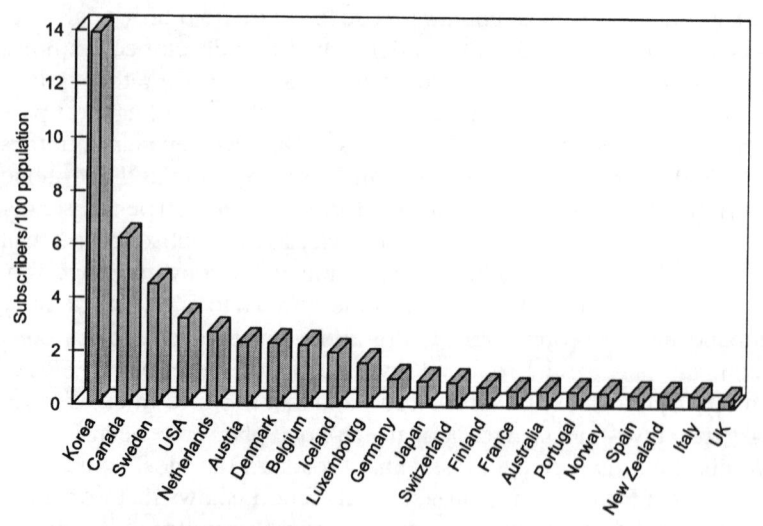

*Source*: OECD, unpublished data.

cost of the subscriber modem, and the cost of installation in the subscriber's premises – but not the continuing cost of providing the service.

A universal service policy for broadband could take a variety of forms. First, regulators could require carriers to offer the new broadband service at the same rate across their entire jurisdictions and to roll out the service at the same rate in low-density and high-density areas and in low-income and high-income areas.[17] Second, the price for broadband service in rural areas or other areas with a low density of prospective subscriptions could be priced below cost and the subsidy funded through a direct charge on all other telecommunications services.[18] Third, a subsidy for broadband service deployment could be defrayed from general tax revenues.

All these alternatives have severe drawbacks, even in a monopoly environment. If broadband is self-supporting, but priced the same across all regions, it is necessarily priced above long-run incremental cost in the high-density areas where early adopters are most likely to reside. This slows the diffusion of the service at a time when it is most needed to generate a critical mass of subscribers. A subsidy to rural or low-income areas, funded by a tax on all telecommunications services, is a better alternative, but such a tax is still very inefficient (see Crandall and Waverman 2000, Chapter 8, and Hausman 1998). A tax based on general tax revenues is the most efficient 'universal service' policy, but the program then competes with all other government programs, reducing its prospects for passage by the legislature.

A subsidy could be paid directly to the service providers for new subscriber enrolments or to the household subscribers. As with most such programs, it is necessary to protect against fraud or unnecessary 'churning' of subscribers. The evidence from traditional telephone services suggests that subsidising the initial cost of connecting a subscriber is likely to be more cost-effective than subsidising the monthly subscription fee (see Crandall and Waverman 2000; Garbacz and Thompson 1997; and Kaserman et al. 1990). In developed countries, a subsidy for connecting telephone subscribers is much more likely to be directed at poorer households than is a broadband connection subsidy. The political support for using tax moneys to induce wealthier households to subscribe to this new service is likely to be difficult to muster.

All discussion of 'universal service' policies for broadband internet connections must now take account of the newly competitive era in telecommunications. If incumbent telephone companies are saddled with uniform rates and 'roll-out' requirements, will the entrants be similarly encumbered? Extending such regulation to new entrants would discourage them from investing in or even entering the market. Building new facilities to compete with incumbent telecommunications carriers is very risky, as the last eighteen months have demonstrated (see Crandall 2001). The new US 'competitive local exchange carriers' (CLECs) have suffered large declines in their stock market values, and many have entered bankruptcy. These companies cannot and should not be saddled with a 'universal service' responsibility. Nor should the existing carriers be so encumbered.

Taxing telecommunications services or drawing on general revenues to fund subsidies for low-density or low-income areas, even if it is welfare enhancing, is surely problematic in a rivalrous or openly competitive telecommunications market. Are incumbents and entrants to receive the same subsidies? If so, how are they to be calibrated – by miles of plant, homes passed, or subscribers enrolled? Targeting subsidies to needy areas is difficult, but deciding on the metric for paying the subsidy is doubly difficult. For this reason, direct subsidies have made almost no inroads in the United States since the passage of the 1996 Telecommunications Act, except for the program aimed at schools and libraries (see Crandall and Waverman 2000, Chapter 8).

Another problem raised by broadband subsidies is the identification of the technologies to be subsidised. Presumably, incumbent telephone companies' DSL services are eligible for such universal-service approaches. But are cable television services eligible for similar subsidies – even if their systems are not now as heavily built out in rural areas? And how does the subsidy program deal with various fixed wireless or satellite services? Are all of these untried and untested services eligible for subsidies in providing services to rural areas even if some may have cost functions that are similar in urban and rural areas?

Finally, and perhaps even more problematic, the politics of any subsidy program such as 'universal service' make it difficult to curtail the program when it is shown to be no longer necessary or even a mistake ab initio. Most developed countries do not need new subsidy programs that become permanent fixtures in their national budgets; most developing countries probably cannot afford them.

## IS BROADBAND ESSENTIAL FOR THE DEVELOPMENT OF E-COMMERCE?

The United States clearly has the most developed e-commerce in the world. It has more internet hosts than any other region of the world and it enjoys greater household internet penetration than Europe or Asia, although the gap appears to be closing. Many of the innovative B2C (business-to-consumer)[19] ideas – such as eBay, Travelocity or Amazon.com – developed in the United States. Yet in 2000, this e-commerce accounted for only 1 per cent of retail sales there (US Department of Commerce 2001). Clearly, e-commerce is not yet a major force, even in the United States.

In a recent study (Crandall and Jackson 2001b), Charles Jackson and I suggested that widespread diffusion of broadband connectivity could generate enormous value to US consumers, but that this value will derive from much more than e-commerce as it is currently understood. I review our findings in this section.

Broadband connectivity is important not only for timely downloads of large files through the internet, but also for its 'always-on' capability. Most households have never experienced the internet without first dialling through a slow modem on an ordinary voice line, waiting for the connection to be made and then deploying an internet browser. The connection is generally terminated by the user or by the ISP after some idle period so as not to tie up the telephone line and, in many countries, accrue telephone calling charges. This process takes so much time that it discourages members of the household from using the internet for routine search functions, such as ascertaining the spelling of a word, finding the address of a local business, or looking up a recent newspaper review. A wide variety of more sophisticated internet applications, such as remote monitoring of household functions, await the widespread diffusion of broadband always-on connections.

No one can foresee the range of such applications today because they are simply impossible without always-on capability, and few subscribers have such capability. Without this information about prospective uses of broadband, one may easily conclude that most individuals have little need for broadband. Teenagers eager to download MP-3 files to create their own audio compact discs (CDs) of favourite popular music may have a need for broadband today, but what about ordinary adults? Until new services develop, many will simply forgo the additional $30 per month required for an always-on broadband connection. Unfortunately, these new services require more broadband households to make them profitable. We are therefore left to conjecture whether these network effects will occur.

The most likely and foreseeable household uses of broadband internet services are home shopping (e-commerce), IP voice telephony, entertainment, telecommuting, telemedicine, education, and household monitoring. E-commerce and entertainment services are the most likely to generate benefits through higher-speed connections, although in the long term such connections could prove to be extremely important for the remote monitoring of chronically ill patients. I will now discuss these categories in more detail. I omit education because there is insufficient information available to even hazard an estimate of the likely effect of broadband on the delivery of educational services to the home.

## Broadband uses

### Home shopping

The internet has become an important shopping medium for books, CDs, travel, home electronics, and a few other standardised products. It is extremely limited as a medium for displaying unique products or differentiated products that are not easily described in writing or by physical characteristics. If one needs to see a variety of views of one or more products to make a purchase decision, the internet becomes extremely cumbersome because of the time required to download most picture formats at speeds of 24 to 56 kbps. Clearly, broadband will make such 'comparison shopping' much easier and more rapid.

### Voice telephony

In the most developed OECD countries, the average household spends about 2 per cent of its total budget on telephone services. Though not a significant burden, most of this expenditure will simply disappear with the full development of universal broadband. The monthly broadband charge, currently $40 to $50 in the United States (including the ISP charge), will pay for the use of all or part of the access line, the electronic transformation and multiplexing of the signal, and the packet switching of inbound and outbound internet communications. Voice calls will simply become a minor component of a household's internet packets. The monthly voice bill will all but disappear.

### Entertainment

The Napster experience provides a glimpse of the potential that broadband video streaming could offer at least the most ardent videophiles. The transition from ordinary television broadcasting to cable/satellite multi-channel video distribution has created enormous value for US households (Crandall and Furchtgott-Roth 1996). Once limited to the choice of programming from three or four channels, US households now receive at least 50 channels from cable television and as many as 200–300 channels from direct broadcast satellites. Fewer than 20 per cent of US homes choose to subsist on the limited off-air choices; 65 per cent have cable television and more than 15 per cent subscribe to one of the two high-powered satellite services. Broadband internet services will expand this choice even more, allowing households to interact with video programs, order their desired programs when they want them, play video clips with other services, such as shopping services, and otherwise customise their video entertainment.[20]

## Telecommuting

In a society dominated by urban sprawl, such as the United States, commuting distances are growing longer and the time required increasing. Clearly, some of us already telecommute to a certain extent. Broadband connections could accelerate this trend by allowing data-intensive professional services to be performed at home or from other remote locations.

## Telemedicine

A great deal has been written about the possibility of delivering medical diagnoses and therapies from remote locations over high-speed internet connections. But most major medical facilities already have access to high-speed services at DS-1 speeds or higher.[21] A more important contribution of broadband may derive from using communications rather than labour to monitor chronically ill patients at home or in dispersed extended-care facilities. A very large share of medical services is consumed by a small number of chronically ill and ageing patients who require some form of 24-hour monitoring and supervision. An always-on high-speed connection may permit remote monitoring of many of these patients, and even an improvement in the quality of care through reduced response time to medical emergencies.

## Household monitoring

Once a household has access to always-on broadband internet services, it will begin to acquire additional PCs or other internet devices so that the service is not limited to one household member's participation at one time. Once there are two or more PCs in a home and perhaps a television set-top box, households will find that establishing wire-based or wireless local area networks (LANs) in the home provides greater user flexibility. The networks may make it easier to install household monitoring devices for security, energy management, or emergency response. Many of these functions will not require broadband speed, but always-on connections at zero marginal cost will surely provide an incentive for developing these services that is simply not possible with costly dial-up connections.

## Estimating the potential value of broadband to consumers

We do not know what the household demand function for broadband will look like in ten or twenty years. Current estimates of the price elasticity of household for broadband in the United States are in the range of –1.0 to –1.5 (Rappaport et al. 2001). By contrast, the demand for narrowband voice connections is in the range of −0.01 to −0.03. Broadband is a luxury today; voice telephony is a necessity.

The potential value to consumers of widespread broadband connections may be estimated through the conventional measure of consumer surplus. The consumer surplus generated by broadband connections today is relatively modest. For instance, let us assume that the demand curve for the service is linear. The annual consumer surplus under this demand curve is currently equal to $S_t^*P/2$ if the demand elasticity is −1.0 and $S_t^*P/3$ if the demand elasticity is −1.5, where $S_t$ is today's number of subscribers and P is the annual subscription cost. For the United States, with about 8.4 million subscribers at $480 per year, the service generates between $1.3 billion and $2.0 billion in annual consumer surplus. Obviously, these are rather small numbers.

What would the value of broadband be if everyone subscribed? I have calculated this on the assumption that a linear demand curve shifts outward at a constant slope. Under this assumption, the estimated consumer surplus rises to $(S_T/St)^*S_T^*P/2$ if the demand curve originally had a price elasticity of −1.0 and $S_T$ is the ultimate number of subscribers. If broadband were to become as common as voice telephony, reaching 94 per cent of US households, the consumer surplus would be $278 billion per year, assuming no household growth. This is obviously a large number. If the price elasticity of demand is currently −1.5, the consumer surplus for universal broadband service would be $(S_T/St)^*S_T^*P/3$, or $186 billion per year, still a large number.[22]

These estimates translate into a consumer surplus of $1,884 to $2,817 per subscriber per year. Personal consumption expenditures in the United States are now approximately $7 trillion per year; therefore, the above estimates suggest that broadband could eventually generate a consumer surplus of between 2.5 and 4 per cent of total consumption expenditures, ignoring the effects of economic growth.

Is such a large value reasonable? Nearly twenty years ago, when fewer than 90 per cent of all US households subscribed to telephone services, Lewis Perl estimated the access externality of voice telephony at $4 per month, or $48 per year (Perl 1983). At current prices, this estimate rises to $72 per month. Perl found that the externality was much larger for smaller communities – in other words, that the value of an additional subscriber is greater to existing subscribers when the number of people on the network is small. For instance, increasing the calling area from 25,000 to 50,000 households would increase the value by $96 per year in 1983 dollars for each household on the network ($144 in today's dollars). It is important to note that Perl's estimates of the access externality were for a service of given quality. Increasing the size of the network in 1983 allowed one to simply talk to more people, not to connect to an expanding array of new and innovative services.

Today, broadband exists in very few homes, making the development of new services tailored specifically for high-speed connections much less profitable than it would be at 25 per cent or 50 per cent household penetration. As subscribers increase, the variety of available services that may be accessed through broadband connections increases. Therefore, the network 'externality' is likely to be quite large at current broadband penetration.

Surely, it would be unreasonable to assume that broadband demand will remain price elastic when everyone has broadband. If the price elasticity does not change from its current level and the demand curve is linear, the network effect is zero. Consumer surplus would simply be a constant $238 per year times the number of subscribers. This would mean that no one would be willing to pay more for broadband when it is universal than the average subscriber is willing to pay today when virtually no one has it and the available services are quite primitive. That assumption seems quite unreasonable.

The assumption that the current demand curve shifts out at its current slope to 'universal' penetration results in a price elasticity of demand of between –0.085 to –0.127, far higher (in absolute value) than the current price elasticity of demand for ordinary telephone services.[23] Surely, this is a reasonable result. However, a linear demand curve with 94 per cent household penetration with that slope implies that *someone* would be willing to pay as much as $510 or $353 per month, respectively, for the service.[24] This may seem less reasonable, given today's use of the broadband network. Ultimately, however, one can surely imagine that Bill Gates, Rupert Murdoch or Warren Buffett might pay this much.

Charles Jackson and I made an attempt to buttress these estimates by building the ultimate consumer surplus estimates from informed guesses about the contribution of each of the sources of value reflected in the above section on 'Is broadband essential for the development of e-commerce?'. Table 4.2 shows the results of our labours.

These estimates, admittedly based on arbitrary assumptions, comport with the above calculations. We are consistent in our optimism.

*Table 4.2* Estimated contribution of eventual universal broadband connectivity to US consumer surplus (billion $/year)

| Service | Estimated benefit Low | Estimated benefit High |
|---|---|---|
| Home shopping | 74 | 257 |
| Voice telephony | 51 | 51 |
| Entertainment | 77 | 142 |
| Telecommuting | 30 | 30 |
| Telemedicine | 40 | 40 |
| Total | 272 | 520 |

*Source*: Crandall and Jackson (2001b).

## The value of many previous technologies was not foreseen

Before dismissing me as a cockeyed optimist, one might reflect on the pessimism that greeted a variety of innovations that now pervade our culture. I review a few of these here.

*Cable television.* To estimate future demand for broadband service without knowing what it may bring in the future would be much like trying to predict the demand for cable television services in 1974, when the only viewing alternatives that cable could provide were distant broadcast stations. At that time, pessimistic economists were predicting that cable would spread to only 30–40 per cent of households in major markets. I concluded at the time that these predictions were decidedly too low, suggesting that 65 per cent cable penetration was possible – almost precisely today's level (Crandall and Fray 1974).

*The photocopier.* When Chester Carlson developed a process for dry photocopying, he searched for years for a corporation that would commercialise his invention. He approached, among others, General Electric, RCA, IBM, and Remington Rand. IBM hired the technology consulting firm Arthur D. Little (ADL) to assess the technology. After receiving ADL's pessimistic report about the likely demand for copiers, IBM chose not to pursue the technology further. A small firm, Haloid Company, subsequently decided to license the technology. After the product began to take off, the firm changed its name to Xerox.

*Wireless telephony.* Many people vastly underestimated the rate at which consumers would adopt wireless (cellular) telephone or personal communications services (PCS). For example, in 1978 a group of researchers at Cornell attempted to forecast the impacts of the yet-to-be deployed cellular technology (Bowers et al. 1978). Their analysis considered three scenarios of likely mobile telephone service penetration. Their medium scenario envisioned only 500,000 to one million US mobile telephone subscribers over a 15–20-year horizon; their most optimistic scenario envisioned no more than 10 million subscribers. By January 2000, there were 86 million actual US subscribers; even their most optimistic scenario had been exceeded, by a factor of almost 9.[25] The Cornell authors were not alone. The forecasts by all parties to the Federal Communications Commission (FCC) rulemakings for cellular service fell far short of actual demand.

*Computer communications.* The pioneers of packet communications and the internet faced great scepticism from the established communications industry. In 1978, Larry Roberts, one of the pioneers of the Arpanet/internet, wrote, 'AT&T and its research organisation, Bell Laboratories, have never to my knowledge published any research on packet switching' (Roberts 1978). He also described how the Advanced Research Projects Agency (ARPA), in an attempt to commercialise the Arpanet, approached AT&T about taking over the Arpanet. AT&T declined the opportunity. The Arpanet ultimately became today's internet.

The relative pessimism that surrounded each of these seminal technologies before they became widely used can be found in discussions of broadband today. It is, of course, possible that these pessimists could be correct. On the other hand, I think that there are sufficient shreds of evidence to suggest an opposite conclusion.

## THE EFFECT OF TELECOM REGULATION ON INVESTMENT INCENTIVES FOR BUILDING THE MODERN TELECOMMUNICATIONS NETWORK

We are only at the beginning of the digital revolution. Most households and small businesses still connect to the internet over copper wires and circuit switches that were installed by telephone companies to deliver voice communications. Many of these facilities are thirty years old or more and are inadequate for delivering high-speed digital communications. If the potential of a broadband internet world is to be realised, someone has to build the facilities to connect these millions of subscribers in a manner that allows them to communicate at high speed. There are several potential competitors in this race, but their success will depend on the regulatory environment.

### Regulatory disincentives – telephone carriers

Despite the recent passage of 'deregulatory' legislation in most developed countries, traditional telephone carriers are still heavily regulated. Indeed, they are more intensely regulated today than ever before, largely because legislators and regulators have a new religion. Having blocked entry for decades in order to preserve cross-subsidies, they now ardently believe that regulation can be used to accelerate competitive entry. Unfortunately, the regulatory policies left over from the old regime reduce the telephone companies' incentives to deploy broadband facilities in several ways.

*Retail price regulation.* First, the regulation of conventional retail voice/data services discourages the incumbent telephone companies from using the requisite facilities to deliver DSL, because broadband services require an entirely different architecture that will eventually replace the circuit-switched architecture that was designed for yesterday's voice telephony. Once the new architecture – based on packet switching – is widely deployed, it will begin to handle voice calls as well, but at a zero price per conversation minute. One cannot easily distinguish voice packets from data packets, nor is the elapsed time of the voice message relevant to the cost of transmitting the message in packets. Therefore, telephone companies cannot extract per-minute charges for only the voice messages delivered over the internet.

As described above, 'universal service' policies lead regulators to place very high prices on using the network and low prices on residential connections to the network. To recover their non-traffic-sensitive fixed network costs,[26] the regulated incumbent carriers must realise substantial revenues from per-minute charges and from a variety of business services. In the broadband internet world, usage charges largely disappear, even for international communications. If, however, regulators continue to require the incumbent telephone companies to recapture their residential access deficit from per-minute charges, the incumbents are likely to weigh the loss of revenues from such charges in their decision to deploy broadband facilities. Because broadband deployment will accelerate the implicit arbitrage of substituting internet telephony for circuit-switched telephony, the profitability of such deployment to the incumbent telephone company is reduced, perhaps substantially.

This opportunity for arbitrage extends to business services as well. Regulators in most countries allow telephone carriers to recover part of their access deficits in connecting residential subscribers by charging extremely high rates for high-speed business services, generally known as DS–1 (1.5 Mbps) or DS–3 (45 Mbps). If these regulated telephone companies offer a new DSL service that is priced to appeal to small subscribers, large businesses may substitute DSL for their much higher priced traditional high-speed services. The potential effect upon the regulated carriers' profitability is obvious.

Finally, any attempt to extend cost-based regulation of retail rates to broadband services will have obvious disincentives for investment. A broadband carrier faces a substantial risk that its deployment of new facilities may face early obsolescence due to the enormous technological progress that is occurring throughout the information technology sector. DSL services may prove to be less flexible and even slower than services delivered over new generations of wireless, satellite, or direct fibre-optic technologies.[27] Cost-based regulation of DSL services, particularly when regulators base the rates on the traditionally long useful lives of older generation equipment, clearly discourages the incumbent carriers from deploying the facilities required to deliver this new service.

*Wholesale regulation.* In many countries, cost-based wholesale regulation, either through 'network unbundling' or resale, is being imposed on incumbent telephone carriers as part of a process of opening telecommunications markets to competition. This type of regulation is relatively new, having appeared in full force first in the United States. The 1996 US Telecommunications Act that opened the local access/exchange markets to competition requires incumbent carriers to 'open' their networks to entrants by leasing 'unbundled' network facilities or entire services at cost-based rates. For instance, an entrant is allowed to lease the line from the subscriber to a telephone switching centre or a transmission path between switching centres rather than building its own facilities. Alternatively, the entrant may simply buy the entire service at a wholesale rate and resell the service to its subscribers.

The World Trade Organization (WTO) is now pressing countries to use wholesale unbundling requirements to liberalise telecommunications markets, even though there is very little evidence that such unbundling regimes are an effective mechanism for promoting competition.[28] Incumbent telephone companies are required to make their transmission, switching, signalling and subscriber connection or 'local loop' facilities available to entrants at cost-based, regulated rates. Moreover, the costs are estimated on the basis of the economic life of the facility, but the entrant is generally allowed to lease the facilities at these rates on a month-to-month basis.

In practice, this unbundling requirement often reduces to just the final loops from the telephone switching centre to the subscribers. Indeed, the unbundling requirement in the EU is often referred to as 'local loop unbundling'. However, there is little experience with this requirement in the EU as yet. The United States has nearly 200 million switched access lines, but, after five years, only about five million subscriber lines are served by entrants via unbundled facilities (US FCC 2001).

The unbundling policy is generally defended as providing access to bottleneck or essential facilities that entrants would have difficulty in replicating economically. However, in the United States, *everything* in the incumbents' networks is now available on an unbundled basis, essential or not. Moreover, entrants can ask for virtually any configuration of facilities that is technically feasible, by simply arguing that they would be impaired in their entry if these facilities were unavailable. Essentially, incumbents must provide interconnection to their networks in a manner that is convenient to entrants, creating numerous technical problems and substantial uncertainty over network design and management.

The unbundling concept might be more defensible if it were limited to only essential facilities in the existing incumbent networks. This is the route that Canada has taken.[29] However, in the United States and the EU, the concept has now been extended to new facilities required for the delivery of broadband DSL services. Incumbents are even required to split the frequencies on their local subscriber lines and to offer the high-speed frequencies to entrants for high-speed DSL services, leaving the incumbents with only the non-remunerative voice services for most residential subscribers. This 'line sharing' obligation creates enormous opportunities for controversy between the incumbent and entrants, each of whom may desire a different technical approach for connecting to part of the incumbent's line.

These wholesaling requirements are likely to be a major deterrent to investment in broadband telecommunications networks wherever they are applied.[30] Telecommunications is not a static technology like railroad or pipeline transportation. One can envision such unbundling requirements for simple technologies, such as railroads or natural gas delivery, because the technology for connecting to the rail line or pipe does not change very much over decades. However, modern telecommunications networks are extremely complex and constantly changing. As higher speed becomes necessary, fibre-optic cables replace copper wires. The fibre must eventually be connected to the subscriber line and part of the signal converted to a digital bit stream while the remainder is delivered as an analogue signal. Complex multiplexers are inserted in the network to allow multiple signals to move rapidly down the same transmission path. Bursts of data from one line must be combined with data from other lines and delivered to a packet switch that is designed to deliver data packets to the internet backbone.

Enormous investments are required in most incumbent telephone networks just to allow the delivery of a high-speed DSL service, and this investment is often tailored to deliver one form of DSL. Even the more advanced national telecommunications networks require such investments, because they have been delivering an older integrated services digital network (ISDN) service that does not provide sufficient speed for widespread residential internet connections to dispersed subscribers. In many cases, the conversion is difficult and fraught with unanticipated technical problems. If rivals may now intervene through the regulatory process to demand that the conversion be tailored to their needs, the investments are likely to be delayed and may even be shelved.

The essential facilities argument surely cannot be used to justify unbundling facilities that do not yet exist for delivering a service that will compete with similar services offered over other networks, such as cable television or, eventually, wireless and satellite facilities of various kinds. The current regulatory rhetoric describes the need for incumbents to open their networks to competitors, but such policies are more likely to delay the development of these networks and prevent incumbents and entrants alike from using them to deliver any DSL service.[31]

*Vertical integration.* In yet another attempt to encourage entry, regulators in the United States are requiring incumbents' competitive services to be offered through separate subsidiaries. In some cases, such as Pennsylvania, regulators are even entertaining notions of requiring incumbents to offer all retail services through a subsidiary that is separate from the basic network company. The network company would offer only wholesale services – to its retail affiliate and to new entrants. Such vertical separation would be designed to prevent the incumbent from favouring its own retail services in its network operations and from employing price 'squeezes' on its new rivals.[32]

Any attempt to require separate retail and wholesale organisations within a complex telecommunications company breaks the connection between service design and deployment and the investment in new or modified network facilities. Losing these advantages of vertical coordination in investment decisions could slow innovation and investment. For this reason, few other countries have followed the United States' 1984 decision to break AT&T's long-distance operations from its local network operations. Indeed, long-distance competition evolved much more rapidly in Canada than in the United States despite the fact that Canada's incumbent telephone companies remain vertically integrated (Crandall and Hausman 2000).

## Cable television

In many countries, such as Japan and Italy, cable television facilities have not been widely deployed. In those countries where they have, such as Canada,

Germany and the United States, cable modem service is being offered as a major source of broadband internet service (see Figure 4.1). However, such deployment requires large investments in updating a cable infrastructure that was originally designed to provide a limited one-way video distribution service. Cable television services have rarely been subject to the detailed cost-based regulation that has been applied to traditional telephone services; therefore, new cable modem services are generally not subject to as detailed regulation as DSL services.

In the United States, however, cable television systems have been regulated, deregulated, re-regulated, and deregulated again over the past 25 years.[33] At present, US distributive video service rates are largely unregulated, but there is considerable uncertainty over the potential regulation of cable modem services because of recent court rulings. The FCC may impose new regulatory rules on cable modem service, because such service has been defined by one appellate court as an interstate 'telecommunications service'.[34]

Obviously, if cable modem service rates are subjected to cost-based regulation, there will be less incentive to invest in modifying cable networks. In the United States, the more likely form of regulation is one that mandates 'open access' to cable modems for competing ISPs or suppliers of content. Unlike telephone companies, cable systems have not traditionally had common carrier obligations; they have therefore developed their broadband services in a manner that allows them to control the choice of ISP. In the United States, some form of mandated access may eventually be imposed on them, thereby reducing their ability to control content and realise economic profits from such control. Given that these systems must compete with other technologies, such as DSL and wireless systems, in the market for high-speed connections, this regulation may be unnecessary and may unduly reduce the incentive to develop new services.

### Implications for future policy

It is still very early in the development of the broadband internet service to make predictions, but it is unlikely that one technology will so dominate the delivery of this service and be subject to scale economies of such a degree that countries will have to design new regulatory systems to protect consumers from monopoly abuse. Therefore, imposing common carriage or cost-based regulation on this new service is surely premature.

Equally important, however, is the need to reform traditional telecommunications policy so as to provide greater incentives for incumbent telephone companies to invest in the requisite facilities. Soon, internet telephony will begin to replace traditional circuit-switched services, but this substitution will be resisted by telephone companies if current universal service policies that require large usage charges for traditional telephone service remain. Once large numbers of consumers have an 'always-on' broadband connection, internet telephony will simply become a component of packet-switched internet services. If regulators wish to encourage the rapid deployment of broadband, they will have to allow telephone companies to recover their costs through much higher fixed monthly charges.

A very large share of telecommunications costs are fixed, non-traffic-sensitive costs that should be recovered through flat fees. Yet most countries continue to recover a large share of their network costs from variable per-minute charges. Figure 4.3 shows the total telecommunications revenues per line and the fixed local rate per line (an average of business and residential rates) for several countries. Note that in Korea, China and Singapore, telephone carriers receive a very small share of their revenues from these flat monthly rates compared to the United States and Japan. Yet even in the United States, per-minute national and regional calls are priced far above incremental cost. These pricing practices are unsustainable in a broadband world.

Finally, this paper sounds an alarm against the unnecessary extension of regulation into the wholesale unbundling of incumbent telephone networks that is occurring around the world, with US leadership. Such unbundling discourages much needed investment in broadband facilities by incumbents and entrants alike. At most, it should be limited to essential facilities that cannot be readily duplicated, such as subscriber lines in rural areas and numbering databases, and even then for only a limited number of years. Technological change may make any 'essential facility' designation obsolete in just a few years. Rather than attempting to regulate existing carriers' new services, countries should focus on opening entry to new players. Encouraging the deployment of advanced technologies by new competitive carriers using fibre-optic cable, coaxial cable or wireless facilities is more important than trying to regulate the older, bureaucratic incumbent telephone companies.

## IMPLICATIONS FOR DEVELOPING COUNTRIES

Much of the discussion in this paper has focused on telecommunications policies in developed economies that already have widespread voice telephony and substantial internet use. It might be argued that developing countries face a very different set of problems. Without widespread deployment of ordinary voice telephone services, many of these countries cannot possibly extend the benefits of internet use to their households, much less provide the dazzling services potentially available from broadband. For most residences in these countries, even narrow-band voice connections are a distant hope.

Nevertheless, policy-makers in developing countries can learn from the mistakes made by their more developed brethren, and they even have the advantage of deploying modern technology that is more suitable for modern internet applications. First, even developing countries should be wary of trying to subsidise residential telephone connections through high local and national calling rates. Any such subsidies should be funded from general revenues rather than through taxing telephone network usage. High local calling rates clearly discourage internet use in developing and developed economies alike.

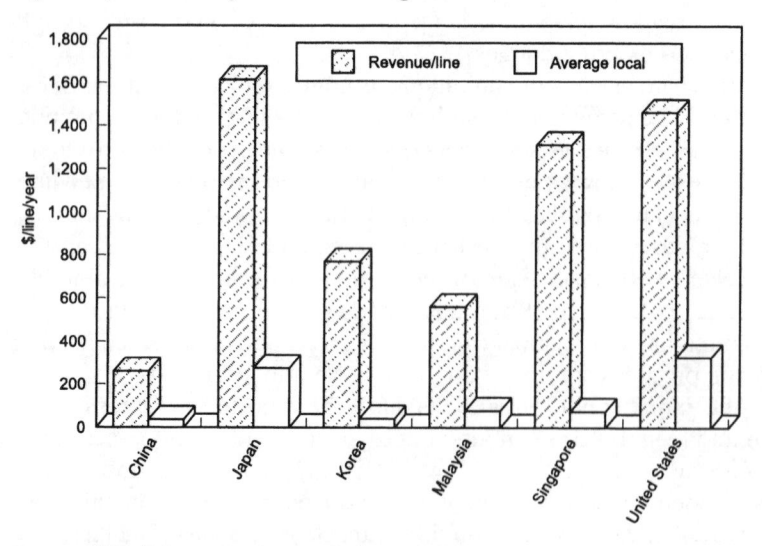

*Figure 4.3* Revenues per line vs average local rate

*Source*: ITU, Telecommunication Indicators, 2000.
*Note*: All data are in US dollars, converted at current exchange rates.

Second, developing countries should promote telecommunications and video services with an eye to developing competition between the suppliers of each. As discussed above, competition between cable television, telephone companies and (eventually) wireless providers is likely to yield greater rewards than any attempt to saddle telephone carriers with onerous regulation.

Third, to reduce the burden of extending the telephone network into high-cost rural areas, developing countries can encourage the deployment of various wireless technologies. These technologies are still developing and may not be competitive with the facilities that are already in place in developed economies, but in developing countries they may be a superior alternative to new copper-wire facilities.

Fourth, it is important to reduce entry barriers. New entry into fixed and mobile wireless services will place competitive pressures on incumbent telephone companies. Competition from cable television systems is likely to be important in the development of broadband services. However, policy-makers should promote facilities-based competition, not the intrusive requirement that existing carriers provide their facilities to competitors at low wholesale rates. Such attempts to subsidise entry risk a substantial slowdown in investment in the very telecommunications infrastructure needed to deliver the services for the new economy.

Competition among ISPs is likely to lead to improved services, lower rates and integration with broadband telecommunications services. Moreover, ISP competition is conducive to the development of new e-commerce applications in each country. Attempts to limit ISP competition are likely to reflect a government's desire to control internet content – surely not a policy that is conducive to the development of e-commerce and other new internet-related services.

Finally, some developing countries may be able to bypass the old circuit-switched technologies that now dominate the developed countries' telephone networks and move directly to packet switching for all telecommunications. There are, after all, some advantages to not being the first mover.

## CONCLUSION

The broadband revolution is following close on the heels of the internet revolution. Already, 30–40 per cent of all individuals in Japan, Korea, Singapore, North America, Europe and Australia use the internet. Broadband connections are necessary for the next step in the development of e-commerce and a myriad of other services that can be delivered over the internet. The principal impediments to the deployment of broadband are technical, commercial and regulatory.

As with many other new services in the information technology sector, there are enormous risks involved in deploying any given technology. The rate of technical progress is so rapid that today's investment may become much less valuable tomorrow because of a new technical breakthrough. In the case of broadband, wireless or satellite technologies could easily advance so rapidly as to render DSL and cable-modem investments much less attractive.

Equally important are the network effects that generate the demand for new products and services. With only 4 or 5 per cent of households using broadband, many investments in broadband content and applications may not be very attractive. Once, say, 50 per cent of all households have the service, such investments will appear much more attractive. Yet, without these new services, many households will not choose to subscribe to the broadband service in the first place. This 'chicken and egg' problem is not easily resolved unless the broadband service providers vertically integrate into the provision of content or subsidise new customer connections.[35] However, it is less important for many Asian countries, which will be able to exploit the benefits created by the widespread deployment of broadband elsewhere.

The largest barrier to broadband deployment in many countries may be their telecommunications regulatory policies. The traditional regulatory approach to voice telephony forces carriers to realise a large share of their revenues from per-minute charges and to charge very high rates for high-speed business services. These policies reduce the incentives for incumbent telephone companies to deploy broadband services. Moreover, the recent trend towards requiring these incumbents to offer pieces of their networks to entrants at regulated wholesale rates reduces their incentives and the incentives of entrants

84  *The New Economy in East Asia and the Pacific*

to invest in new facilities. Pacific Basin countries would be well advised not to follow in the footsteps of the EU and the United States in adopting such policies.

## NOTES

I am grateful to Wendy Dobson and Henry Ergas for comments on an earlier draft of this paper.

1. US Bureau of the Census, *Current Population Survey*.
2. Long-distance transmission was confined to copper wires until after World War II, when microwave (radio) transmission replaced the copper wires for most long-haul services. Thereafter, satellites began to replace microwave circuits in certain locations, and ultimately fibre-optics replaced both microwave and satellite circuits for virtually all long-distance transmission of telecommunications services.
3. This discussion is taken largely from Crandall and Waverman (2000). See also Garbacz and Thompson (1997).
4. Crandall and Waverman (2000) estimated that this loss in the United States is between $1.1 billion and $3.6 billion per year just from overpriced residential long-distance calls. The total loss in consumer welfare from all pricing distortions caused by the universal service policy is likely to be much greater.
5. I use the term 'subsidy' to denote the burden placed on other rates by the policy of keeping residential line rates low. This may or may not be an economic subsidy, because these services are joint products of the telecommunications network and all rates may still be between incremental cost and stand-alone cost.
6. I assume that the incremental cost of circuit switching capacity is no more than US$0.002.
7. The old voice network transmitted calls in an analogue format over dedicated circuits. The internet, by contrast, requires all communications to be digital and divided into discrete packets. These packets are then sent over the network, sharing lines with other packets. At their destination, the packets are reassembled and delivered to the addressee. The transmission and delivery of these packets permits much more efficient use of transmission and switching facilities.
8. In the United States, the Department of Commerce's National Telecommunications and Information Agency has been in the forefront of documenting a digital divide (see US Department of Commerce 1999). Recently, the World Bank has entered the rhetorical fray through its information and communication technology (ICT) group, which says: 'ICT presents unprecedented opportunities to combat poverty by increasing income, opening markets and providing a channel through which the voices of the poor can be heard. Our main challenge is to broaden the reach of ICT to reach those who risk being left behind. This is the essence of the "Digital Divide"'. Accessed 23 July 2001 at http://wbln0018.worldbank.org/external/lac/lac.nsf/sectors/inftelecoms/175bfef0e678f649852569ad00018365?opendocument.
9. Broadband services are those that deliver much more information per unit of time. A video signal requires the transmission of as much as 500 times as much information per second as an ordinary voice telephone call. Therefore, a much broader pipe or transmission path is required to deliver video images over the internet than to deliver simple text or voice messages. The latter can be accommodated over a simple dial-up telephone connection through a modem at 56 kilobits per second; the former require speeds of 1 megabit per second or more, depending on the degree to which the video signal is compressed by modern electronics technology.

10 Internet protocol (IP) telephony is the transmission of voice telephone calls in packet form over the internet. This technology is still being perfected, but its quality is already sufficient to provide a substitute for high-price international and national calls in many countries.
11 This is not to say that these services do not currently compete for customers in any area. Moreover, further technical progress, such as the development of spot-beam geostationary satellites, will undoubtedly make them more important competitors in a few years.
12 As this is being written, even the more modest '2.5G' systems are being postponed.
13 Crandall and Jackson (2001a) provide a brief review in Appendix A of their paper.
14 Broadband service comes in different flavours and speeds, but most definitions specify that the service must deliver signals at a minimum of 200 kbps in one direction.
15 See 'Estimating the potential value of broadband to consumers', below in this paper.
16 The latest is Paul Rappaport et al. (2001).
17 This approach is taken in legislation currently being debated in the US House of Representatives, HR 1542, which will require telephone companies to have a broadband service available in 100 per cent of their central offices within five years.
18 This is the implicit requirement of the 1996 US Telecommunications Act.
19 'Business-to-consumer' internet services are those that are developed by businesses to communicate with and sell to consumers online. By contrast 'business to business' applications are those that are developed by businesses to contact suppliers or their downstream business customers.
20 Apparently, this process has already begun in a video-starved Korea. The demand for greater variety in television programming helps to explain Korea's stunning current lead in household broadband subscription.
21 DS–1, 2, 3, etc. is the term used for standard digital transmission rates. 'DS' stands for 'digital signal'.
22 Note that this estimate is based on consumer demand for broadband services, which reflects the reduced cost of goods and services due to e-commerce. It does not, however, reflect the savings that narrowband or broadband connections are creating for all types of organisations, including manufacturers, health care providers, educational institutions and governments. Two colleagues, Robert Litan and Alice Rivlin, have completed an edited volume that addresses some of these other benefits of the internet (Litan and Rivlin 2001). They concluded that the internet could contribute between $100 billion and $230 billion in total cost savings, even without any savings in retailing and education.
23 This calculation is based on demand increasing from its current level of 8 per cent of US households to 94 per cent of households, with no growth in households.
24 These are the zero-output intercepts of the demand curves. See Hausman (1997).
25 See Cellular Communications and Internet Association (CTIA), 'Frequently asked questions and fast facts' (downloaded from: http://www.wow-com.com/consumer/faq/articles.cfm?ID=101).
26 These are the costs that do not vary with the intensity of using the network. Included are the terminal equipment (handset or personal computer) and the copper line connecting the subscriber to the closest switching point. These costs are a large share of the total cost of the telecommunications network.
27 In Japan, NTT Communications is beginning to provide to the subscriber fibre that is capable of delivering up to 100 Mbps.

28  For a critique of these requirements, see Crandall and Hausman (2000). See also Alleman and Noam (1999).
29  Even Canada is reconsidering its 1997 decision now. The Canadian regulator – the Canadian Radio-television and Telecommunications Commission (CRTC) – is under pressure from the new local entrants to extend unbundling and reduce wholesale rates.
30  Henry Ergas suggested in his comments on this paper that the unbundling requirements could be improved if the rates charged for unbundled facilities reflected a charge for universal service obligations and if the contract period more closely reflected the useful life of the facility. Under such conditions, I believe that there would be virtually no demand for unbundled facilities by new carriers, and the policy would simply become irrelevant.
31  There is a developing empirical literature on this phenomenon. For a paper that demonstrates that low unbundled loop rates restrain competitors' investment in their own facilities, see Eisner and Lehman (2001). For a contrary conclusion, see Jamison (2001).
32  A form of this type of separation was imposed on AT&T in the United States to settle a case that was brought by the US government charging AT&T with violating the Sherman act, the landmark US competition statute. No other country has pursued such a policy. This author and Alfred Kahn supported a proposal for separation of retail and wholesale operations proposed by the Rochester (New York) Telephone Company in the 1990s. This proposal was a second-best solution to a regulatory dilemma faced by Rochester under New York's jurisdiction.
33  See Crandall and Furchtgott-Roth (1996) for a discussion of the history of cable regulation in the United States.
34  AT&T Corp. v. City of Portland, 216 F.3d 871.
35  US cable television companies overcame this problem by jointly investing in scores of cable television programming services. Similarly, wireless companies routinely subsidise handset purchases to overcome this problem.

## REFERENCES

Alleman, J. and Noam, E. (eds) (1999) *The New Investment Theory of Real Options and Its Implications for Telecommunications Economics*. Boston: Kluwer,

Bowers, R. et al. (eds) (1978) *Communications for a Mobile Society: An Assessment of New Technology*. Beverly Hills CA: Sage Publications.

Cairncross, F. (2001) *The Death of Distance 2.0: How the communications revolution will change our lives,* Boston: Harvard Business School.

Commission of the European Communities (2000) *Sixth Report on the Implementation of the Telecommunications Regulatory Package*. Brussels: Commmission of the European Communities, 7 December, p.65.

Crandall, R.W. (2001) *An Assessment of the Competitive Local Exchange Carriers Five Years After the Passage of the Telecommunications Act*, Report prepared for SBC Communications, June, available at http://www.criterioneconomics.com/documents/Crandall%20CLEC.pdf.

Crandall, R.W. and Fray, L.L. (1974) 'A reexamination of the prophecy of doom for cable television', *The Bell Journal of Economics and Management Science*, 5, 1, Spring: 264–287.

Crandall, R.W. and Furchtgott-Roth, H. (1996) *Cable TV: competition or regulation?*, Washington DC: Brookings Institution.

Crandall, R.W. and Hausman, J. (2000) 'Competition in U.S. telecommunications services: the effect of the 1996 legislation,' in Peltzman, Sam and Winston,

Clifford, *Deregulation of Network Industries: what's next?* Washington DC: Brookings Institution.

Crandall, R.W. and Jackson, C.L. (2001a) (March) 'Regulatory barriers to innovation: The case of high-speed internet access,' unpublished manuscript, Washington DC: Brookings Institution.

Crandall, R.W. and Jackson, C.L. (2001b) *The $500 Billion Opportunity: The potential economic benefit of widespread diffusion of broadband internet access.* Report prepared for Verizon Communications, July, available at http://www.criterion economics.com/.

Crandall, R.W. and Waverman, L. (2000) *Who Pays for Universal Service? When telephone subsidies become transparent*, Washington DC: Brookings Institution.

Eisner, J. and Lehman, D.E. (2001) 'Regulatory behavior and competitive entry', Paper presented at the 14th Annual Western Conference, Center for Research in Regulated Industries, 28 June.

Garbacz, C.C. and Thompson, H.G. Jr (1997) 'Assessing the impact of FCC lifeline and link-up programs', *Journal of Regulatory Economics,* 11: 67–78.

Hausman, J. (1997) 'Valuing the effect of regulation on new services in telecommunications', *Brookings Papers on Economic Activity, Microeconomics*: 1–38.

—— (1998) 'Taxation by telecommunications regulation', *Tax Policy and the Economy*, 12: 29–48.

ITU (International Telecommunication Union) (2001) *Telecommunications Indicators '01 (Stars* database), Geneva.

Jamison, M.A. (2001) 'Incumbent and entrant incentives with network interconnection: the case of U.S. telecommunications', unpublished manuscript, University of Florida.

Kaserman, D.L., Mayo, J.W. and Flynn, J.E. (1990) 'Cross subsidization in telecommunications: beyond the universal service fairy tale', *Journal of Regulatory Economics*, 2: 231–49.

Litan, R.E. and Rivlin, A.M. (eds) (2001) *The Economic Payoff from the Internet Revolution*, Washington DC: Brookings Institution.

Perl, L.J. (1983) *Residential Demand for Telephone Service, 1983*, Study prepared for the Central Services Organization, Inc. of the Bell Operating Companies, December.

Rappaport, P., Taylor, L.D. and Kridel, D. (2001) 'Residential demand for access to the internet', Temple University, unpublished manuscript.

Roberts, L. (1978) 'The evolution of packet switching', Institute of Electrical and Electronics Engineers, *Proceedings of the IEEE,* 66, 11, November.

US Department of Commerce (2001) 'Estimated U.S. retail e-commerce sales: 4th quarter 1999 – 1st quarter 2001', Washington DC. Downloaded from www.census.gov/mrts/www/mrts.html.

US Department of Commerce, National Telecommunications and Information Administration (1999) (July) *Falling Through the Net: Defining the Digital Divide: A Report on the Telecommunications and Information Gap in America*, Washington DC: US Department of Commerce.

US FCC (Federal Communications Commission) (2001) (May) *Local Telephone Competition: Status as of December 2000.* Washington DC: US FCC.

# 5 Factors affecting growth in the region: R&D and productivity

*Shandre M. Thangavelu and Toh Mun Heng*

## INTRODUCTION

The term 'new economy' seems to mean different things to different researchers,[1] but, no matter how the new economy is defined, growth in the new economy will be driven by technological innovation and scientific progress. Developments in information and communication technologies (ICTs) are clearly the most important 'new' elements of recent economic growth in the 1990s, and have led to other innovations such as the greater role of networking in the new economy. ICTs are seen as important developments in the new economy, but some of their impacts might not have occurred without broad changes in national innovation systems[2] in the world (OECD 2000).

Based on the most recent data, the most important trend in the new economy is the rising share of investment in research and development (R&D). In most Organisation for Economic Co-operation and Development (OECD) countries, overall investment in R&D rose in the 1990s. In 1997, overall R&D expenditure[3] in OECD countries was nearly US$500 billion, nearly 2.2 per cent of the overall OECD gross domestic product (GDP) (OECD 2000). In addition, the composition of the R&D expenditure is changing towards more business-funded and market-oriented research: the business-funded R&D expenditure relative to overall OECD GDP was nearly 1.6 per cent in 1998 (OECD 2000).

Like OECD countries, East Asian countries are redefining their economic structure, particularly in terms of innovation and technological development, which are seen as being of paramount importance for long-term growth in these countries. The key elements in the new economy are the creation, absorption and dissemination of knowledge; these elements are clearly emphasised in the economic structures of the East Asian countries as they face new internal and external challenges in the global economy.

The new economy has engulfed not only the OECD countries, but also the East Asian countries. The 'miracle growth' driven by trade and foreign direct investment (FDI) in the 1970s and 1980s raised the standard of living in these countries and integrated their economic structures in the global economy.

However, in the 1990s the East Asian countries faced important new external challenges in their ability to sustain the 'miracle growth' of the 1980s. As the 'miracle growth' from adopting foreign technology decreases, and as the participation of China and India intensify the competition to attract FDI in the Asian region, the East Asian countries are trying to restructure their economies towards more technology and knowledge-intensive economic structures to maintain their competitiveness in their outward-oriented growth engines.

In the East Asian region, the key to long-term sustainable growth in the new economy lies in the contribution of knowledge to output growth. Indigenous R&D is increasingly important as the region as a whole struggles to attract more FDI, as rapidly emerging large countries like China and India tend to divert the attention of multinational corporations (MNCs). Given their large domestic market, abundant resources and untapped human capital, China and India will be key growth poles for MNCs. The smaller East Asian 'tigers' of Korea, Singapore and Taiwan tend to lack the lure of the large domestic market and must rely heavily on external markets for export growth and for augmenting domestic capital and technology.

The East Asian countries are moving into higher-value-added production as their economies and industrial structures mature. Increasingly, the East Asian tigers are forced to compete with developed countries in higher-value-added exports as countries with abundant labour and resources in Asia increase their presence in the global marketplace. In the new economy, the creation of indigenous knowledge is the key for the East Asian tigers to maintain their export competitiveness and to attract MNCs. But there is not only strong competition to maintain global export market share; the economic fundamentals in East Asian countries are also depreciating much more quickly in the current global environment due to rapid technological developments and thus shorter technology cycles. To maintain long-term competitiveness, there is a concerted effort by the East Asian countries to increase indigenous knowledge capital through R&D expenditure.

The importance of R&D for technological change, innovation and long-term growth is clearly emphasised by theoretical and empirical studies by Griliches (1999), Fagerberg (1994), Grossman and Helpman (1991), Jones (1995) and Stokey (1995). In fact, R&D provides an important contribution to total factor productivity (TFP) growth and hence to long-run output growth in the economy. In the new economy, R&D investments not only provide broad changes in domestic innovation, but also provide important linkages for technology transfer from abroad. In this chapter, we will study the impact of R&D expenditure on the productive performance of Korea, Singapore, and Taiwan; we will use a neoclassical cost function to study the impact of R&D on the aggregate productive performance of these East Asian countries.

In the following section, we discuss the traditional measure of multifactor productivity growth (MFPG) across the Asian region. We then provide a summary of innovations and technological developments in East Asian countries. In subsequent sections, we discuss the methodological framework to measure the productive impact of R&D capital on output growth, and provide relevant data. We then present the results from the model and discuss the implications for policy.

## THE TRADITIONAL MEASURE OF MULTIFACTOR PRODUCTIVITY GROWTH

In every stage of growth, productivity is considered to be the most important source of long-term growth. Traditionally, productivity is defined as the residual of output growth after accounting for input growth. The idea of long-term growth driven by productivity growth is given by the neoclassical assumption of diminishing returns in increasing the factor inputs. Given that increases in inputs are subject to diminishing marginal returns, long-term growth must be achieved from productivity growth, not input growth. The traditional MFPG measures the combined productivity of all factors of production; the methodology is derived from the neoclassical framework of competitive markets and constant returns to scale. MFPG is estimated by subtracting the weighted growth rates of factor inputs from the growth rate of value-added output at constant prices. The weights are the income shares for factor inputs.

Table 5.1 shows recent estimates of MFPG. In 1991–97, MFPG in Indonesia and Singapore was higher than in neighbouring countries. In Korea, India, Japan, Malaysia, Pakistan, the Philippines, Taiwan and Thailand, there was a trend for MFPG to fall from 1986–90 to 1991–97. There is no observable MFPG pattern across the Asian countries, except for Indonesia, which experienced an average growth rate of 5 per cent from 1978 to 1997.

There are several weaknesses in the traditional measure of MFPG, given its strong assumption of competitive markets and constant returns to scale

*Table 5.1* Multifactor productivity growth for selected Asian countries, 1978–97 (per cent)

|  | 1978–85 | 1986–90 | 1991–97 |
|---|---|---|---|
| Korea | 1.12 | 2.80 | 2.32 |
| Japan | 2.38 | 2.63 | 0.69 |
| India | 2.25 | 2.92 | 1.04 |
| Indonesia | 5.76 | 5.35 | 5.94 |
| Malaysia | –0.72 | 2.80 | –0.87 |
| Pakistan | 1.82 | 1.92 | –1.00 |
| Philippines | –3.65 | 1.2 | –0.87 |
| Singapore | 0.41 | 3.33 | 3.29 |
| Taiwan | 3.47 | 5.02 | 1.75 |
| Thailand | 0.64 | 3.47 | 0.85 |

*Source*: APO (2001).

*Factors affecting growth in the region* 91

(Hall 1986, 1988). For example, if there are returns to scale in the economy, the traditional measure will be biased and overstate the contribution of pure technical change. Moreover, since traditional MFPG is derived as a 'Solow residual' in the production function, it does not provide any information on the sources of productivity growth in the economy. In particular, based on the residual, it is difficult to extract any information on the economy's capacity for innovation or technological development.

In contrast to the standard neoclassical growth accounting framework based on the production function, several authors have used the cost function approach to account for sources of MFPG.[4] In this paper, we use the dual measure of MFPG (growth accounting using the cost function) to study the sources of MFPG in Korea, Singapore and Taiwan; in particular, we try to account for innovation capabilities in these economies.

## INNOVATION AND TECHNOLOGICAL DEVELOPMENT

R&D expenditure in Korea, Singapore and Taiwan increased rapidly in the 1990s (see Figure 5.1). In 1978, the ratio of R&D expenditure to GDP was only 0.6 per cent for Korea, 0.3 per cent for Singapore, and 0.4 per cent for

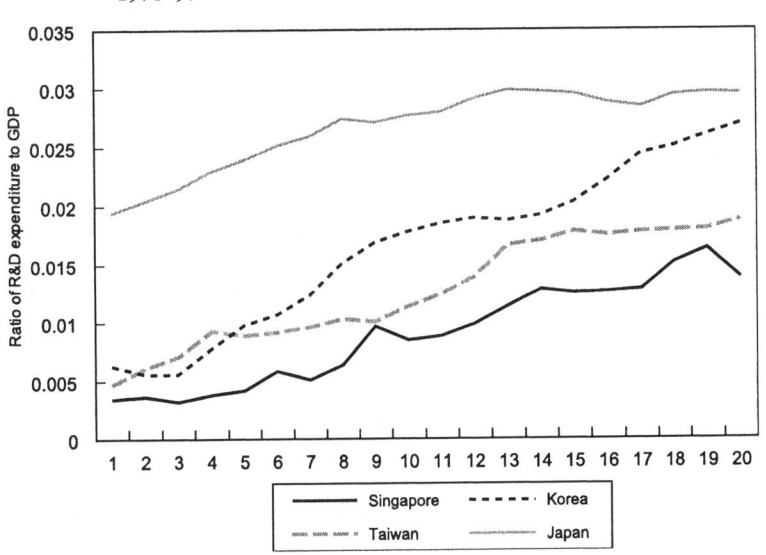

*Figure 5.1* R&D expenditure to GDP ratios for Korea, Singapore and Taiwan, 1978–97

*Source*: Drawn from data obtained from statistical yearbooks (Korea and Taiwan), the National Survey on R&D in Singapore, National Science and Technology Board, various issues (Singapore) and the Japan Statistical Yearbook, various issues (Japan).

*Table 5.2* Government and private R&D expenditure in Korea, Singapore and Taiwan, 1978–97 (per cent)

|  | Government expenditure | Private expenditure |
|---|---|---|
| *Korea* | | |
| 1978–85 | 42.36 | 57.64 |
| 1985–90 | 21.79 | 78.21 |
| 1991–97 | 19.10 | 80.90 |
| *Singapore* | | |
| 1978–85 | 47.82 | 52.18 |
| 1985–90 | 42.82 | 57.18 |
| 1991–97 | 38.06 | 61.94 |
| *Taiwan* | | |
| 1978–85 | 59.18 | 40.82 |
| 1985–90 | 52.28 | 47.72 |
| 1991–97 | 47.16 | 52.84 |

*Source*: For Korea and Taiwan, statistical yearbooks, various issues; for Singapore, National Science and Technology Board, various issues.

Taiwan. However, in 1997 the ratio increased to 2.7 per cent for Korea, 1.4 per cent for Singapore, and 1.9 per cent for Taiwan.

Table 5.2 shows the share of government and private R&D expenditure in Korea, Singapore and Taiwan. The role of the government in R&D development is clearly evident in these countries. In 1978–85, government expenditure accounted for more than 40 per cent of R&D expenditure for all three countries; as in OECD countries, the share of private R&D expenditure is gradually rising. In Korea, government R&D expenditure was about 42 per cent of the total in 1978–85, but declined to less than 20 per cent in the early 1990s. In contrast, the government R&D expenditure in Singapore and Taiwan was still about 40 per cent of the total R&D expenditure in the 1990s, reflecting the government's important role in promoting and developing R&D activities in the domestic economy.

The output of innovation processes is reflected by changes in patenting activities. Table 5.3 shows the change in the number of patents granted from 1995 to 1998 for residents and non-residents in East Asian and selected OECD countries. It is clear that most Asian countries rely heavily on non-residents, particularly foreign MNCs, for most of their technology developments. For example, in Korea and Singapore, the number of patents granted to non-residents is much larger than the patents granted to residents. The only exception is Taiwan, where local residents make a substantial contribution to technological developments in the domestic economy.

*Table 5.3* Technology indicators for selected countries

|  | Δ Patents (1995–98) Residents | Non-residents | R&D expenditure (% GNP) (1987–97) |
|---|---|---|---|
| Japan | 25,277 | 23,168 | 2.8 |
| Korea | −8,535 | 33,729 | 2.8 |
| China | 3,938 | 36,578 | 0.6 |
| Hong Kong | 51 | 12,601 | – |
| Malaysia | 38 | 2,361 | 0.24 |
| Singapore | 301 | 32,766 | 1.13 |
| Taiwan[a] | 6,988 | 3,343 | – |
| United States | 13,866 | 13,481 | 2.6 |

*Source*: Data for Taiwan are from the Taiwan Statistical Yearbook; data for other countries are from the World Development Report (various issues).

*Note*
a   In Taiwan, the change in the number of patents was calculated from 1994 to 1997. Dashes indicate that data were not available.

## THEORETICAL FRAMEWORK AND METHODOLOGY

In this section, we discuss the theoretical framework to decompose the impact of R&D capital on the performance of the aggregate economy based on a cost function. Morrison and Schwartz (1994) and Morrison and Siegel (1997) provide a more detailed theoretical analysis using the production–cost approach. Morrison and Siegel (1997) provide the theoretical and empirical framework to account for the quasi-fixity of private capital, long-run (internal) scale economies, and external 'knowledge' factors such as investment in R&D and education based on the cost function framework. As in their paper, we will assume that the knowledge factors are exogenous and create positive spillovers on the overall economic performance of the economy.[5] The assumption of exogeneity of R&D capital could be relaxed and introduced as one of the inputs in the production function independent of disembodied technological developments, in which case the price of R&D should be included in the cost function. However, the difficulty of obtaining a price for R&D capital tends to limit this application in most studies. In addition, the public or social nature of R&D capital justifies the idea of exogeneity of knowledge capital to production activities, particularly in terms of the proprietary nature and spillovers of R&D investments. It should be noted that the social nature of R&D capital is represented only by government investment in R&D, so private investment in R&D capital should be excluded from the measurement of R&D capital. However, in most East Asian countries, the government still plays an important role in inducing more R&D activities and hence increasing domestic R&D investments. To account for the role of the government the current study includes both government and private R&D investments in the calculation of R&D capital.

The cost function is given as $C(k_r, w_i, Y, t)$, where $k_r$ is a vector of $r$ exogenous knowledge factors (R&D capital), $w_i$ is the vector of $I$ variable inputs, $Y$ is the level of output, and $t$ is the state of technology arising from disembodied technological change. The cost function defined above is a long-run cost function where all the factors are variable except for R&D capital, which is considered to be a fixed exogenous variable. Further, there are no restrictions on the cost function in terms of constant returns to scale. The cost function in the long run captures cost minimisation and reflects the technology that is dual to the production function. In addition, R&D capital is included as a fixed input to allow its impact to shift the cost function and hence accommodate external economies of scale. Like Morrison and Siegel (1997), we will assume that R&D capital is exogenous and that its availability is not affected by the payments made by the firm.

Differentiating the cost function with respect to time and letting $\varepsilon_{ct} = \frac{\partial C}{\partial t} \frac{t}{C}$ and $\varepsilon_{cy}^L = \frac{\partial C}{\partial Y} \frac{Y}{C}$, we have the following formula:

$$\frac{\dot{C}}{C} = \varepsilon_{ct} + \varepsilon_{cy}^L \frac{\dot{Y}}{Y} + \sum_i \frac{v_i w_i}{C} \frac{\dot{w_i}}{w_i} - \sum_r \frac{Z_r K_r}{C} \frac{\dot{K_r}}{K_r} \tag{1}$$

The term $\varepsilon_{ct}$ is the elasticity measure representing MFPG; it reflects the reduction in total cost due to technical progress. The expression $Z_r$ reflects the shadow value or marginal benefits to firms from having an additional unit of R&D capital.[6] The firms will benefit from an additional unit of R&D capital as long as the shadow value of R&D capital is positive ($Z_r > 0$). A positive shadow value of R&D capital suggests that further investments in R&D will lead to benefits in terms of cost savings and hence improvement in the productive performance of the economy.

The term $\varepsilon_{cy}^L$ represents the inverse of the long-term returns to scale (that is, the changes in long-run cost function to changes in output). The full effect of the returns to scale is given as follows:[7]

$$\varepsilon_{cy}^L = \varepsilon_{cy} + \sum_r s_{k_r} \varepsilon_{k_r y} \tag{2}$$

The first term in equation (2), $\varepsilon_{cy}$, is the direct effect of changes in output on the cost function; the second term is the indirect effect of the R&D

Factors affecting growth in the region    95

capital on the cost function through its effect on output. Therefore the indirect effect of R&D capital consists of two components: $s_{k_r} = \dfrac{z g k_r}{C}$ is the elasticity of cost with respect to R&D capital and $\varepsilon_{k_r y} = \dfrac{\partial k_r}{\partial Y} \dfrac{Y}{k_r}$ is the long-run elasticity of R&D capital with respect to output.

In a perfectly competitive environment with constant returns to scale ($\varepsilon_{cy} = 1$) and with no R&D investment, a traditional measure of MFPG ($\varepsilon^*_{ct}$) emerges from the total cost function:

$$\dfrac{\dot{C}}{C} = \varepsilon^*_{ct} + \dfrac{\dot{Y}}{Y} + \sum_i \dfrac{v_i w_i}{C} \dfrac{\dot{w_i}}{w_i} \qquad (3)$$

Using (3) and (1), we can rewrite the MFPG measure as the error biases in the traditional measure:[8]

$$-\varepsilon^*_{ct} = -\varepsilon_{ct} - (\varepsilon_{cy} - 1)\dfrac{\dot{Y}}{Y} - \sum_r s_{k_r} \varepsilon_{k_r y} \dfrac{\dot{Y}}{Y} + \sum_r s_{k_r} \dfrac{\dot{K_r}}{K_r} \qquad (4)$$

Based on equation (4), the impact of R&D capital on productivity growth of the industries depends on the direct effect ($s_{k_r} \dfrac{\dot{k_r}}{k_r}$) and the indirect effect ($s_{k_r} \varepsilon_{k_r y} \dfrac{\dot{Y}}{Y}$)

The impact of R&D capital on productive performance of the economy is given by the direct and indirect effect of R&D capital. The direct effect of R&D capital on productivity growth is positive if both the growth of R&D capital, $\dfrac{\dot{k_r}}{k_r}$, and the shadow value of R&D capital, $z_{k_r}$, are positive. In this case, the traditional measure of productivity growth is overstated by the growth of R&D capital. The indirect effect of public capital is negative when both the shadow value of R&D capital and output growth are positive. This suggests that increasing output without increasing R&D investment will understate the traditional measure of MFPG. Therefore the net effect of R&D capital, which takes both effects into account, suggests that, if the elasticity of cost with respect to R&D capital is positive ($s_{k_r} > 0$), the rate of growth of

R&D capital must at least be greater than the growth rate of output, weighted by $\varepsilon_{k_r}y$, for R&D capital to have a positive contribution to traditional MFPG.

## EMPIRICAL IMPLEMENTATION

The empirical implementation of the above theoretical model is undertaken with a generalised Leontief cost function with three variable inputs (capital, labour and energy) and one fixed input (R&D capital stock). The generalised Leontief cost function is given as:

$$\begin{aligned}C(K_g, w_i, Y, t) = &Y[\alpha_{ll}P_l + 2\alpha_{lk}P_l^{0.5}P_k^{0.5} + \alpha_{kk}P_k + \delta_{ly}P_l Y^{0.5} + \delta_{lt}P_l t^{0.5} + \\
&\alpha_{ee}P_e + 2\alpha_{le}P_e^{0.5}P_l^{0.5} + 2\alpha_{ek}P_e^{0.5}P_k^{0.5} + \delta_{re}P_e Y^{0.5} + \delta_{et}P_e t^{0.5} \\
&\delta_{ky}P_k Y^{0.5} + \delta_{kt}P_k t^{0.5} + (P_k + P_e + P_l)(\gamma_{yy}Y + 2\gamma_{ty}Y^{0.5}t^{0.5} + \\
&\gamma_{tt}t)] + Y^{0.5}[\delta_{lk_r}P_l K_r^{0.5} + \delta_{kk_r}P_k K_r^{0.5} + \delta_{er}P_e K_r^{0.5} + (P_l + P_k + P_e) \\
&(\gamma_{yk_r}Y^{0.5}K_r^{0.5} + \gamma_{tk_r}t^{0.5}K_r^{0.5})] \\
&+ (P_l + P_k + P_e)(\gamma_{k_r k_r}K_r) \end{aligned} \quad (5)$$

where $P_e$, $P_l$ and $P_k$ are the price of energy, labour and capital respectively, $Y$ is the industry manufactured output, $t$ is the state of technology, and $k_r$ is the R&D capital stock.

Sheppard's lemma can be used to derive the factor demand equations from the cost function; the model is estimated as a system of factor demand equations by appending an error term to each of the above equations. Since the estimating system is not simultaneous,[9] a seemingly unrelated (SUR) system technique is used to estimate the above model.[10]

### Data

The data on GDP at factor cost, GDP deflator, electricity consumption (in kilowatt hours per capita),[11] electricity price index, interest rate, deflator for gross capital formation, number of workers, and labour remuneration were obtained from the respective statistical yearbooks of Korea, Singapore and Taiwan. The capital stock data for Korea and Taiwan are available in APO (2001). In the survey reported in APO (2001), the capital stock series for Taiwan was calculated from data from the National Wealth Censor, Industry and Commercial Censor and Directorate General of Budget (various issues) (see Chi-Yuan Liang in APO 2001). The capital stock series for Korea was obtained from Pyo (1998) (see Byoungki Lee in APO 2001). The capital stock series for Singapore was obtained mainly from Rao and Lee (1995); the series after 1995 was calculated by perpetual inventory method using their depreciation rates. The price of capital was constructed as described by Morrison and Schwartz (1994).[12] All data in the study were debased to 1995 prices.

The data on R&D expenditure were obtained from statistical yearbooks for Korea and Taiwan. The R&D data for Singapore were obtained from the National Survey on R&D in Singapore (National Science and Technology Board, Singapore, various issues). The complete series of data for Singapore was available only from 1990, and it was given at three-yearly intervals from 1978 to 1990. To produce a compete series of data, the missing data were extrapolated from the available data. The R&D capital stock was constructed by the perpetual inventory method with a depreciation rate of 10 per cent (Nadiri and Mamuneas 1994). Based on Griliches (1980), the initial R&D capital stock ($K_{r0}$) was estimated by:

$$K_{r0} = \frac{R_0}{(g+\delta)}, \qquad (6)$$

where $R_0$ is R&D expenditure in period 0, $g$ is the annual logarithmic growth rate of R&D expenditure for which data are available and $d$ represents the depreciation of knowledge capital.

## RESULTS

We estimated the cost function for the three East Asian countries from 1978 to 1997. Given that we estimated the model with a system of factor demand equations, we could increase the degrees of freedom from the sample size of only twenty observations. If we observed auto-correlation in our estimation, we assumed a first-order autoregressive form (AR1) to correct it. Further, in order to establish robustness in our results, we used a Wald test to test the model for the join insignificance of the key coefficients in the estimated elasticities. The results are given in Table 5.4.[13]

The test results show that the coefficients of the cost elasticity of R&D capital ($S_{kr}$) for the three countries are statistically significant at the 5 per cent level, rejecting the null hypothesis that all R&D capital stock coefficients are insignificant. The coefficients in the cost elasticity of MFPG ($\varepsilon_{ct}$) and output (the inverse of long-run scale economies) ($\varepsilon_{cy}^L$) are also statistically significant at the 5 per cent level for the three countries, rejecting the null hypothesis that key coefficients in the measured elasticity are jointly insignificant.

Table 5.5 shows the elasticity of cost to R&D capital for Korea, Taiwan and Singapore. The elasticity is positive for all three countries, suggesting that the shadow values are positive but less than one. Since the cost elasticity of R&D capital is less than one for all three countries, the decrease in aggregate cost will be less than the unit increase in the R&D capital. However, the positive shadow values suggest that R&D capital has a positive impact on the productive

*Table 5.4* Wald test on the insignificance of elasticities for R&D capital stock, scale effects and technological change

|  | Chi-square statistic | | |
| --- | --- | --- | --- |
|  | Korea | Singapore | Taiwan |
| Effects of R&D capital: $s_{k_r}$ $H_0: \delta_{lk_r} = \delta_{klk_r} = \gamma_{lk_r} = \gamma_{k_r k_r} = \gamma_{k_r k_r} = 0$ | 69.11 | 22.86 | 37.01 |
| Effects of technological change: $\varepsilon_{ct}$ $H_0: \delta_{lt} = \delta_{kt} = \delta_{et} = \gamma_{ty} = \gamma_{k_r k_r} = \gamma_{tt} = 0$ | 196.37 | 30.21 | 47.82 |
| Effects of economies of scale: $\varepsilon_{cy}^L$ $H_0: \delta_{ly} = \delta_{kky} = \delta_{ey} = \gamma_{yy} = \gamma_{yk_r} = \gamma_{ty} = 0$ | 33.28 | 68.87 | 481.14 |

*Note*: Critical chi-square value at 5 per cent level of significance with 6 degrees of freedom =12.59.

performance of the East Asian countries and that an additional unit of R&D capital will increase the productive performance of the economy.

In the 1990s, Korea tended to have higher cost elasticity to R&D capital than Singapore and Taiwan, suggesting that the Korean economy is in a slightly better position to absorb new innovations and technologies in order to increase its productive performance. As compared to Singapore and Taiwan, the cost elasticity is rising for Korea, suggesting that the productive impact of R&D capital on the economy is improving over time.

Table 5.6 shows the results for the direct and indirect effects[14] of the R&D capital on production performance. The indirect effect reveals the impact of increasing output for a given stock of R&D capital and thus the contribution of R&D capital through the scale effects. The indirect effect is positive across the three countries, suggesting that increasing output for a given R&D capital stock will tend to reduce productivity growth of the economy through its

*Table 5.5* Elasticity of cost to R&D capital stock $s_{k_r} = -\dfrac{\partial C}{\partial k_r} \dfrac{k_r}{C}$

|  | 1978–85 | 1986–90 | 1990–97 |
| --- | --- | --- | --- |
| Korea | 0.015 | 0.042 | 0.081 |
| Singapore | 0.026 | 0.011 | 0.021 |
| Taiwan | 0.0087 | 0.0024 | 0.001 |

*Table 5.6* The indirect, direct and total effects of R&D capital on the productive performance of Korea, Singapore and Taiwan, 1978–97

|  | 1978–85 | 1986–90 | 1991–97 |
|---|---|---|---|
| Indirect effects $\left(-s_{k_r}(\frac{\partial k_r}{\partial Y}\frac{Y}{k_r})\frac{\dot{Y}}{Y}\right)$ | | | |
| Korea | 0.0009 | 0.0033 | 0.0044 |
| Singapore | 0.0016 | 0.0008 | 0.0017 |
| Taiwan | 0.0012 | 0.0003 | 0.0012 |
| Direct effects $\left[s_{k_r}\frac{\dot{k_r}}{k_r}\right]$ | | | |
| Korea | 0.0077 | 0.0145 | 0.0244 |
| Singapore | 0.0043 | 0.0021 | 0.0035 |
| Taiwan | 0.0039 | 0.0017 | 0.0014 |
| Total = direct − indirect | | | |
| Korea | 0.0068 | 0.0112 | 0.0200 |
| Singapore | 0.0027 | 0.0013 | 0.0018 |
| Taiwan | 0.0027 | 0.0014 | 0.0002 |

effect on the cost of production in the economy. This suggests that there is little contribution from R&D capital through the economies of scale effect across the three countries.

On the other hand, the direct effect of R&D capital shows the direct impact of R&D investment on the productive performance of the economy. Table 5.6 shows the direct effect of R&D capital on each of the East Asian economies. The direct effect of R&D capital on productive performance is positive and higher than the indirect effect for all the countries. Table 5.6 also shows the total effect of R&D capital, which takes both the direct and indirect effects of R&D capital into account. Since the positive direct effects are greater than the indirect effects for all three countries, there is a positive contribution from R&D capital on productivity growth in terms of direct gains from R&D investments. Among the three countries, Korea tends to have the highest contribution from R&D capital to productive performance of the economy.

Table 5.7 shows the cost elasticity on the long-run economies of scale. The figures suggest that the cost elasticity to output is less than one for all three countries, suggesting that there are significant increasing returns to scale in these countries. In fact, the impact of increasing returns to scale was more pronounced in Taiwan (the elasticity averaged about 0.152 for 1978–97) than in Singapore or Korea. Our findings of increasing returns to scale are comparable to those of Park and Kwon (1995), who also found increasing returns to scale in the Korean manufacturing industries.

*Table 5.7* The cost elasticity of output and contribution of scale to productive performance of Korea, Taiwan and Singapore, 1978–97

|  | 1978–85 | 1986–90 | 1991–97 |
|---|---|---|---|
| $\varepsilon_{cy}$ (effect of changes in output on the cost function) | | | |
| Korea | 0.748 | 0.611 | 0.410 |
| Singapore | 1.03 | 0.824 | 0.591 |
| Taiwan | 0.151 | 0.196 | 0.160 |
| $(\varepsilon_{cy} - 1)\dfrac{\dot{Y}}{Y}$ (contribution from scale effects) | | | |
| Korea | 0.013 | 0.037 | 0.039 |
| Singapore | –0.0038 | 0.014 | 0.026 |
| Taiwan | 0.057 | 0.068 | 0.058 |

Table 5.8 shows that, after adjusting for effects from economies of scale and R&D capital, the MFPG ($\varepsilon_{ct}$) is much lower than the traditional measure of MFPG ($\varepsilon_{ct}^{*}$) for Taiwan and Korea but not for Singapore. In the case of Singapore, the adjusted MFPG ($\varepsilon_{ct}$) is positive, which suggests positive technological progress even after adjusting for the economies of scale and R&D effects. On the other hand, the adjusted MFPG ($\varepsilon_{ct}$) is negligible for Taiwan and negative for Korea. The traditional measure of MFPG could be

*Table 5.8* Multifactor productivity growth for Korea, Singapore and Taiwan, 1978–97

|  | 1978–85 | 1986–90 | 1991–97 |
|---|---|---|---|
| $\varepsilon_{ct}$ (%) | | | |
| Korea | –3.11 | –2.64 | –2.21 |
| Singapore | 3.2 | 5.6 | 4.5 |
| Taiwan | 0.57 | 0.39 | 0.28 |
| $(\varepsilon_{ct}^{*})$ (%) | | | |
| Korea | –1.1 | 2.2 | 3.7 |
| Singapore | 3.1 | 7.1 | 7.2 |
| Taiwan | 6.5 | 7.3 | 6.1 |

derived from equation (4); it is shown in Table 5.8. The traditional measure of MFPG is quite significant for Taiwan, and is at an average of 6 per cent for 1978–97, which is much higher than for Korea or Singapore. When we compare the direct effects in Table 5.6 and the contribution from the scale effects in Table 5.7 across the three countries, we see that the scale effects tend to overstate the contribution of traditional TFP compared with R&D investment. Conversely, the results suggest that the scale effects are a more important contributor to output growth than the total R&D investments in the aggregate economy in all three countries.

## CONCLUSIONS

This paper considers the impact of R&D capital on the productive performance of the East Asian countries of Korea, Singapore and Taiwan. The results indicate that the shadow value and cost elasticity of R&D capital is quite low but positive across all three countries. All three countries are experiencing positive effects from R&D capital and economies of scale, but the contribution from economies of scale is much larger than from R&D capital to output growth. Given the rapid accumulation of R&D capital in Korea, the results indicate that Korea tends to enjoy a more positive effect of R&D capital on the productive performance of its economy than do Singapore or Taiwan.

The low shadow values could be attributed to several factors. There are complementary effects between human capital investments and R&D capital, since relatively educated and skilled workers can absorb and unbundle knowledge much faster than can unskilled workers. The R&D employment data in Table 5.9 show that the rising R&D expenditures in the three countries studied is not matched by the increase in R&D employment. Table 5.9 also shows the ratio of R&D employment to total employment. In Korea, people engaged in R&D comprised only 0.26 per cent of the total labour force in 1978–85, but this increased to 0.82 per cent in 1991–97. In Singapore and Taiwan, the ratios of R&D employment to total employment were only 0.26 per cent and 0.44 per cent, respectively, in 1978–85. In 1991–97, this ratio increased to 0.67 per cent for Singapore and 1 per cent for Taiwan, but it was still very insignificant.

The key component for long-term growth in the knowledge-based economy depends on the 'absorptive capacity' of the economy. The most important aspect of absorptive capacity is skill and education of the workforce. Educated workers have the capacity and basic foundations to absorb and implement new technologies faster than uneducated workers. Therefore policies should aim to increase educational levels in the domestic economy; in particular, they should focus on tertiary education to increase the level of general education in the domestic economy. The general infrastructure in the domestic economy is another important component of absorptive capacity. General infrastructure is becoming very important hardware for creating network economies within and across countries. The impact of R&D capital will be

Table 5.9  R&D employment in Japan, Korea, Singapore and Taiwan, 1978–97 (per cent)

|  | Ratio of R&D manpower to total employment | Share of the number of researchers to total R&D employment | Share of the number of technicians to R&D employment |
|---|---|---|---|
| *Korea* | | | |
| 1978–85 | 0.26 | 0.686 | 0.314 |
| 1986–90 | 0.49 | 0.705 | 0.295 |
| 1991–97 | 0.82 | 0.678 | 0.322 |
| *Singapore* | | | |
| 1978–85 | 0.26 | 0.472 | 0.528 |
| 1986–90 | 0.46 | 0.591 | 0.409 |
| 1991–97 | 0.68 | 0.790 | 0.210 |
| *Taiwan* | | | |
| 1978–85 | 0.44 | 0.763 | 0.237 |
| 1986–90 | 0.66 | 0.677 | 0.323 |
| 1991–97 | 1.00 | 0.698 | 0.302 |
| *Japan* | | | |
| 1978–85 | 1.00 | 0.579 | 0.421 |
| 1986–90 | 1.20 | 0.615 | 0.385 |
| 1991–97 | 1.30 | 0.664 | 0.336 |

*Source*: For Korea and Taiwan, statistical yearbooks, various issues; for Singapore, National Science and Technology Board, various issues; and for Japan, Japan Statistical Yearbook, various issues.

felt on the domestic economy only if the absorptive capacity of the domestic economy is high enough to absorb and implement the new technologies created from R&D investments.

The private sector must undertake more R&D activities in the new economy. There is strong evidence that technology cycles are getting shorter; this will have important implications for R&D activities. In shorter technology cycles, the benefits of R&D activities are reaped more rapidly by the profit-based private sector than by the government sector. There is a greater need for the private sector to invest more in R&D than does the government, because it is easier for profit-based corporations to translate research and innovations into successful products than it is for government-sponsored laboratories.

The role of the private sector in financing risky ventures in R&D activities is equally important for productive returns from R&D investments. There should be more support from financial systems, in particular venture capital, which is better able to evaluate and monitor high-risk innovating firms in the domestic economy. This screening and monitoring increases the probability that well-defined, market-based R&D activities will be undertaken, increasing the returns from R&D activities in the domestic economy.

R&D investments may not have a significant impact on the aggregate economy across the three countries, but there may be positive spillover effects and externalities at the disaggregated industry and firm level. Thus it is important to carry out studies at a more disaggregated level.

It clear that in the new economy the government's role is not to 'pick winners', but to facilitate R&D activities by providing proper institutional arrangements and public capital in terms of infrastructure and human capital development. This will become a pertinent policy issue in the new economy.

## NOTES

1. See Lipsey, R., 'The new economy: theory and measurement', in these proceedings.
2. The national innovation system is defined as the link between the institutional framework and technological developments in the domestic economy.
3. R&D expenditure includes expenditures and investments in R&D activities by private, government, and non-profit organisations (OECD 2000).
4. See Good et al. (1999) on the survey of the MFPG literature.
5. Moreover, the impacts of R&D on technology developments and spillovers have been more critical for productivity growth than as an input in the production function (Griliches 1999).
6. As given in Morrison and Siegel (1997), the shadow value of R&D capital is

    $$z_r = -\frac{\partial C}{\partial K_r}.$$

7. See Morrison and Schwartz (1994) for the full derivation.
8. Since the primal side of the multifactor productivity growth is equal to the negative of the dual, we need to multiply the above measure by a negative sign to give a positive MFPG measure (see Morrison and Schwartz 1994 for more details).
9. Also, the model was estimated by applying cross-equation restriction to the factor demand equations.
10. The cost function used in the analysis is a standard neoclassical cost function that must satisfy all the regularity and concavity conditions. The fitted cost function is positive and the second derivative of the hessian matrix is negative semi-definite for all years. These results are available from the authors.
11. The electricity consumption variable is used as a proxy for energy consumption in the domestic economy.
12. Unlike Morrison and Schwartz (1994), we excluded the effective rate of corporate tax from our construction due to lack of data. The depreciation rate of 6.6 per cent for Korea (Pyo 1998) and 9.05 per cent for Singapore and Taiwan (average interest given in Hulten and Wykoff 1980 and used by Rao and Lee 1995) is used to construct the price of capital.
13. The full results of the above model, including estimated coefficients and test statistics, are available from the authors. Due to limited space, we excluded the full results in the text.

14 The long-run elasticity of R&D capital to output, $\varepsilon_{k_r,y}$, is estimated by assuming the following linear relationship between R&D capital and output: $LnK_r = a_0 + LnY_i + e$.

## REFERENCES

APO (Asian Productivity Organization) (2001) 'Measuring total factor productivity: survey report', Tokyo: APO.

Directorate-General of Budget, Accounting and Statistics, Executive Yuan, Republic of China (various issues), Statistical Yearbook of the Republic of China, Taipei, Republic of China.

Economic Planning Council, Executive Yuan, Taiwan Statistical Data Book, Taipei, Republic of China (various issues).

Fagerberg, J. (1994) 'Technology and international differences in growth rates', *Journal of Economic Literature*, 32: 1147–1175.

Good, D., Nadiri, M.I. and Sickles, R.C. (1999) 'Index number and factor demand approaches to the estimation of productivity', in Pesaran, M.H. and Schmidt, P. (eds), *Handbook of Applied Econometrics*, Oxford: Blackwell Publishers.

Griliches, Z. (1980) 'R&D and productivity slowdown', *American Economic Review*, 70, 2: 343–348.

Griliches, Z. (1999) R&D and Productivity: The Econometric Evidence, Chicago: University of Chicago Press.

Grossman, G.M. and Helpman, E. (1991) Innovation and Growth in the Global Economy. Cambridge MA: MIT Press.

Hall, R.E. (1986), 'Market structure and macroeconomic fluctuations,' Brookings Papers on Economic Activity, vol. 2.

Hall, R.E. (1988) 'The relation between price and marginal cost in U.S. industry', *Journal of Political Economy*, 96, 5: 921–947.

Hulten, C. R. and Wykoff, F.C. (1980) 'The measurement of economic depreciation', in Hulten, Charles (ed.), Depreciation, Inflation and Taxation of Income from Taxation, Washington: Urban Institute Press, pp. 81–125.

Jones, C.I. (1995) 'R&D based models of economic growth', *Journal of Political Economy*, 103: 759–784.

Korea National Statistical Office, Korean Statistical Yearbook, Korea, various issues

Morrison, C. and Schwartz, A.E. (1994) 'Distinguishing external from internal scale effects: the case of public infrastructure', *Journal of Productivity Analysis*, 5, 3: 249–270.

Morrison, C. and Siegel, D. (1997) 'External capital factors and increasing returns in U.S. manufacturing', *Review of Economics and Statistics*, 79, 4 (November): 647–654.

Nadiri, M.I. and Mamuneas, T.P. (1994) 'The effects of public infrastructure and R&D capital on cost structure and performance of U.S. manufacturing industries,' *Review of Economics and Statistics*, 76: 22–37.

National Science and Technology Board. National Survey of R&D in Singapore, Singapore, various issues.

National Science and Technology Board. Annual Report, Singapore (various issues).

OECD (Organisation for Economic Co-operation and Development (2000) 'A new economy? The changing role of innovation and information technology in growth', Paris: OECD.

Park, Seung-Rok and Kwon, Jene K. (1995) 'Rapid economic growth with increasing returns to scale and little or no productivity growth', *Review of Economics and Statistics*, 77, 2: 332–351.

Pyo, Hak-Kil (1998) Estimates of Fixed Reproducible Tangible Assets in Republic of Korea: 1954–1996, Korean Institute of Public Finance.

Rao, V.V. Bhanoji and Lee, C. (1995), 'Sources of growth in the Singapore economy and its manufacturing and service sectors', *Singapore Economic Review*, 40, 1: 83–115.

Singapore Department of Statistics, Yearbook of Statistics, Singapore, various issues.

Stokey, N.L. (1995) 'R&D and economic growth', *Review of Economic Studies*, 62: 469–489.

# 6.1 Beyond Silicon Valley: the regional spread of innovation

*Juan J. Palacios*

One of the most distinctive features of the Silicon Valley model is the continuous creation and development of high technology firms within a particular locale. High technology firms are those dedicated to the design, development and production of new products or processes through the systematic application of scientific knowledge and the most advanced technologies, and which thus have both a high proportion of scientists and engineers and a large expenditure in research and development (R&D) operations.

The continuous generation of synergy that drives this process of firm creation is often described as the Silicon Valley phenomenon. It is the result of a complex interaction of social, institutional, organisational and economic factors and territorial structures that together make up an enabling milieu for technological innovation. 'Silicon Valley' is essentially an innovative milieu, a place where innovation is generated from the sheer flow of information over production and communication networks connecting scientifically oriented individuals across a wide spectrum of public and private entities (Castells and Hall 1994).

The critical ingredients in the emergence of the Silicon Valley cluster in California were originally the boom of aeronautics and space industries in Southern California after World War II, the presence of a first-rank research university (Stanford), the creation of one of the first industrial and technology parks in the United States, the existence of an abundant supply of cheap migrant labour, the availability of an abundant supply of venture capital from both private firms and military institutions, and a pleasant natural and cultural environment.

Silicon Valley is not, of course, the only kind of innovative milieu in history. Innovative milieux have developed in old, large cities of the world such as London, Paris, Tokyo, Berlin and New York. Nor, according to Castells and Hall (1994), is the Silicon Valley model necessarily replicable. It is a model in the sense of an experience that illustrates the essential elements underlying the formation of leading technological milieux, as well as the forms and sequencing of their combination (Castells and Hall 1994).

Going beyond Silicon Valley implies both replicating the Silicon Valley phenomenon elsewhere and transcending it, in a conceptual sense. There have been many deliberate attempts to emulate Silicon Valley, both in the United States and elsewhere. Even in the United States, experience shows that it takes more than building a 'research park' and luring a huge semiconductor factory to the locality. Solid investments in schools and universities, as well as local and state government policies that support and foster the process are also needed. A strong academic scene seems to be a necessary condition, but it is not sufficient on its own. As Michael Malone (1996) has pointed out, creating a local company incubator is harder than it looks. It requires a special combination of entrepreneurial personalities, venture capital, a supportive business community, and 'a matrix of suppliers, consultants and service companies'.

Major high-tech clusters exist in a number of cities throughout the Asia Pacific region. There has been a geographic spread to Asia of both innovation capacity and high-tech production sites, which has led to an extensive growth in cross-border production networks throughout the region.

To date, those developments would seem to have been more a product of the investment and location strategies of foreign multinational corporations, especially from the United States, Europe and Japan, than the outcome of deliberate efforts to replicate the Silicon Valley model.

In the Asia Pacific region, the geographic spread of innovation has been largely a function of the extent of the de-verticalisation of production processes and the resulting formation of inter-plant production networks in localities where multinational corporations have set foot. Production networks of this nature in localised territorial complexes (Scott 1987) resemble the ones which developed spontaneously in Silicon Valley in California, which came to constitute the nestling framework for the spurt of innovation and production potential in that region of southwestern United States.

The continued creation of small-scale 'proto-Silicon Valleys' (Palacios 1994), and the direction and pace of this diffusion of innovation and production capability, will largely depend on the investment strategies multinational corporations adopt in the future. And, as knowledge is the main precondition for innovation diffusion, its continued geographic spread across the Asia Pacific region is likely to depend increasingly on the development of the internet and other telecommunications networks as a key facilitating force.

## REFERENCES

Castells, Manuel and Hall, Peter (1994) *Technopoles of the World. The Making of 21$^{st}$ Century Industrial Complexes*, London: Routledge.

Malone, Michael S. (1996) 'Which way to the Silicon Forest?' www.UpsideToday.com, 29 February.

Palacios, Juan J. (1994) 'Foreign direct investment and technology transfer in the Pacific Rim: the case of the electronics industry in two proto-Silicon Valleys'. *Canada Asia Pacific Research Initiatives Special Series* No. 4. University of British Columbia, Vancouver, Canada.

Scott, Allen J. (1987) 'The semiconductor industry in South East Asia', *Regional Studies* 21: 143–160.

# 6.2 The Indian experience

*Mangesh G. Korgaonker*

## INTRODUCTION

The 'Silicon Valley phenomenon' is characterised by the continuous creation and development of high technology firms (those dedicated to the design and production of new products or processes through the systematic application of scientific knowledge and the most advanced technologies). The firms have a high proportion of scientists and engineers (at least 40 per cent) and a large expenditure in research and development (R&D). The Silicon Valley phenomenon is founded on certain critical ingredients: a nearby first-rank university (Stanford); the creation of one of the first industrial and technology parks in the United States; having an abundant supply of cheap migrant labour; having an abundant supply of venture capital from private capital investment firms and military institutions; and having a pleasant natural and cultural environment.

Is Silicon Valley a replicable model? Arguably, it is the most successful innovative environment in history, but even in the United States replication has not been at all easy or assured. Many Silicon Valley clones have discovered that creating a local company incubator is harder than it looks. It takes a unique combination of entrepreneurial personalities, venture capital, smart investors, and a business community that supports entrepreneurship through a matrix of suppliers, consultants or service companies. These ingredients are not available in many technological enclaves.

In the Asia Pacific region, there is a also a need to distinguish between the universal spread of innovative capacity, the spread of high-tech production sites and the ensuing growth of cross-border production networks throughout the region. It could be argued that the spread of production sites will depend on the 'de-verticalisation' of production processes and the resulting formation of inter-plant networks in localities where multinational corporations (MNCs) have an established presence. These local networks, often dubbed 'small-scale Silicon valleys', could resemble the networks that spontaneously developed in Silicon Valley and constituted the framework for the spurt in the area's innovation and production potential.

In India, the upsurge of innovative activity seems to have had more to do with the advent of liberalisation than with the Silicon Valley phenomenon. I shall therefore focus on the nature of innovative activity in India, the extent of diffusion across key sectors of the economy, the major drivers of innovation, and the impacts it is producing, particularly in the post-liberalisation period.

### Economic liberalisation: a key innovation driver in India

Liberalisation has dramatically changed the market and supply conditions, from being shortage and seller driven to being buyer and competition driven. To survive and grow, firms have to focus on improving their competitiveness. They are realising that the real source of industrial competition today lies in innovation and the rapid technological change taking place throughout the world. Technology is now a key determinant of strategic change in Indian firms. Industrial development based on indigenous technology development is still an elusive dream, but the 'process' of technology acquisition and assimilation is now very much a strategic process, aligned with firms' need to build competencies.

Liberalisation has stimulated the rapid growth of innovation-driven industries such as information technology (IT), communications technology, biotechnology and pharmaceutical industries. This has led to a new type of business enterprise known as the 'knowledge enterprise' and a new sector of the economy known as the 'knowledge economy'. This sector is now a significant component of the national economy and accounts for a large portion of economic growth.

Liberalisation created renewed interest in innovative entrepreneurship as a key driver for the rapid diffusion of innovation in business and industry. Entrepreneurship occupies centre stage in the wealth creation process in the knowledge economy.

It is beyond the scope of this paper to review all these developments in detail. Rather, I review some major recent findings, particularly in manufacturing and IT, which have undergone particularly dramatic change. The manufacturing sector is important for India for two reasons: the industrial economy is substantially dependent on it, and it is the major engine for the growth of new, knowledge-based industries. In this paper, I begin by reviewing innovation and technical change in the manufacturing sector. I then consider the spread of innovation in the IT industry and related industries.

## TECHNOLOGY, QUALITY AND MANUFACTURING BEFORE LIBERALISATION

### Technology

Before liberalisation, the Indian approach to industrial technology and innovation management lacked a strategic vision designed to provide a competitive edge in world markets. The national effort in science and technology (S&T) focused on setting up a large S&T infrastructure in areas such as defence, atomic energy, agriculture, space, selected industrial activities and medicine. There are nearly 1,250 recognised industrial R&D units, as well as government-owned laboratories. Infrastructure is extensive, but not well developed; moreover, because it is under government control, it is poorly linked to user industries. Before 1983, the technology import policy sought to regulate rather than promote technology diffusion. The technology policy of 1983 tried to create some balance between technology acquisition, generation and diffusion, but India's inward orientation, restricted access to capital goods, lack of incentives for innovation and use of non-equity forms of technology imports effectively resulted in ageing, obsolescent products and technology.

In India, the central government funds much S&T infrastructure. For example, during 1985–95, India's R&D expenditure averaged around 1 per cent of gross national product (GNP), of which 80 per cent was in the government sector. Private sector R&D expenditure accounted for a mere 13 per cent of the total, and industrial R&D expenditure for just 23 per cent of the total (Bowonder 1998; Korgaonker 1998). Moreover, expenditure was confined to a handful of industries such as electrical equipment, defence, metallurgy, drugs and pharmaceuticals, transportation, fuels and chemicals (Table 6.2.1).

With little equity participation, there was heavy dependence on technology licensing as a source of technology acquisition. From 1948 to 1986, about 20 per cent of technology acquisition was through equity-based collaborations, the remainder through technology licences. The collaborations were mainly from the United States, the United Kingdom and West Germany. Nearly 63 per cent of the technology payments were in the form of royalties, or royalties and lump sum payments. Unfortunately Indian firms were unable or unwilling to master the basic knowledge and processes for technical development, so

*Table 6.2.1* R&D expenditure in selected industries in India, 1990–91 (per cent of total)

| Industry | R&D expenditure | Industry | R&D expenditure |
|---|---|---|---|
| Electrical equipment | 15.74 | Telecommunications | 5.66 |
| Defence | 11.86 | Industrial machinery | 4.37 |
| Metallurgical | 9.79 | Fertilisers | 2.64 |
| Drugs and pharmaceuticals | 8.03 | Machine tools | 1.26 |
| Transportation | 7.66 | Agricultural machinery | 1.13 |
| Fuels | 7.63 | Others | 16.80 |
| Chemical | 7.63 | | |

*Source:* Bowonder (1998) and Korgaonker (1998).

innovation suffered. The Indian model, instead of being 'assimilate and innovate', degenerated into 'adopt and obsolete'.

As regards the search for technological services, the interactions of industrial firms and technology institutions are also very limited. Table 6.2.2 shows the findings of a recent World Bank study on the technological capabilities of industrial firms in India. The study showed that information services and contract R&D (product and process engineering) were the services most sought from technology institutions; the institutions were not providing services such as problem solving, testing, education and training, collaborative R&D or commercial advice. Further, large firms are more likely to use the services of technology institutions than are small and medium firms. Clearly, firms have rather low technological dynamism. Technological capabilities have been hampered by problems such as government policy and the availability of financial resources, equipment and machinery, skill or knowledge-based human resources and knowledge of current technological trends.

## Manufacturing and quality

Manufacturing in India has lacked competitive focus and strategy. There has been little strategic emphasis, no stimulus for innovation and a lack of focus on infrastructure decisions such as the choice of manufacturing software systems. However, there has been overwhelming emphasis on structural decisions such as capacity planning, and making or buying. Measures such as capacity utilisation, utilisation of men and machines, labour productivity and manufacturing cost have dominated. Most decisions have been based on

Table 6.2.2 Technological services used in India and the role of technology institutions

| Service | Importance (/ 5) | Beneficial (rank) | Extent of use (/ 5) | Obtained from TI (/ 5) | Overall Score [a] |
|---|---|---|---|---|---|
| Information | 3.77 | 2 | 2.7 | 2.04 | 10.38 |
| Problem solving/ trouble shooting | 3.90 | 6 | 3.4 | 1.27 | 2.65 |
| Standards/testing | 3.98 | 4 | 3.4 | 1.56 | 5.28 |
| Education and training | 3.65 | 5 | 3.6 | 1.64 | 4.31 |
| Contract R&D | 3.17 | 1 | 3.1 | 1.27 | 12.48 |
| Collaborative R&D | 3.89 | 3 | 2.4 | 1.28 | 3.98 |
| Commercial advice | 3.53 | 7 | 2.1 | 1.08 | 1.14 |
| Technological networks | 3.7 | 3.8 | n/a | n/a | n/a |

Source: Korgaonker (1998), based on data from Goldman et al. (1997).

Note
a Overall score = [(Importance rank * Extent of use rank * Obtained from TI rank)/Beneficial rank].

'play safe' attitudes, whether they have been hardware decisions such as the choice of technology or software decisions such as inventory control or production planning. In most industries competition has been limited or absent. The policies of reserving some items for manufacture by small companies and of industrial capacity licensing have prevented economies of scale. Import procedures have caused long delays and high costs associated with large inventories of raw materials. Excessive overtime and absenteeism have become symptoms of a decaying work culture. This has contrasted with a culture elsewhere of manufacturing excellence, the building of long-term capabilities, benchmarking with global standards, improvements in supplier capabilities, the installation of quality assurance systems, collaborative and teamwork processes, and integrated problem solving. The capital productivity ratio in the manufacturing industry had deteriorated to a level of around 0.25 in 1997. In summary, manufacturing has been largely an extension of colonial practice.

## TECHNOLOGY, QUALITY AND MANUFACTURING AFTER LIBERALISATION: A RENAISSANCE

### How liberalisation affected the manufacturing sector

Several key reforms in India have affected the manufacturing sector. The scope of industrial licensing and the number of areas reserved for the public sector and the small-scale sector have been reduced. Rules, regulations and procedures have been simplified. Duties have been rationalised and reduced. Funding for selected public sector undertakings has been discontinued. Foreign financial institutions have been permitted to operate and invest in India. Foreign equity limits have been increased. Foreign technology collaborations with no equity participation and small lump sum payments have been automatically approved. Trade and exchange rate policies have been liberalised. Requirements of the World Trade Organization (WTO) have been accepted and national policies aligned with these requirements. Investment norms for Indian firms investing in other countries have been relaxed. 'Sick' industries have been required to be referred to the Bureau of Industrial Finance and Reconstruction. A National Renewal Fund has been created to protect the interests of workers affected by downsizing. Finally, there has been a much greater thrust on new, knowledge-based industries such as IT, communications technology and biotechnology.

### Innovation and technological change

Liberalisation has changed Indian firms' perception of the competitive role of innovation and technology. Firms are now investing in technological R&D. Table 6.2.3 shows the composition of technological collaborations since liberalisation. Much foreign technology investment occurs through foreign direct investment (FDI). Although technology acquisition and diffusion are still predominant, there is a clear shift from licensing to joint ventures as the

*Table 6.2.3*  Foreign collaborations in India, 1991–94

| Year | Number of collaborations | Purely technical | Equity involved |
|---|---|---|---|
| 1991 | 950 | 661 (69) | 289 (31) |
| 1992 | 1,520 | 828 (55) | 692 (45) |
| 1993 | 1,476 | 691 (47) | 785 (53) |
| 1994 (up to August) | 1,061 | 409 (39) | 652 (61) |

*Source*: Chadha (1996).

*Note*: Figures in brackets are per cent of total.

dominant mode of technology acquisition in India. There is cooperative R&D between Indian firms and MNCs; R&D centres are being built in India by foreign firms; and knowledge-based industrial activity is emerging. In the discussion below, I outline several interesting developments.

*Foreign direct investment.* FDI has moved up sharply in recent years. By 1995–96, approved FDI had climbed to Rs 371.13 billion, and actual FDI to Rs 67.5 billion. This contrasts with figures of Rs 11.7 billion and Rs 1.58 billion in 1991–92. More FDI means a greater inflow of foreign technology investment.

*R&D by global firms.* Some foreign firms are setting up R&D units in India. They include Eli Lily, SGS Thompson, Alfa Laval, Texas Instruments, Motorola, Philips, Siemens, Proctor and Gamble, Hoechst, IBM, Unilever, Hewlett Packard and Daimler Benz. The companies are investing in areas such as drug development, electronics design, very large-scale integration design, pagers, manufacturing processes, software, herbal products and food processing.

*R&D alliances with Indian firms.* There is a growing trend to establish cooperative research activities between Indian and foreign firms in selected areas.

*Outsourcing by transnationals.* Transnational companies are now obtaining components from India and even undertaking product development (software) there.

*Technical collaboration.* There is a sharp increase in joint ventures and foreign technical collaborations involving FDI – from 20 per cent to nearly 65 per cent. Most collaborations are in technology-intensive areas such as telecommunications, power and petroleum, chemicals, metallurgy, consumer durables, transportation, food processing and textiles.

*Stimulation of technological innovation.* The government has launched special funding schemes, set up a Technology Development Fund with a Technology Development Board, and introduced fiscal incentives and support measures to stimulate industrial innovation. Financial and development institutions have launched similar schemes.

*Technology parks and technology business incubators.* New technology parks and technology business incubators have been set up. The initiative for these has come as much from the government as from private enterprise.

## Innovation and quality

Indian firms have begun to adopt total quality management (TQM) principles to foster a quality culture and follow it up with innovations for product, process and service quality improvement. Over 5,000 firms are believed to have received International Organization for Standardization (ISO) certification in quality. Several firms have adopted key quality practices in the following areas: employee empowerment; training and retraining of managers, supervisors and workers; process analysis and documentation; quality benchmarking; customer satisfaction; cross-functional teamwork; an internal customer approach; waste elimination; colour coding of raw materials; use of process capability measures and statistical analysis; and policy deployment. One study (Korgaonker 1999a, b, c) focused on quality improvement in six selected industry categories: electrical, light engineering, heavy engineering, automobiles, chemicals and textiles. Table 6.2.4 shows the findings of this study. The overall quality rating was highest for the heavy engineering industry and lowest for textiles. The average quality rating for Indian industry was 6.25 out of 10 in 1998, which reflects an average quality improvement of about 5 per cent a year. Among quality parameters, the lowest rating was for application of TQM techniques, and this applied uniformly across all industries.

## A renaissance in the new millennium?

The idea of the 'new economy', driven by information, communications and entertainment, is sweeping the world. India, with one billion people, and growing, needs to be a strong manufacturing nation. The manufacturing sector is the backbone of industrial growth and the prime generator of demand for industries such as IT, telecommunications, power, logistics and services. Even employment opportunities are substantially driven by the manufacturing sector. In India there is an urgent need to enhance national capability and competitiveness in manufacturing, but the sector is in crisis, with sluggish or negative growth in several industries, falling investments, falling market capitalisation of leading companies, falling profit to earning (p/e) ratios and shareholder returns and stagnant capital productivity and total productivity ratios.

Between 1997 and 2000, private sector investment in the manufacturing sector fell by 48 per cent. The sector's share in gross domestic product (GDP) fell sharply to 24.7 per cent. By contrast, the service sector share in GDP rose to 46.1 per cent. The manufacturing sector's problems have resulted from optimistic projections, higher real interest rates, overcapacity and cheaper imports. The new economy, driven by knowledge-based information, communications and entertainment industries seeking high growth and profitability, may also have played a role. How have the Indian manufacturing

*Table 6.2.4* Quality ratings for selected industries, 1998

| Quality parameters | Electronic | Light eng | Heavy eng | Auto-mobiles | Chemicals | Tex-tiles |
|---|---|---|---|---|---|---|
| *Product design and development* | 6.81 | 7.29 | 8.02 | 7.34 | 7.09 | 6.20 |
| *Materials quality* | | | | | | |
| Vendor control | 5.94 | 7.04 | 7.05 | 6.50 | 6.50 | 5.50 |
| Control of purchased material | 6.71 | 6.54 | 7.75 | 7.20 | 7.34 | 6.12 |
| Total materials quality | 12.65 | 13.58 | 14.80 | 13.70 | 13.84 | 11.62 |
| *Process quality* | | | | | | |
| Process economic value added | 5.79 | 6.58 | 7.55 | 7.04 | 6.91 | 4.80 |
| Process control | 5.68 | 5.91 | 6.72 | 6.59 | 6.59 | 4.50 |
| Maintenance and control of production equipment | 6.50 | 6.25 | 7.52 | 6.81 | 6.95 | 6.25 |
| Measurement quality | 6.02 | 6.10 | 7.15 | 6.54 | 6.97 | 5.92 |
| Non-conformity | 5.83 | 6.25 | 7.20 | 6.52 | 7.20 | 5.75 |
| Total process quality | 29.82 | 31.09 | 36.14 | 33.50 | 34.62 | 27.22 |
| *Distribution and logistics quality* | | | | | | |
| Handling, storage, packaging preservation, and delivery | 6.39 | 6.97 | 7.22 | 7.77 | 7.70 | 6.70 |
| *Customer service quality* | | | | | | |
| Product Installation and servicing | 6.58 | 7.06 | 7.52 | 7.56 | n.a. | n.a. |
| Customer feedback, servicing satisfaction, product liability, quality, cost | 6.29 | 6.46 | 7.52 | 7.09 | 6.32 | 5.22 |
| Total customer quality | 12.87 | 13.52 | 15.04 | 14.65 | 6.32 | 5.22 |
| *Quality systems implementation* | | | | | | |
| Quality documentation, audits, records | 4.71 | 5.34 | 8.29 | 5.75 | 5.27 | 3.56 |
| Strategic quality planning | 4.89 | 5.69 | 7.17 | 6.95 | 4.86 | 4.90 |
| Training and humanistic aspects of quality | 5.46 | 6.33 | 7.50 | 7.25 | 4.57 | 5.21 |
| Use of TQM techniques | 2.64 | 2.60 | 3.67 | 4.54 | 2.39 | 2.40 |
| Total system quality | 17.70 | 19.96 | 26.63 | 24.49 | 17.09 | 16.07 |
| Total score | 86.24 | 92.41 | 107.85 | 101.45 | 86.66 | 73.03 |

*Source*: Korgaonker (1999).

*Notes*: All parameters are rated on a scale of 0 to 10, 10 being the highest; TQM = total quality management; n.a. = not applicable.

enterprises been responding to the challenge of competition? What are the realities? To address this question, a nationwide survey has been conducted in association with the Confederation of Indian Industries (CII), the peak industry association in India. I briefly discuss the survey findings below.

## The manufacturing survey

The survey sought information on the background and profile of the participating firm; critical success factors for manufacturing; attributes of the competitive strategies of the firms; manufacturing technologies and innovations adopted by the firms; performance measurement of the firms; and manufacturing enhancement programs. A major interest was to find out how Indian firms were rating their own efforts and strengths vis-à-vis the challenges and requirements of the new competitive environment and how the priorities and capabilities to remain competitive were likely to shift over the next three years. The survey sought to identify the key manufacturing technologies and innovations currently adopted by the firms and how the firms rated their own performance. Finally the survey attempted to identify the major manufacturing enhancement programs being implemented by the firms.

### *Background of the participating firms*

A total of 150 firms participated in the survey. About 67 per cent of respondents were senior managers (chief executive officer, chief operating officer, chairman, managing director, president, vice-president or director).

### *Company profiles*

Korgaonker (2000) has assessed company profiles in terms of manufacturing competitiveness. Most firms were organised as process shops. They belonged to a wide array of industries, with about 40 per cent in electrical, electronic, automobile and related areas, and metal working machinery. Only 7 per cent were in export processing zones. Twenty-seven per cent of firms were in backward area locations. The average annual sales volume was Rs 1.03 billion, indicating that most firms were of medium size. The average number of employees was 1,068, giving sales per employee of Rs 0.96 million per year, just 5.5 per cent of the global benchmark of Rs 17.5 million per employee per year. This indicates that Indian firms are a long way from global size or global productivity benchmarks. The average return on assets was 9.8 per cent, which was low given the interest rates and cost of capital. The average number of inventory turnovers per year was 6.08 (i.e. two months inventory), far from the global benchmark of twenty turnovers. The average market share was 26 per cent, implying that the overall market size for the product/market segments was Rs 4 billion, nowhere near global size. The average ratio of profit before tax to sales was 7.8 per cent. The sales to asset ratio was just 1.26, so the asset utilisation was low. Nearly 80 per cent of companies reported profits, with only 9 per cent indicating a loss. The capital investment in advanced process technology was just 4 per cent of sales and in information systems only 1.98 per cent of sales. These figures indicate a very low level of expenditure on process technology and IT upgrade. Indian firms must guard against falling into the vicious trap of not investing enough for a more secure future.

*Major conclusions*

The survey led to four key findings.

*Critical success factors.* The study identified six critical success factors: global capabilities, overall customer service, overall product pricing and cost, overall technological leadership, overall flexibility and responsiveness, and overall product quality. On a scale of 0–5, Indian firms rated 3.11 on critical success factors, ranging from 3.6 for overall product quality to 1.5 for global capability. Firms will require higher ratings in the next three years (3.9, ranging from 2.75 for global capabilities to 4.25 for overall product pricing and cost). The gap between the capability required and the current rating is approximately 25 per cent. However, the rating for product quality level achieved is quite high and the perceived gap is minimal (14 per cent). The rating was above average for all other critical success factors except global capabilities (Korgaonker 2001). The most serious concern is with respect to global capabilities, where both the current lack of strength and the perceived gaps are alarming. This is due to the local geographical focus of Indian manufacturing companies. Despite the fact that the United States is the largest market for Indian software firms, developed markets such as the United States, Japan and Canada do not tend to rank highly in the overall business strategies of Indian companies.

*Manufacturing technologies.* Indian manufacturing firms are implementing five main groups of manufacturing technologies: information technologies, manufacturing planning and control, production process, expert systems and artificial intelligence, and product/process design. The firms surveyed had little experience in the application of advanced technology, with a rating of 2.08 (Korgaonker 2000); the contribution of such applications to the firm's success was also low, with a rating of 2.7 (Korgaonker 2000), indicating a low technological dynamism. Technologies where both the experience and contribution to success are relatively higher are shown in Table 6.2.5. Technologies used in relation to expert systems and artificial intelligence, and manufacturing planning and control, contributed relatively more to the firm's success.

There were significant gaps between current experience and the necessary competence. The overall gap was 35 per cent, ranging from 24 per cent (IT) to 42 per cent (manufacturing planning and control). There was extremely low diffusion of computer-integrated manufacturing. Over 90 per cent of firms either lacked experience with computer-integrated manufacturing or were experimenting with it in parts of their enterprises. The barriers to computer-integrated manufacturing were economic or market uncertainties, capital expenditure, insufficient in-house technical support capability, inadequate training and education of production workers, technical staff and management, and short-term profit incentives (Korgaonker 2000). The main benefits from advanced technologies (Table 6.2.6) were increased product

*Table 6.2.5* Manufacturing technologies in India: implementation and success

| Technologies implemented | | Contribution of technologies to firm's success | |
|---|---|---|---|
| *Information technologies* | | | |
| Integration of information systems across functions | (2.8) | Integration of information systems across functions | (3.8) |
| Internet/intranet | (3.4) | internet/intranet | (4.1) |
| Electronic mail | (3.9) | Electronic mail | (4.3) |
| Personal computers | (3.8) | Personal computers | (4.4) |
| Local area networks | (3.4) | Local area networks | (3.9) |
| *Manufacturing planning and control* | | | |
| Shop floor data collection | (2.9) | Shop floor data collection | (3.6) |
| Predictive maintenance | (2.5) | Total productive maintenance | (3.4) |
| Enterprise resource planning | (2.5) | Predictive maintenance | (3.5) |
| Materials requirement planning | (2.7) | Supply chain management | (3.0) |
| | | Enterprise resource planning | (3.4) |
| | | Manufacturing resources planning | (3.1) |
| *Production process* | | | |
| Programmable logic controllers | (2.7) | Programmable logic controllers | (3.2) |
| *Expert systems/artificial intelligence* | | | |
| Customer service | (2.6) | Customer service | (3.5) |
| Machine maintenance | (2.5) | Machine maintenance | (3.3) |
| Process control/improvement | (2.8) | Process control / improvement | (3.5) |
| Total quality management | (2.5) | Total quality management | (3.0) |
| | | | (3.4) |
| *Product/process design* | | | |
| Value analysis/engineering | (2.5) | Value analysis / engineering | (3.1) |
| Computer aided design | (2.9) | Computer aided design | (3.2) |

*Source*: Korgaonker (2001).

*Note*: Figures in brackets are ratings assigned on a scale of 0–5, with 0 implying the lowest score and 5 the highest score. They are the averages derived from the ratings assigned by the individual company respondents.

quality (as measured by reduced customer returns) and labour productivity, improved on-time delivery and reduced inventory requirements.

Technologies have yielded impressive improvements wherever they have been applied. Local area networks, wide area networks (WANs), TQM, ISO 9000 compliance, computer numerical control and computer-aided design (CAD) are among the most beneficial technologies. The Indian firms have adopted a cluster approach to manufacturing technologies. There are three frequently mentioned clusters: Cluster I includes CAD, computer-aided manufacturing and computer numerical control; Cluster II includes 'just in time', TQM and manufacturing cells; Cluster III includes materials requirement planning, manufacturing resources planning and enterprise resource planning.

*Table 6.2.6* Improvements in manufacturing performance from the introduction of advanced technologies, 1997–2000

| Parameter | Improvement (%) | Parameter | Improvement (%) |
|---|---|---|---|
| Labour productivity | 38 | Finished goods inventory | 14 |
| Customer returns | 38 | Market share | 12 |
| Profitability | 27 | Changeover times | 11 |
| First pass yield | 23 | Procurement lead times | 10 |
| On time delivery | 19 | Raw material inventory | 9 |
| Manufacturing to design | 14 | Work in progress inventory | 8 |
| Manufacturing cycle time | 14 | Raw material defect rate | 7 |
| Speed of new product development | 14 | Timely completion of new product | 7 |
| Delivery lead time | 14 | Average unit production cost | 5 |

Source: Korgaonker (1999a).

*Manufacturing performance.* The following parameters were used to assess manufacturing performance: sales and profitability, cost structure, inventory, quality, utilisation/efficiency/effectiveness, customer service, organisation and human assets. Firms rated the importance of performance parameters at 3.36 now and 4.17 over the next three years, an increase of 25 per cent, indicating the need to improve performance substantially across all dimensions in the next three years. Firms assigned high importance to sales volume, overall perceived and conformance product quality, customer satisfaction, on-time delivery, customer complaints, material costs and total manufacturing costs, vendor quality level, labour productivity, and the existence of a knowledge-based, flexible, cross-functional workforce.

*Manufacturing enhancement programs.* Six categories of important manufacturing enhancement programs were being undertaken by the firms surveyed: customer service, facilities, supply chain, organisation, human resources and product/process. Firms rated managerial commitment to manufacturing enhancement programs as 2.7 in the past three years (rather low), and 3.4 over the next three years, an increase of 25 per cent. Even this may be less than adequate. The firms indicated that in future 'supply chain' factors would receive the highest increase in emphasis (33 per cent); facilities and human resource factors were expected to have the lowest increase in emphasis (22.5 per cent). Firms indicated that some specific programs would receive high managerial commitment. The selective emphasis on manufacturing enhancement programs is welcome, but manufacturing enhancement programs based on technological facilities are conspicuous by their absence. This does not augur well for Indian firms becoming globally competitive or competitive enough to take on the world's leading manufacturing firms in India.

Indian manufacturing firms must speedily address a number of areas. They must take up automation and advanced technologies at a greater intensity. They must show greater competence in the design and development of products and in building and managing world-scale facilities. They must create a knowledge-based workforce and make a concerted effort to improve infrastructure. Finally, they must take quick action to deal with old plants, machinery and workforces.

Indian promoters in the manufacturing sector will have to restructure their operations, including outright sale, mergers, closures or redefinition of their roles. India has the potential to be a global hub for some finished and intermediate products. This potential is being exploited by the world's leading manufacturers, through greater FDI and through global benchmarking of capacity, technology, manufacturing excellence, workforce and management capabilities. Products with strong domestic markets and factor advantages are witnessing multinational interest. In the WTO environment, candidates for manufacturing in India include high technology, knowledge-based products protected by intellectual property rights but requiring global manufacturing locations due to cost imperatives, logistics, global scales and volumes. Examples include food items and agricultural inputs and the cement and steel required for infrastructure projects. Food and agricultural inputs have large markets. Cement production and steel production are capital intensive and require economies of scale to attract investment. A specific illustration is the ambitious takeover and expansion plan of the French firm Lafarge in the cement industry. Other candidates are consumer goods, pharmaceuticals, automobiles and white goods, whose production is categorised as knowledge and technology intensive. The trend to new manufacturing priorities will by no means spell the death of Indian companies. Some will emerge as direct competitors; some will become vendors, sources and service providers; and some will diversify into the support services required for global competitiveness. Low-cost producers will also gain entry into generally price-sensitive Indian markets.

## TECHNOLOGY AND INNOVATION IN THE INFORMATION TECHNOLOGY INDUSTRY

Innovation and technological change in the manufacturing industry reflect an epic struggle for Indian firms to remain competitive. In stark contrast, India is gaining a strong presence in the IT industry. The IT revolution is the new mantra for economic growth, with its high potential to generate wealth, foreign exchange and employment. The government, public and private sectors are increasingly using IT. In the last decade, software engineering, web-based services and e-commerce have become the new jewels of the Indian economy. The country has been quick to ride the innovation and technology wave in the IT industry. The fact that the industry is characterised by knowledge intensity rather than capital intensity is an important plus. This factor is also behind the rapid geographical diffusion of innovation. What have been important contributing factors and accomplishments?

## The knowledge and human resources advantage

India, along with Israel, Taiwan and the United States, is known for its globally competitive human capital. The success achieved in software and IT in India is against the odds, given inadequate infrastructure, expensive hardware, restricted access to foreign resources and limited domestic demand. The number of scientific and technical personnel in India is the second largest in the English-speaking world. Some Indian management and technology institutes are globally known as centres of excellence. Every year thousands of young people complete specialised diploma courses in computers and other technical areas. Hundreds of thousands of engineers and managers graduate every year. Given this quality and magnitude of human capital, India's potential to create enterprises is unlimited. In Silicon Valley, there are many success stories concerning Indians who have reached their potential and achieved prominence in terms of wealth creation and credibility. At least 30 per cent of the start-up enterprises in Silicon Valley are started or backed by Indians (Mehta Dewang 2001; Mukherjea 2001). Indian software exports are growing by an estimated 50 per cent per annum (NASSCOM 2001a). The market capitalisation of listed software companies is about 25 per cent of total market capitalisation.

## Quality and cost advantage

An IT software and services (ITSS) company must deliver a product or service that meets customer requirements, is delivered on time, is able to work right first time, is user friendly, and is technically well designed and developed. This needs talented professional staff, customer focus, a well-defined methodology, and the best available development and test tools. There are many IT quality standards worldwide, including SEI CMM (the Software Engineering Institute capability maturity model), PCMM (the people capability maturity model), ISO 9000, Bootstrap and SPICE (software process improvement and capability determination). The costs of developing hardware or software while adhering to a quality standard depend on the level at which processes are already prescribed and documented.

The Indian ITSS sector has relentlessly pursued the goal of acquiring the highest standards of quality, setting in place processes for offering world class IT software products and services. In October 2001, India had thirty-two companies at SEI CMM level 5 assessment, compared to only forty-nine organisations world-wide. Of the top 400 IT firms in India, more than 250 have acquired ISO 9000 certification. There is growing adoption of PCMM by the Indian software industry. For India – with its large asset of skilled human resources – the relevance of PCMM needs no emphasis.

In January 2001, the National Association of Software and Service Companies (NASSCOM) – the peak body of the software, e-commerce and IT services industry in India – conducted a survey to find out how many ITSS companies in India had adopted international quality standards. The study suggested that firms have laid a strong foundation to extend the value proposition created by high quality skilled manpower in ITSS at low cost, with growing vendor sophistication: more than 195 companies are quality accredited and serve the needs of over 185 Fortune 500 companies. By 2001, more than 285 ITSS companies in India were expected to be quality accredited. The key components of India's value proposition are high-quality workers, attractive price performance and the sophisticated capabilities of Indian vendors.

## The IT industry performance

In 2000–01 the ITSS industry in India had revenues of Rs 377.60 billion (US$8.26 billion) and a growth rate of 55 per cent rate (NASSCOM 2001a). Gross software exports amounted to Rs 283.5 billion and the domestic software market fetched a total of Rs 94.10 billion. In recent times, many US companies have shown increased interest in the Indian software industry. Du Pont, Deutsche Bank, Intel, Hewlett Packard and others have announced an increase in IT investment and outsourcing to India. Firms in Europe and Japan are also increasing their outsourcing to India. E-commerce software solutions are a major growth area of the Indian ITSS industry. By 2008, e-business solutions for India are expected to generate revenues of US$10 billion (NASSCOM 2001e). In 2000–01, Indian e-commerce software solutions were worth US$1.2 billion; this is expected to amount to US$1.8 billion in 2001–02.

## Software

India's innovation-led software sector is mainly export driven, with over 60 per cent annual growth. ITSS exports account for 14 per cent of India's total exports, compared to only 2.5 per cent a few years ago (NASSCOM 2001a). By 2008, they are expected to reach 35 per cent of total exports. The ITSS industry accounts for almost 2 per cent of India's GDP and will be 7.7 per cent of GDP by 2008 (NASSCOM 2001a). In 2000–01, almost 30 software companies exported more than Rs 2 billion worth of ITSS; 75 companies exported more than Rs 500 million worth of ITSS. India now exports software and services to 102 countries. Of these, almost 62 per cent go to the United States and Canada; 24 per cent to Europe; 4 per cent to Japan; and 10 per cent to the rest of the world (NASSCOM 2001a). One out of every four global giants outsourced their 'mission critical software requirements' to India.

The United States is India's largest export destination; more than 270 Indian companies have set up offices, subsidiaries and marketing alliances in the United States. The slowdown in the US economy is pushing Indian firms to expand into newer markets in Europe and Asia. In 2000–01, total Indian software exports to Europe increased by 68 per cent, the United Kingdom being the most favoured destination. European firms are fast recognising the competitive advantage of working with Indian software companies, making

Indian ITSS firms expand their base and reassess their strategies. Software and services exports to Japan amount to US$250 million and are expected to exceed US $500 million in 2002–03. The industry is growing exponentially and moving up the value chain: from staffing to software development to integration and IT business consulting. However, it needs to move faster, to be more involved in strategic consulting, brand management and R&D, to be more web based and to provide more interactive e-commerce services to customers. R&D spending in the software industry has increased from 2.5 per cent in 1997–98 to over 4 per cent during 2000–01. It will soon increase to 10 per cent of total spending (NASSCOM 2001a).

*Domestic market*

The growth in the domestic software market is slower than that for exports, at 30 per cent. However, growth will increase with the proliferation of internet, e-business and WAP-enabled technologies (those relying on wide-area paging technologies) focusing on e-governance and e-banking. The drop in software piracy from 89 per cent in 1993 to the current figure of 60 per cent is a plus. The losses due to software piracy are estimated to be Rs 11 billion in India. The industry target is to reduce software piracy levels to below 25 per cent by 2005. Local software companies and overseas companies launched, respectively, over 92 and 152 new software products and/or upgrades in the domestic market during 2000–01.

*Offshore software development*

Offshore services are becoming more dominant in the software exports sector, which is another important move up the value chain. Offshore services increased to about 44 per cent of total exports, and on-site services contributed 56 per cent of export revenues in 2000–01. In 1991–92, offshore services were only 5 per cent of total software exports, with on-site services representing 95 per cent (NASSCOM 2001a).

## Emerging sectors in the IT software and service industry

Increased bandwidth and universal networking standards are allowing multinationals to move many parts of their business systems. By 2008, global remote-service operations could amount to half a trillion dollars around the world and represent every element of the value chain: research and design, procurement and logistics, marketing and sales, accounting, IT, and human resource management.

IT-enabled service applications in India include customer interaction services, business process outsourcing and management, back office operations, insurance claim processing, medical transcription, legal databases, digital content, online education, data digitisation and geographic information systems, payroll and human resources services, and web site services. By outsourcing IT-enabled services to India, large overseas firms, including

Fortune 500 companies and existing overseas service providers, realise benefits in cost, quality and time and create platforms for new businesses. Leading IT-enabled service hubs like Ireland and Singapore now 'back-end' their operations in India, since skilled labour is a scarce resource in these countries. Several firms in the financial service sector have saved at least 50–60 per cent of their process costs by operating in India. The use of IT-enabled services also allows companies to capitalise on time zone differences and provide round-the-clock services. For example, doctors in the United States can have their notes transcribed overnight in India. IT-enabled services also provide new business opportunities. GE Capital, one of the largest IT-enabled service operations, is extending its services beyond financial services to other GE group companies and external customers. It aims to expand its IT-enabled services in India to over 10,000 employees.

A large number of smaller Indian companies also provide IT-enabled services. In 2000–01, IT-enabled service revenue in India in 2000–01 amounted to Rs 41 billion. The IT-enabled service sector was expected to achieve growth of 54 per cent in 2001. Customer interaction services, including call centres and customer support centres, are the highest-growing segment, with a growth rate of 112 per cent, and revenue of Rs 8.5 billion in 2000–01. Back office operations grew by 42.1 per cent, to Rs 13.5 billion, in 2000–2001. IT-enabled services are expected to generate revenues of US$17 billion and provide employment for 1.1 million people in the next eight years. In India, the industry currently employs 70,000 people and accounts for 10.6 per cent of the total ITSS revenues world-wide (NASSCOM 2001b).

## Internet spread

The internet is virtually a household name in India, with an increasing number of private internet service providers (ISPs) offering access at competitive prices. The internet is viewed as a vital medium for information, entertainment and communication and as the sole means for electronic commerce. Internet usage in India is still less than in China, Japan and Taiwan, but there are already over two million subscribers (and more than 5.5 million users) and the increasing pace of growth is expected to result in an estimated 50 million users by December 2003.

In January 2001, a survey was conducted in 68 cities and towns in India (accounting for over 92 per cent of the total internet users in the country) to gain information about internet usage trends (NASSCOM 2001e). In-depth studies carried out during the survey using nation-wide data revealed that more than 200 cities and towns in India had internet connectivity; that more than 3.7 million of a total five million personal computers (PCs) have Pentium I or better processors (that is, could be effectively used for the internet); and that over 81 per cent of PCs sold now are purchased to allow people to use the internet. The survey found that by June 2001 more than 120 private ISPs and 12 private international internet gateways were operational. More than

86 per cent of the top 100 corporate companies that responded to the survey had endorsed the internet and e-commerce as an integral part of their corporate strategic framework for the next year. The survey also found that 91 per cent of India's corporate web sites were located overseas.

### E-commerce

India is in the midst of an e-commerce revolution. With the arrival of the internet and the rapid growth of web applications, there is strong government support for e-commerce, both business-to-business (B2B) and business-to-consumer (B2C). In 2000–01, the value of e-commerce transactions in India amounted to Rs 23 billion. A 1999 study (see NASSCOM 2001e) has suggested that by 2008 India could earn US$10 billion from e-business solutions. Indian firms can increase their share of revenue from e-business by increasing their reach, establishing deep relationships, and expanding the scope of businesses they cover. Two particular areas have been identified as important for increasing India's share of e-commerce and e-business software exports (NASSCOM 2001e). The first includes supply chain management and customer relationship management, which will be the strongest drivers of the global e-commerce solutions market. More than 72 per cent of Indian software houses have strong expertise in supply chain management and customer relationship management. The second area is e-commerce services such as legacy application integration, internet application integration, electronic data interchange (EDI), migration to web-based services, new IT frameworks, integration with business strategies and e-commerce training services.

The Indian government has taken some important initiatives to increase the country's revenues from IT. First, in 2000, it introduced the Information Technology Act, heralding a 'cyberlaw regime' in the country. Second, it has introduced a policy to permit the entry of private ISPs, allowed private ISPs to set up international gateways and allowed internet access through cable television infrastructure. Third, the government has initiated the concept of the 'national internet backbone' and announced that the national long-distance service will be extended to private operators. Fourth, the government has endorsed policies under which there can be no monopoly of undersea fibre connectivity for ISPs and under which there will be a 'free right of way facility', to allow service providers to lay optical fibre networks along national highways, state highways and other roads. Fifth, the government has endorsed the interconnectivity of government and closed user group networks and permitted the establishment of public 'tele info centres' with multimedia capabilities. Finally, the government has permitted 100 per cent FDI in B2B e-commerce.

### VENTURE CAPITAL

Venture capital is having a growing influence on innovation and technology. Venture creation has four stages: idea generation, start-up, growth, and exit.

Venture capital financing involves funding innovative ideas with the potential for high growth but inherent uncertainties. Thus it is a high-risk, high-return investment. Venture capitalists also provide networking, management and marketing support. This blend of risk-financing and holding the hand of entrepreneurs is suitable for the 200,000 or so engineering graduates coming out of government and private colleges in India every year.

At one stage in India, entrepreneurs depended on private placements, public offerings and funding from financial institutions. In 1988, the government allowed banks and other financial institutions to set up venture capital funds and provided guidelines for doing so. In September 1995, the government issued guidelines for overseas venture capital investment in India and for tax exemption purposes. In 1996, the Securities and Exchange Board of India framed regulations for venture capital funds. The venture capital industry is still nascent in India. According to the Indian Venture Capital Association (IVCA), domestic venture capital funds and offshore funds (IVCA members) provide funding of about Rs 22 billion, 52 per cent of which is directed to technology firms. Between 1996 and 2000, venture capital and 'angel' investments in high-tech firms grew more than thirty-fold. By 2008, they are expected to reach Rs 500 billion. In 2000–01, more than 38 funds were registered in India. Given the vast potential for IT, software, biotechnology, telecommunications, media, entertainment, medicine and health, the venture capital industry will play a catalyst's role.

One venture capital fund in India is known as the 'National Venture Fund for Software and IT Industry'. It is a ten-year fund set up mainly to help small ventures to achieve rapid growth rates and maintain global competitiveness. The fund seeks to develop international networking, obtain co-investments from international venture capitalists, and obtain a listing with foreign stock markets (Nasdaq). Part of the fund is used for incubation projects for software products. In India, many state governments have already set up venture capital funds for the IT sector in partnership with local and national financial institutions.

## Critical success factors for venture capital in India

There are six critical factors for the success of venture capital industry in India. First, the regulatory, tax and legal environment needs to play an enabling role and to provide a level playing field between domestic and offshore venture capital investors. Second, there must be flexibility for investment, management and exit, in line with global trends. For example, initial public offerings, mergers and acquisitions must be permitted globally, not just within India. Third, there is a need to increase domestic funding for venture capital investment and to increase opportunities for Indian venture capital firms and venture capital-financed enterprises to invest abroad. In addition, the venture capital industry must be 'professionalised'. Fourth, high-level educational institutions must encourage entrepreneurship, ventures and incubators. Fifth, infrastructure creation must be prioritised, with government support and private management. For example, there is a need to create technology and knowledge incubators to support innovation and to promote R&D by private organisations. Finally, there is a need to encourage the growth of risk capital by creating Yozma-like funds,[1] ensuring fair return on the risk capital, allowing provident funds, pension funds and banks to participate, and supporting venture investment fairs.

## CYBERLAWS IN INDIA

Rapid innovation, technical change and sustained high growth of the cyber sector call for legal and regulatory guidelines for orderly development. The Information Technology Act 2000 has created this framework by enacting 'cyberlaws'. The act aims to provide the legal infrastructure for e-commerce in India, and to accord legal sanctity to electronic records and activities carried out by electronic means. The act states that contracts accepted by electronic communication will have legal validity and enforceability.

The act includes a number of important provisions. First, it allows authentication of an electronic record by affixing a digital signature; allows an electronic record to be verified by use of a public key of the subscriber; and makes provision for electronic governance and digital signature certificates. Second, the act provides that a Controller of Certifying Authorities will supervise the activities of certifying authorities and specify the form and content of digital signature certificates. Third, the act deals with the penalty for damage to computers, computer systems and the like, fixing compensation for damage at an amount not exceeding Rs 10 million. Fourth, the act establishes the Cyber Regulations Appellate Tribunal. Fifth, the act deals with the investigation of offences by a police officer not below the rank of Deputy Superintendent of Police. Offences include tampering with computer source documents, publishing obscene information and hacking. Finally, the act provides the constitution of the Cyber Regulations Advisory Committee.

## HARDWARE DEVELOPMENT

India has set a target of US$87 billion for software exports and domestic sales in 2008. To achieve this target, it is important to determine the hardware requirement.[2] Domestic hardware manufacture must proceed rapidly if imports are not to place an intolerable burden on the economy.

## CONCLUSION

This chapter has dealt at some length with innovation and technological change in India in the aftermath of the economic liberalisation which began in the early 1990s. My analysis dealt with two major sectors: the longstanding manufacturing sector and the newly emerged and rapidly growing IT sector, including software development, IT-enabled services, internet services and e-business. Innovation is not restricted to these sectors alone – biotechnology, telecommunications, media and entertainment are also witnessing strong innovative and entrepreneurial transformations – but a complete analysis of all sectors is beyond the scope of the present paper.

As measured by indigenous innovation, technical change and impact on the economy, the manufacturing and IT sectors are by far the most significant in India. Moreover, these two sectors present studies in contrast. The innovations in the manufacturing sector are mainly to improve national competitiveness and capabilities and are therefore primarily driven by instincts of survival rather than gaining a pre-eminent position in the global economy. These innovations are not closely linked to the Silicon Valley phenomenon, but are extremely important for a large nation like India, which needs a strong and vibrant manufacturing sector because of the size of its population and workforce. On the other hand, innovations in the IT sector have some linkages with the Silicon Valley and are to some extent stimulated by partnerships between Indian people working in Silicon Valley and those back home. From the very beginning, innovation in the IT sector has been driven by sophisticated customers in places like the United States. In other words, innovations in the IT sector are mainly export driven. They promise to create for India a substantial share in the global marketplace.

If the main purpose of innovation is to introduce new opportunities for the creation of wealth, value and employment, IT innovation in India is unmistakably successful. However, the nation's infrastructure, which has so often been the major stumbling block for rapid growth even in the manufacturing sector, needs to be expanded and upgraded if high levels of growth are to be sustained. Furthermore, it is time for the government to take steps to stimulate internal demand. Recent experience clearly shows that exclusive reliance on exports, particularly in sectors where technology and innovation are changing fast, can cause deep vulnerability and threaten to erode the gains which have been so painstakingly achieved after long years of hard and smart work.

## NOTES

The author wishes to acknowledge many helpful comments received from the discussants of the paper at the conference, particularly those from Professor Juan J. Palacios, University of Guadalajara, Mexico, and Professor Raghbendra Jha, from the Australian National University. The paper has been revised to incorporate some of their comments and suggestions.

1   Yozma Group is the pioneer of Israel's venture capital industry. Yozma makes equity investments in technology companies, in fields where Israel has demonstrated world leadership. The group targets high growth companies in the communications, information technology and life sciences sectors. Yozma created the Israeli venture capital market in 1993 through its venture fund Yozma I. In three years, the group established ten drop-down funds, each capitalised at US$20 million. In parallel, Yozma made direct investments in start-up companies, leading to a professionally managed venture capital market in Israel. Yozma launched its second fund, Yozma II, in 1998, continuing the strategy of direct investments, and also playing the role of value-adding investor, by recruiting

senior managers, formulating business strategies, raising additional capital rounds, and attracting strategic and financial investors to its portfolio companies.
2 One estimate puts it at $160 billion, according to Professor R. Jha, discussant for the paper.

## REFERENCES

Bowonder, B. (1998) 'Industrialization and economic growth of India: interactions of indigenous and foreign technology', *International Journal of Technology Management*, 15, 6 /7: 622–645.

Chadha, Vikram (1996) 'India's foreign technical collaborations – pattern and growth', *Productivity*, 37, 3: 491–502.

Department of Scientific and Industrial Research (2000) *Outstanding in house R&D achievements*, New Delhi: Government of India, November.

Goldman, M., Ergas, H., Ralph, E. and Felker, G. (1997) 'Technology institutions and policies: their role in developing technological capability in industry, Washington DC: World Bank Technical Paper 388.

Korgaonker, M.G. (1998) 'Strategy– the driving force in technology acquisition – Part I', *MM – The Industry Magazine*, December: 63–70.

—— (1999a) 'Technology, quality and manufacturing (TQM – II) in a competitive environment', paper presented at All India Seminar on Switchgear: Recent Development and Technology Management, Academy of Resource Management, 27–28 August, Mumbai.

—— (1999b) 'Dynamic world class manufacturing: a paradigm for the new millennium – Part I', *MM – The Industry Magazine*, July.

—— (1999c) 'Dynamic world class manufacturing – a paradigm for the new millennium – Part II', *MM – The Industry Magazine*, September.

—— (2000) 'Competitiveness of Indian manufacturing enterprises', *MM – The Industry Magazine*, 2nd anniversary issue, December: 25–36.

—— (2001) 'No simple answers', *Business World*, 29 January: 14–15.

Mehta, Dewang (2001) 'How the west was won – Indians in USA', *Business India*, 22 January – 4 February 2001, Mumbai.

Mukherjea, D.N. (2001) 'Essay – the celebration of the magnificent obsession', *Business World*, January 2001: 14–19.

NASSCOM (National Association of Software and Service Companies) (2001a) *Indian Software Industry Survey – a report*, New Delhi: NASSCOM, August.

NASSCOM (2001b) *Spectrum of IT Enabled Services – a featured article*, August 2001, New Delhi.

—— (2001c) *Venture Funding in India – a report*, August 2001, New Delhi.

NASSCOM (2001d) *Quality of Indian Software – a report*, August 2001, New Delhi.

NASSCOM (2001e) *Internet and E-commerce – a report*, August 2001, New Delhi

Srinivas, Alam, Dubey, Rajeev and Saxena, Vishal (2000) 'For software's big boys: which is the real face?', *Business World*, 30 October 2000, New Delhi: 22–28.

# 6.3 The Chinese experience

*Dong Yuntin*

## INTRODUCTION

The information technology (IT) industry is a pillar of China's national economy. It has developed rapidly since the market opening and reform policy was introduced. The IT industry contributes more than any other industry to the growth of the national economy as measured by output value and sales revenue. IT is one of forty industries in the industrial sector; its market share increased from 3.1 per cent in 1990 to 8 per cent in 1999. The IT industry has developed more rapidly than any other Chinese industry and has the largest output, best profits and largest export turnover. In this chapter, I will discuss the development of the IT industry in China, focusing on its rapid development, and drawing attention to some structural and other problems.[1] I will discuss the impact of the US economic slowdown and how China's entry into the World Trade Organization (WTO) will affect the industry. I will limit my discussion to the manufacturing and software industry, although the industry also includes telecommunications.

## RAPID DEVELOPMENT

### Industry size

From 1990 to 1999, the Chinese IT industry grew by 32.1 per cent per annum, compared with a total industry figure of 14.2 per cent and national economy growth of 9.7 per cent. In 2000, the output of the IT industry exceeded US$120 billion, sales revenue increased by 34 per cent to US$70 billion, and profits and taxes grew by 66 per cent to US$6.7 billion. Exports reached US$55.1 billion, up by 41.2 per cent from 1999. In 2000, the IT industry created value added of US$16.7 billion, which accounted for 1.54 per cent of gross domestic product (GDP), compared with 1.29 per cent in 1999. Table 6.3.1 compares value added of the IT industry for 1995–2000 with GDP growth over the same period.

At the end of 2000, there were 229 million telephone subscribers in China, 85.26 million of whom were mobile phone subscribers. At the end of July 2001, there were 160 million fixed-line telephone subscribers and 120.6 million

*Table 6.3.1* Comparison of GDP and value added of the IT industry, 1995–2000 (US$ billion)

| Year | GDP | Value added of IT industry | Share of GDP (per cent) |
| --- | --- | --- | --- |
| 1995 | 706.43 | 5.42 | 0.77 |
| 1996 | 820.06 | 6.44 | 0.78 |
| 1997 | 899.52 | 7.85 | 0.87 |
| 1998 | 959.12 | 9.75 | 1.01 |
| 1999 | 991.23 | 12.83 | 1.29 |
| 2000 | 1084.42 | 16.7 | 1.54 |

Source: Yearbook of China Economy 2001, Yearbook of China Electronic Industry 2001.

mobile phone subscribers. This represents more than 50 per cent growth and accounts for 37 per cent of the total Asia Pacific cellular market. China has the largest mobile phone network in Asia, exceeding even that of Japan; this further highlights the fact that China is the biggest and fastest growing network in the region. At the end of 2000, there were nineteen personal computers (PCs) per 1,000 people, compared with only four PCs per 1,000 people in 1996 (*Yearbook of China Electronic Industry* 2001). In 2000 China produced more PCs, mobile telephones, colour television sets, telephones, audio devices, video compact disks (VCDs) and magnetic heads than any other country in the world. Table 6.3.2 shows the increased production of major electronic products between 1995 and 2000.

## Information projects

In China, information projects have made great progress and have facilitated the development of the IT manufacturing industry. Country-wide, the number of computers increased from 500,000 sets in 1990 to 24 million sets in 2000;

*Table 6.3.2* The production of major electronic products in China, 1995–2000

| Product | 1995 | 2000 |
| --- | --- | --- |
| PCs | 449,000 sets | 8.6 million sets |
| Switches | 11.23 million lines | 62.37 million lines |
| ICs | 515 million pieces | 5 billion pieces |
| Software | US$82.1 million | US$2.72 billion |
| Colour TVs | 19.12 million sets | 37.42 million sets |
| Mobile phones | 13,000 sets | 52.1 million sets |

Source: Yearbook of China Electronic Industry, 1996, 2001.
Note: PC = personal computer; IC = integrated circuitry.

the output value of the computer industry increased from US$845 million in 1991 to US$33.8 billion in 2000. The output value of software and information services increased from US$51 million in 1990 to US$6.64 billion in 2000.

At the end of 2000, fixed-line and mobile telephone networks and subscriber bases were the second largest in the world. The data and multimedia telecommunications network has reached all cities and most counties. About 108 computer information systems have been established through public networks in the finance, customs, taxes and trade sectors. More than seventy-one other economies have installed direct lines into China, with mobile telecommunications in fifty-nine countries and areas. The first bankcard came out in 1985; by May 2001, 277 million bankcards and 700 million non-bank cards had been issued. In 1998, the value of card transactions reached US$25.6 billion.

Since 1996, the number of internet users has grown exponentially, doubling every six months. At the end of June 2001, there were 26.5 million internet users – a figure exceeded only by the United States and Japan, but accounting for only 2 per cent of China's population.

### Improved technology

Since the 1990s, more large and medium-sized electronic enterprises have established their own research and development (R&D) centres and injected more capital into R&D. There are about 2,000 major research achievements every year, which have brought about breakthroughs in fields as diverse as local area digital switching, Galaxy large-scale computing, a Chinese electronic publication system, a large-scale software development platform, an air traffic control system, a 32-bit central processing unit of 0.25 microns, and thin film transistor (TFT) industrial manufacturing technology.

### Economic structure

The industrial structure of the information-related electronic product manufacturing industry has gradually been rationalised and the economic structure is being adjusted continuously. The ratio of capital equipment (including telecommunications equipment and computers) to consumer products and components is 48:25:27.

The top 100 IT companies in China include thirteen companies with a turnover of RMB 10 billion (US$1.21 billion) and seventy-five companies with a turnover of RMB 1 billion (US$120 million). The largest firm had a turnover of RMB 46.5 billion (US$5.6 billion) in 2000.

### Exports

From 1991 to 1999, exports of electronic products increased by 31.6 per cent per annum, 15.7 percentage points faster than total exports. Moreover, the share of IT exports in national exports increased from 6.0 per cent in 1991 to 20.6 per cent in 1999. In 2000, IT exports leapt to US$55.1 billion, an increase of 41.2 per cent from the previous year.

China has become the world's largest exporter of some IT products. It produces 40 per cent of the world output of recorders, telephones, loudspeakers and magnetic heads. Its output of television, digital video disks (DVDs), monitors and printed circuit boards ranks third in the world.

Enterprises with foreign investment or cooperation accounted for 46 per cent of the IT enterprise structure, and private enterprises for about 10 per cent in 2000.

## E-commerce

E-commerce first appeared in China in 1993, but it did not make significant progress until 1997. By the end of 2000, there were more than 1,500 e-commerce sites, including 600 online sales sites, 100 auction sites, 180 distance education sites and twenty distance medical sites, and involving the industries of finance, airlines, railways, IT, domestic electrical appliances, tourism, toys, food, cars (or parts), books and flowers.

There are 370 business-to-business (B2B) sites and 667 business-to-consumer (B2C) sites. In 2000, e-commerce transactions were valued at US$9.34 billion, of which US$9.29 billion was from B2B and US$47.2 million from B2C. The development of business sites has been paralleled by the development of national economic and social information projects. The implementation of modern digital information projects in customs, banking, taxation and trade has established the hardware foundation for the development of e-commerce in China.

Applied software and databases for networks form the software basis for the development of e-commerce. Currently, there are 15,000 Chinese sites, of which 620 sites are of internet service providers (ISPs) and more than 1,000 of internet content providers (ICPs). In 2000, the internet service market was valued at US$641.6 million, of which US$592 million came from the access market (US$343.8 million from dial-up access and US$118.6 million from special-line access).

## GOVERNMENT POLICY

The rapid growth of the IT industry in China can be attributed to the government's emphasis on technology innovation. Technology and innovation are important drivers of economic growth, especially for high technology industry, including the IT industry.

Government support for and investment in the IT industry are fundamental guarantees for its development. In the 1980s, the Chinese government created special development funds for integrated circuits (ICs), computers, software and switches. In 1992, the IT industry was listed as a pillar industry, and later IT products were emphasised as the new drivers for economic growth. Recently, the government launched special programs for internet construction, high-definition television (HDTV), mobile communications and software.

By the end of 2000, the Chinese government had pumped at least RMB 20 billion (US$2.42 billion) into the IC industry and allocated RMB 2.2 billion

(US$266 million) to an electronic development fund to support ICs, computers and software. The fund has supported 616 projects, of which 184 were in the IC industry, 279 in computing and software, and 51 in telecommunications. Such funding has allowed many Chinese IT companies to grow rapidly to be important players in the world arena.

ICs and software are critical for the IT industry. At present, China imports 90 per cent of the ICs it uses. There are over 10,000 software enterprises in China, with more than 200,000 employees. However, in 2000 their sales revenue reached only US$2.72 billion. This figure accounted for less than 1 per cent of the world total. IC and software have become bottlenecks to further development of the IT industry. The government has therefore launched policies for supporting the IC and software industries, with numerous preferential policies in finance, taxation, exports, skilled people and intellectual property.

## Social environment

The Chinese government is nurturing the social environment for technology innovation. Excellent information network systems have been established in China and the government has launched a series of policies for technology innovation. More and more high technology industrial parks and development zones have been set up. The government has established market standards; sustained fair market competition; prevented unjust competition behaviour; protected legal rights and interests for manufacturers, operators and consumers; set up normal market order; and guaranteed the healthy development of the IT industry. The government is improving legal protection for intellectual property rights and patent technology. A just and fair market system is being set up.

An innovation fund totalling RMB 1.2 billion (US$145 million) has been created for the high technology industry, especially for small and medium-sized enterprises. The fast-growing stock market helps high-tech companies to raise more money, because more and more high-tech companies are allowed to issue stock to raise money. By the end of 2000, about eighty IT companies had issued stock, raising RMB 30 billion (US$3.63 billion), accounting for 10 per cent of initial public offering (IPO) companies. In addition, 50 per cent of IPO companies have invested in the IT industry or are ready to do so. This will provide ample capital for the IT industry.

## Foreign investment and exports

By the first half of 2001, actual foreign investment in the IT industry had reached US$12.1 billion, of an intended figure of US$16.3 billion. It occurred in high-tech fields such as switches, telecommunications, transmission equipment, computers, IC and software. China introduced technology, management and skilled people as well as capital and products. The technology and quality of many products reached international advanced levels. Self-development capability was strengthened. Expanding exports and the tapping of foreign markets have been other important factors for rapid development of the IT industry.

### Technology innovation

Most technology innovation occurs in enterprises. Since the 1990s, large and medium-scale enterprises have established R&D centres and pumped capital into R&D. One-fifth of the top 100 IT companies in China inject more than 5 per cent of their sales revenues into R&D, approaching the international level. Three switch companies invest more than 10 per cent of their sales revenue in R&D, so that China develops switches with its own intellectual property in critical technologies.

China has explored a new method of technology innovation. Integration of enterprises, universities and research centres helps universities and research centres to develop new designs and innovative ideas into industrial processes and products. This partnership also helps industries to shorten development cycles and lower production costs. Some large companies have established postdoctoral working stations to help commercialise their research achievements.

Innovation and creativity are reflected in patent applications and authorisation. From 1987 to 2000, about 2,500 patents were authorised in China in high-tech industries, including computers and telecommunications. From 1997 to 1999, domestic companies applied for 3,284 patents related to the IT industry, of which 1,107 related to computers and automation and 422 to telecommunication technologies, together accounting for 46.6 per cent of IT-related patents.

### International collaboration

The Chinese government promotes collaboration with multinational companies specialising in the design and manufacturing of electronic products, wireless technologies, and e-commerce platforms. Microsoft, Intel, IBM, Bell, Lucent and Ericsson have established their own R&D centres in Beijing and Shanghai. This helps to nurture developers and enhances China's R&D capability. In addition, Konka Company has established a strategic alliance with Lucent, and Haier Company has developed a strategic alliance with Dow and Motorola. More and more multinational companies have established joint ventures or agencies in China.

## PROBLEMS

### Sales revenue

There is a large gap between the sales revenue from China's IT industry and that of industrialised countries. For example, the sales revenue of China's IT industry reached US$52 billion in 1999; that of IBM was US$87.5 billion in the same year. The Legend group, China's largest computer firm, recorded sales revenue of US$2.45 billion, only 2.8 per cent that of IBM.

## Quality and profit

Many IT enterprises are still running at a deficit. At the end of 2000, there were 966 deficit enterprises, a 33.1 per cent share of 2,919 electronic enterprises with a total deficit of RMB 310 million (US$37.5 million). Unsold product inventories reached US$5.5 billion, an increase of 9.6 per cent. In 1999, the profit rate of the Legend group was only 2.26 per cent, compared with 24.9 per cent for Intel and 8.8 per cent for IBM.

## Structural problems

China suffers from a number of structural problems, which are outlined in the sections below.

*Industrial structure.* The manufacturing industry and service industry are out of proportion. In 2000, the sales revenue of the service industry (including software) was US$6.65 billion, less than 10 per cent of the total revenue of the IT industry. In addition, the production values of computer hardware and software are out of proportion. In 2000, the ratio of hardware to software in China was 4:1, compared with 41:59 in the United States.

*Product structure.* China's scarcity economy led to duplication and excess production capacity of television and switches. The contrast between product ageing and time-delayed development is increasingly serious. Most high value-added products such as new monitors and network products are imported. In 2000, about 20 billion pieces of IC were imported, accounting for 90 per cent of the domestic demand.

*Enterprise structure.* The average production capability of eighty PC manufacturers is 30,000 sets over one year, but the government defines the rational economic scale as 500,000. Only three companies, including the Legend group, have achieved that scale. Many international companies can produce five million sets, so our enterprises are less competitive in the global market. Only ten of ninety-one television manufacturers reach the rational economic scale of one million sets; fifty-one television manufacturers produce fewer than 100,000 sets a year.

*Geographic distribution.* In 2000, the output, sales revenue and profit of the electronic industry in nine coastal provinces or cities accounted for 84 per cent, 89.5 per cent and 93 per cent of the whole industry, respectively. The ratio of sales revenue in eastern China to that in middle and western China was 13:1:1. The irrational investment system resulted in duplicated construction, overlapping spatial distribution, and a lack of product diversity and complementarity.

*Export structure.* In 2000, 90 per cent of exports of electronic products were derived from processing imported materials; only 8.4 per cent were domestically sourced. Joint ventures accounted for 72 per cent of exports; state-owned enterprises accounted for only 26 per cent. Sixty per cent of exports go to Asia; few go to Europe, Latin America or Africa.

*Investment structure.* R&D expenditure in most high-tech enterprises constituted less than 1 per cent of their sales revenues, resulting in the delayed development of products and limitations of key technologies. From 1996 to 2000, the share of R&D expenditures in China's GDP ranged from 0.6 per cent to 0.75 per cent. The R&D expenditure per person in China was only 0.4 per cent that of the United States and 0.3 per cent that of Japan.

## IMPACT OF THE US ECONOMIC SLOWDOWN

Since the third quarter of 2000, US economic growth has slowed down, with many negative effects on the global economy. What has China's IT industry suffered as a result of this?

In the first six months of 2001, China's IT industry grew by 32.9 per cent compared to the same period in 2000, 20 percentage points more than the average growth rate of industry as a whole. The IT industry contributed 21.4 per cent to the growth of the national economy. IT exports reached US$28.92 billion, with an increase of 21 per cent compared with the same period in the previous year. However, the slowdown of the US economy has affected the export of IT products. In 2000, exports of IT products to the United States reached US$12.58 billion, representing 22.9 per cent of total IT exports; in the first half of 2001, exports to the United States rose by only 7.5 per cent. Moreover, exports to the United States decreased by 11 per cent between April and May. The slowdown's full impact on the IT industry was expected to appear in the second half of 2001. To counteract the impact of the US economic recession, the Chinese government is seeking to expand domestic demand. The government's internet access project, information project, mega-development in western China and upgrading of traditional industries drive the strong demand for IT products.

On the other hand, China is also likely to benefit from the US economic slowdown. International capital might withdraw from the United States, but could be attracted to China, with its stable exchange rate and broad market. Inspired by China's entry into the WTO, foreign companies are pumping more capital into China, especially into high-tech industries. In the first quarter of 2001, the value of foreign direct investment agreements leapt to US$16.3 billion, an increase of 44.3 per cent over the same quarter of the previous year.

Skilled people are required for the high-tech industry. In the first half of 2001, multinational companies such as Microsoft, Lucent, Cisco and Nortel laid off tens of thousands of engineers. Many of these were from China. They are likely to be attracted to China with preferential talent policies and an improved environment. The Chinese government has established several industrial parks for Chinese people who have studied overseas.

## IMPACT OF CHINA'S ENTRY TO THE WORLD TRADE ORGANIZATION

China entered the WTO at the end of 2001, at which time its IT industry faced many opportunities and challenges. As a member of the WTO, China will have a greater opportunity to join international competition and global affairs. This will help to promote the development of the market economy system and the upgrading of enterprises, introduce and keep pace with the top technologies and open up the world market for Chinese products. On the other hand, challenges are first reflected in the ideological sphere, including the concept of markets, competitive attitudes and quality standards. The economic system, national quality, industrial civilisation and urban construction in China do not yet meet the standards of industrialised nations. WTO entry will encourage the Chinese government to open markets and reduce or even cancel tariffs. This will result in a large shock on IT products that have enjoyed long-term protection from tariffs and licences (for example, switches and colour tubes) or that lack technological innovation and intellectual property rights (for example, chips, software and network products).

China's entry into the WTO has prompted the government and related sectors to focus on WTO regulations, accelerating the reform of the economic system and improving national quality levels. To minimise the shock to the IT industry, China planned to take eleven countermeasures. First, it planned to accelerate structural adjustment to strengthen competitiveness in the world market. Second, it planned to maximise China's comparative advantage in processing and assembly, and also to emphasise joint development. Third, it planned to establish technology innovation systems in enterprises and promote the commercialisation of high technology. Fourth, it planned to improve the development environment to attract and retain skilled people, and prevent the loss of senior skilled people. Fifth, it planned to study international regulations as a basis for removing regulations and policies contradictory to WTO regulations and establishing regulations consistent with international regulations. Sixth, it planned to adjust tax structures and supplement important tariffs. Seventh, it planned to implement special policies for products such as software, mobile telecommunications, digital TV and network products, to support their rapid development. Eighth, it planned to enlarge domestic demand to strengthen the capacity to withstand risks. Ninth, it planned to expand exports to mitigate the conflict between supply and demand in the domestic markets. Tenth, it planned to establish and improve the economic, commercial and industrial aspects of the legal system to protect and promote the healthy development of the IT industry by legal means. Finally, it planned to establish independent research institutions to focus on the study of countermeasures for access to the WTO.

## FORECAST

According to the 'tenth five-year plan' of the Ministry of Information Industry, the output value of the IT industry is projected to increase from US$120 billion in 2000 to US$300 billion in 2005, sales revenue from US$70 billion to US$180 billion, value added from US$16.1 billion to US$38 billion, and exports from US$55.1 billion to US$100 billion. E-commerce transactions are projected to reach US$5 billion by 2005. By 2005, it is expected that twenty enterprises in the IT sector will have an annual turnover of more than RMB 20 billion (US$2.4 billion) and that five to ten enterprises will have an annual turnover of RMB 50 billion (US$6 billion).

Key products will rely on self-development and their technology and quality will be developed to meet domestic and foreign needs. The technology and quality of communication products, computers and video and audio products will achieve international advanced levels.

## NOTES

1   The information in this chapter is derived from the following sources: *Calling China*, Asia Business, July 2001, Hong Kong; *Yearbook of China Economy 2000*, 2001, Beijing; Dong Yunting (2000); *Prospects for the IT industry in the tenth five-year plan*, 2000, Beijing; *Regulations of the WTO and the strategy of China*, 2000, Beijing; *World Trade Organization annual report on world trade 2000*; *Yearbook of China Electronic Industry 2000*, 2001, Beijing; and *Yearbook of World Electronic Data 2000*.

## REFERENCES

Dong Yunting (2000) 'Impacts of WTO entry on China information product manufacturing industry and our countermeasures', *Journal of University of Electronic Science and Technology of China*, 2, 1: 9–15.

# 6.4 The Taiwanese experience: impact on production and trade

*Sheng-Cheng Hu and Vei-Lin Chan*

## INTRODUCTION

Accelerated economic growth in the United States in the 1990s has often been claimed to be a 'new economy' phenomenon. The new economy emphasises the importance of information and communication technology (ICT), human capital and innovation for economic growth. However, empirical support for the new-economy phenomenon seems to be mixed. While earlier empirical studies had various explanations for Solow's productivity puzzle,[1] recent evidence supported the presence of a 'new economy' for the United States in the second half of the 1990s.[2]

In contrast, Taiwan's economic growth has been slower in the 1990s than in the 1980s. Partly due to structural changes in the country's economy and the effects of the Asian financial crisis in mid-1997, the average growth rate of real gross national product (GNP) per capita fell from 6.84 per cent in the 1980s to 5.66 per cent over 1990–94 and 4.84 per cent over 1995–99. Since the 1990s, Taiwan has become a leading producer of computer hardware in the world. Taiwan plays an important role in the global production networks of the personal computer (PC) industry. The computer hardware industry contributes to Taiwan's economy not only through production but also through investment, exports and the capital market. This paper discusses Taiwan's recent economic development from three perspectives: ICT, human capital and innovation. In discussing ICT, we will focus on the three channels through which ICT boosts economic growth: the ICT sector itself, ICT investment, and spillover effects (where the use of ICT stimulates technological progress).

Taiwan is a small island with few natural resources. Traditional manufacturing industries played an important role in earlier economic development. In the past decade, the ICT industry has created another 'miracle' for the economy. Global networks of human resources and flexible, network-type industry structures are often regarded as key factors in explaining the rapid development of the computer hardware industry in Taiwan.

Since 1987, Taiwan's economy has experienced a dramatic structural change; the service sector and the ICT-manufacturing sector have grown rapidly. Real GDP of the service sector as a share of total real GDP rose from 47.34

per cent in 1987 to 64.3 per cent in 1999. This ratio is near those of advanced economies. Within the manufacturing sector, technology-intensive manufacturing has grown much faster than traditional and basic manufacturing industries. The electrical and electronic machinery industry is used as a proxy for ICT manufacturing because we cannot separate ICT manufacturing from the electrical and electronic machinery industry using two-digit industry-level data. The GDP of the electrical and electronic machinery industry as a share of manufacturing GDP rose from 11.73 per cent in 1987 to 30.11 per cent in 1999. In fact, the electrical and electronic machinery industry is the only manufacturing industry to enjoy a two-digit growth rate.

Taiwan has benefited from the globalisation of the computer industry and in the past decade has become a leading producer of computer hardware. Since the 1960s, it has relied on light industries such as textile and shoe manufacturing for its rapid economic development. In the 1970s, heavy and chemical industries developed fast. Since the mid-1980s, Taiwan's manufacturing sector (especially the traditional manufacturing industries) has shifted production to countries with low labour costs. On the other hand, Taiwan's computer hardware industry is highly competitive, because it is flexible, well networked, and export oriented.

Taiwan's earlier industrial policy focused on promoting the ICT-producing sector and overlooked the domestic market and the use of ICT. This slowed the adoption of ICT technology. The government has now recognised that the use of computers is important to productivity and hence economic growth. As a result, it has recently begun to promote domestic demand for software and information services by relaxing telecommunications regulations and investing heavily in ICT infrastructure. This process is still going on. Taiwan is one of the top ICT countries in Asia as measured by indicators such as mobile phones, phone lines, computers, internet users, and internet hosts (per 1,000); moreover, Taiwan has better telecommunications networks than many other Asian countries and can thus provide a better environment for e-commerce and e-business. For example, Taiwan had a mobile phone penetration rate over 100 per cent in 2000, the highest in Asia. According to the META Group, Taiwan's potential for global e-commerce ranks tenth among the forty-seven countries covered in 2000.[3]

One significant recent change is that the ICT sector has speeded up the movement of low-end goods production to China. In the mid-1980s, following the limited relaxation of Taiwan's 'three no's' policy and Deng's announcement of the 'one country, two systems' formula, China–Taiwan cross-strait relations began to expand. Indirect trade between them has increased rapidly since 1987. The political 'cold war' dated from Lee Teng-hui's 'private' visit to Cornell University in the United States in June 1995 and followed his announcement of 'special state-to-state relations' with China (the 'two states' theory) in July 1999. Both events caused great tensions across the strait. But China adopted a policy to 'separate politics from economy'. As Lee further adopted the 'go slow' policy on 14 August 1996, cross-strait economic relations remained very strong. Cross-strait investment and indirect trade continued, deeply affecting the development of Taiwan's domestic ICT sector.

Since 1949, Taiwan has banned the three direct links to China (postal, air and shipping services) and direct trade with China. China has pushed for the three direct links since the late 1970s. Taiwan's government resisted because of political and security concerns. Beijing's 'one China' principle is unacceptable to Taiwan's government. The pressure to open postal, air and shipping services forced President Lee to announce the 'go slow' policy. The policy has restricted investment in China by high value-added manufacturing firms; much of Taiwan's investment and trade with China has been done via Hong Kong and other third countries. A considerable amount of investment and trade was conducted directly in defiance of the official ban without being properly recorded. Even so, China, especially Hong Kong, was Taiwan's biggest export market in 2000. Taiwan has had a large trade surplus with Hong Kong and China. In 2000, it was US$27.2 billion, of which almost one-third (US$7.8 billion) was from the trade of electronic products and information and telecommunications products. This almost equalled Taiwan's total trade surplus (US$8.3 billion).

In 2000, electronic and information and telecommunications products generated a total trade surplus of US$12.7 billion, 62 per cent from trade with Hong Kong and China.[4] A close production network between Taiwan and China's ICT industries lies behind these statistics. This production network is just like that of the United States, Japan and the newly industrialised economies (NIEs) of Taiwan, Korea and Singapore: companies in industrial countries looked for low-cost production bases, and underdeveloped countries such as NIEs, southern Asian countries and China looked for chances to develop their own ICT industries. Hence, there is no way to discuss Taiwan's ICT industry without discussing the role of China's ICT industry. Postal, air and shipping services gradually became a hot issue while the manufacturing sector struggled for structural changes due to rising labour costs. There was no plan by the Taiwanese government to change the 'three links' policy before Taiwan entered the World Trade Organization, which does not allow Taiwan to ban these direct links with China. When Taiwan changes this policy, the ICT sectors of Taiwan and China will be further integrated.

ICT is the most important sector in Taiwan as measured by investment, exports, trade surplus, research and development (R&D), foreign direct investment (FDI) and capital market share. In Taiwan, the 'new economy' seems to have been established through the production rather than the use of ICT and its spillover effects in the rest of the economy. Further development in ICT production is likely to require manufacturers to shift production to low-cost countries and upgrade to more advanced products or process technologies.

The following sections of this chapter describe Taiwan's ICT sector and industrial policy and the contribution of ICT to Taiwan's production and trade, before concluding with some remarks about the ICT industry in Taiwan.

## CHARACTERISTICS OF TAIWAN'S COMPUTER INDUSTRY

Roughly speaking, the ICT sector includes both the manufacture and service of office, accounting and computing machinery, computer hardware (including semiconductors and electronic components), computer software and communication equipment.[5] The development of Taiwan's electronics and information industry started in the 1960s. Foreign multinational corporations (MNCs) pursued Taiwan as an original equipment manufacturer of consumer electronics. Domestic Taiwanese firms are major ICT manufacturers. The electronics and information industry initially concentrated in small and medium-sized enterprises (SMEs). SMEs had an advantage in coping with a fast-changing world market but had disadvantages in their reduced ability to be innovative. In the past decade, Taiwan has turned into one of the world's leading producers of computer hardware. Increased efficiency and lowered production costs come from increasing the size of firms, so mergers and acquisitions have become a growing trend in the electronics and information industry.

### Computer hardware industry

Taiwan specialises in the production of electronic data processing and components; since 1995, it has been the third largest computer hardware producer in the world.[6] Value added in Taiwan's ICT industry accounts for a larger share of GDP (8 per cent in 1999) than that of Organisation for Economic Co-operation and Development (OECD) countries, in which levels are between 2.5 per cent and 4.5 per cent (OECD 2000a).

Taiwan's computer hardware industry depends heavily on the United States, Japan and European countries for critical technologies, components and equipment. The key factors of Taiwan's success story are its engineering capabilities and highly flexible manufacturing techniques, not R&D and innovation. Taiwan's computer hardware companies make technological improvements in products that were originally developed by advanced countries. Cooperating with MNCs and licensing technologies are important channels for Taiwan's ICT manufacturers to upgrade their technologies. Another channel is government-affiliated laboratories and research institutions such as the Institute for Technological and Industrial Research (ITIR), which develops new technologies and then transfers them, along with human resources, to the private sector. This channel was especially important in the startup of Taiwan's major integrated circuit (IC) firms such as the United Micro Electronics Corporation (UMC) and the Taiwan Semiconductor Manufacturing Corporation (TSMC). Taiwan's internet industry has grown rapidly and the country is now the second largest supplier of internet products such as hubs, ethernet local area network (LAN) cards and modems in the world, with a production value of US$3.5 billion in 2000. The production value of the telecommunications industry was US$5.2 billion in 2000. 'Thin client'[7] and internet terminal are major products; in 2000, cable communication had a production value of US$1.45 billion, with an annual growth of 224 per cent.

Since the mid-1980s, the increasing globalisation of the computer industry has been reflected in global sourcing (the shifting of computer hardware production to Thailand, Malaysia and China) and the vast amount of FDI in southern Asia and China. At first, Taiwan's ICT manufacturers just sought production bases with lower production costs. In recent years, they have begun to play an important role in global logistics networks.

Table 6.4.1 shows the total, domestic and offshore production value of computer hardware from 1995 to 2000. The production value of offshore manufacturing grew much faster than that of domestic manufacturing. In fact, the proportion of the production value of domestic computer hardware manufacturing to the total production value of Taiwan's computer hardware has fallen by roughly 5 per cent every year. The proportion was 88 per cent in 1992 (MIC, ITIS Project 2000), 72 per cent in 1995 and only 48 per cent in 2000, the first year in which the offshore production value of computer hardware exceeded that of domestic production. The widening gap between export value and export orders also reflects the trend for Taiwanese firms to receive orders from foreign buyers and manufacturers offshore. In 2000, notebook PCs, motherboards, monitors and desktop PCs accounted for 78 per cent of total production value. Some 84 per cent of desktop PCs, 81 per cent of monitors, 47 per cent of motherboards and 6.7 per cent of notebook computers were produced offshore (MIC, IT IS Project 2000).

China has lagged behind Taiwan in ICT technology, but it is Taiwan's leading competitor in the computer hardware industry. On the other hand, because of low labour costs and close cultural backgrounds and networks, China is the largest production base abroad for Taiwan's computer hardware manufacturers. Most factories funded by Taiwan's entrepreneurs are located

*Table 6.4.1* Taiwan's computer hardware production, 1995–2000 (US$ million, per cent)

| Year | Total Production value | Growth rate | Domestic manufacturing Production value | Growth rate | Offshore manufacturing Production value | Growth rate |
|---|---|---|---|---|---|---|
| 1995 | 19,543 | 34.0 | 14,071 | 21.5 | 5,472 | 82.2 |
| 1996 | 25,025 | 28.1 | 16,999 | 20.8 | 8,036 | 46.9 |
| 1997 | 30,174 | 20.5 | 18,889 | 11.1 | 11,285 | 40.4 |
| 1998 | 33,776 | 11.9 | 19,240 | 1.9 | 14,536 | 28.8 |
| 1999 | 39,398 | 18.1 | 21,023 | 9.3 | 18,375 | 26.4 |
| 2000 | 48,076 | 20.5 | 23,209 | 10.4 | 24,867 | 35.3 |

*Source*: MIC IT IS Project, Institute of Information Industry, November 2000.

in the Pearl River delta in Guangdon and in the Yangtze River delta. Over one-third of the production value of Taiwan's computer hardware comes from China, and over 70 per cent of the production value of China's computer hardware comes from factories funded by Taiwan's entrepreneurs (MIC, IT IS Project 2000). This is the main reason that China replaced Taiwan as the third largest computer hardware producer in 2000. When offshore production is included, Taiwan is still one of the world's dominant computer hardware producers.

Recently, China has encouraged computer industrial clusters in special economic zones and coastal open cities. In the past decade, it has set up fifty-four high-tech science-based industrial parks, which have attracted more Taiwanese high-tech manufacturers to shift their production to China. Initially, the Taiwanese manufacturers tended to move labour-intensive ICT production with mature production technologies to China; since 2000, those with higher value-added production and/or more advanced products such as semiconductors and notebook PCs have invested large amounts in China. However, they continue to manufacture high-end products in Taiwan, because China lacks the high-quality human resources and environment for high-tech production. Taiwan need not worry about this offshore movement in the ICT sector in the short run, but in the longer term the government should be concerned about the situation if domestic hardware production keeps falling behind offshore production without manufacturers upgrading to more advanced products or process technologies.

Many Taiwanese computers and components are manufactured for original equipment manufacturers, so most ICT goods are exported, accounting for about one-third of total exports. Taiwan's ICT exports therefore play an important role in the world trade in ICT products. Taiwan's economy relies on international trade, and international trade relies heavily on ICT products. Moreover, a recent survey by the Chung-Hua Institute of Economics showed that foreign-funded economic units produced 80 per cent of China's high-tech exports, most of which are information and electronic products. Some 72 per cent of information and electronic goods exports were from factories funded by Taiwanese entrepreneurs. The economies of Taiwan and China are closely connected through the ICT sector integration.

During the 1990s, China became the largest offshore production base for Taiwan's computer hardware industry and the largest host country for Taiwan's outward investment. According to the official data of China's Ministry of Foreign Trade and Economic Cooperation (MOFTEC), Hong Kong has been the largest source of China's FDI: until 1999, half the cumulative total FDI inflow into China came from Hong Kong. Taiwan is China's fourth largest source of FDI; this investment accounted for only 7.76 per cent of China's cumulative total FDI, but part of Hong Kong's investment in China was actually from Taiwan. There is no way to estimate the actual amount of Taiwanese investment in China, but Taiwan is certainly an important FDI source. Further evidence is provided by reference to the trend of investment into China's electronic and electrical appliances industry. Taiwan's official indirect mainland investment statistics show that the electronic and electrical appliance industry accounted for about 28 per cent of total indirect mainland investment over 1991–2000. In 2000, the electronic and electrical appliance industry was the dominant host industry, accounting for 56.2 per cent of indirect mainland investment. If we count unreported numbers, the statistics should be higher.

**Software industry**

Taiwan's software industry (or information services industry) is much less successful than the computer hardware industry. For example, though the software market has grown steadily, the production value of the software industry is only about US$3 billion, 8 per cent of that of computer hardware in 2000. The underdevelopment of Taiwan's software industry is partly due to the slow adoption of information technology. It also reflects the fact that Taiwan's computer policies have focused more on production than on use and more on exports than on domestic markets. Table 6.4.2 shows the production values and growth rates of product, project and service markets in Taiwan from 1996 to 2000. The recent popularity of e-business and e-commerce can explain the rapid growth rates of the project and service markets. In 2000, packaged software accounted for 86 per cent of total software exports, and antivirus programs (made by Trend Inc) for 41 per cent of package software exports. The project and service markets are mainly domestic. Because the development of software industries has been slow, there were only ten over-the-counter listed software companies in the Taiwan Stock Exchange Corporation.

Table 6.4.2 Taiwan's computer software production, 1996–2000 (NT$ million, per cent)

| Year | Total Production value | Growth rate | Product market[a] Production value | Growth rate | Project market[b] Production value | Growth rate | Service market[c] Production value | Growth rate |
|---|---|---|---|---|---|---|---|---|
| 1996 | 48,893 | – | 25,545 | – | 14,953 | – | 8,395 | – |
| 1997 | 57,818 | 18.34 | 28,596 | 11.94 | 19,646 | 31.39 | 9,576 | 14.07 |
| 1998 | 73,391 | 26.93 | 38,887 | 35.99 | 20,213 | 2.89 | 14,291 | 49.24 |
| 1999 | 95,437 | 30.04 | 48,840 | 25.59 | 26,816 | 32.67 | 19,781 | 38.42 |
| 2000 | 118,728 | 24.40 | 57,119 | 16.95 | 33,520 | 25.00 | 28,089 | 42.00 |

Source: See Table 6.4.1.

Notes
a The product market includes packages software and turn-key systems.
b The project market includes system integration and professional service.
c The service market includes network services and processing services.

## Industrial policy

The concept of the new economy emphasises the role played by R&D and the use of ICT in technological progress. Initially, Taiwan's industrial policies focused on promoting the computer hardware industry and overlooked computer use and the domestic market. In the past few decades, development of Taiwan's ICT industries relied on government initiatives, active participation by the private sector and investments by MNCs. The government established research institutions, and transferred human resources and R&D results to private enterprises for commercialisation. The Ministry of Economic Affairs founded the Industrial Technology Research Institute (ITRI) in 1973 and the Institute for Information Industry (III) in 1979. Both institutes are non-profit R&D organisations. ITRI engages in applied research and technical service. It has recruited many people with PhD and master degrees: 76 per cent of the 13,000 employees who have left ITRI have joined private firms, especially in the ICT sector. There are twelve research divisions. Three divisions – the Electronics Research and Service Organisation (ERSO, founded in 1974), the Computer and Communications Research Laboratories (CCL, founded in 1990), and the Opto-Electronics and Systems Laboratories (OES, founded in 1990) – have had an impact on the development of Taiwan's ICT industries.[8] ERSO has helped to establish spin-off IC companies such as UMC, TSMC and the Vanguard International Semiconductor Corporation, which are now Taiwan's major IC companies. III initially carried out information technology R&D, focusing on innovation and software development, but is now promoting, developing and deploying e-government and e-commerce applications and technologies.

The Taiwanese government has provided subsidies, loans and tax breaks for private R&D efforts. Unlike Japan and Korea, Taiwan's industrial policies benefited many firms rather than just a few large firms or groups. In the early 1990s, R&D expenditure from the public sector accounted for more than half of total R&D expenditure. Recently, R&D from the private sector began to play a more important role. In the late 1980s, private R&D expenditure was about 40 per cent that of the public sector; by 1999, the proportion had risen to 62.1 per cent.[9] The electrical and electronic machinery industry had the highest ratio of R&D expenditure to sales of any manufacturing industry: the proportion was about 1 per cent in the late 1980s and 2.2–2.6 per cent over 1995–99. In 1995–99, information services had a higher ratio of R&D expenditure to sales (lying between 3.4 per cent and 4.1 per cent of sales) than did the manufacturing and non-manufacturing sectors. From 1990 to 1999, no ICT company spent more than 10 per cent of the value of sales on R&D, so Taiwan's ICT industry does not meet the OECD definition of a high-tech industry.[10] Moreover, if we take the small scale of Taiwan's firms, industry and economy into account, the amount of R&D is too small to compete with that of advanced countries.

A closer look at the data shows that almost all R&D expenditure is for applied research and technological development; none is for basic research.

For example, the electrical and electronic machinery industry allocates about 75 per cent of R&D expenditure to technological development and 25 per cent to applied research. The information service industry allocates 50 per cent to technological development and 50 per cent to applied research.[11] These data suggest that Taiwan's ICT industry may not be good at developing new key technologies or new products, but is good at upgrading technologies or process technologies originally developed abroad.

In recent years, industrial policies have switched to promoting ICT use to create domestic demand for software and information services. In 2000, the government published white papers on the electronic industry. It has also promoted the industrial automation and digitisation program by combining three existing programs[12] and in 2000 it established the Nangkang Software Science-Based Park in suburban Taipei. The aim is to build up Taiwan as an international procurement and logistics base.

## ICT'S CONTRIBUTION TO PRODUCTION AND TRADE

Taiwan is a typical small, open economy. The R&D-based endogenous growth theory suggests that economic growth for such economies is driven by technological progress generated by human capital investment, and technology transfer through FDI and international trade (Jones 1998). Small economies cannot afford large R&D expenditures for original invention and innovation; instead, technological progress and economic growth are achieved through the diffusion and transfer of technology.[13]

In the 1990s, ICT played an important role in promoting technological progress in the world. Taiwan was no exception, with new capital in ICT, ICT trade and FDI, and total factor productivity (TFP) growth due to innovation in the production and use of ICT. The changes involved human capital, which is required for learning advanced technology, and FDI and international trade, which are major channels of transmitting and learning new ideas and new technologies. Many empirical studies have supported these channels of promoting economic growth.

In this section, we will discuss Taiwan's recent economic development in the ICT sector. We focus our discussion on the ICT sector itself, ICT investment and spillover effects. The new economy focuses on the use of ICT, represented by the rise of ICT investment and TFP growth over various industries. The TFP growth rate in the ICT sector itself would reflect innovations in ICT production. TFP growth in ICT-using industries would suggest that the use of ICT indeed has spillover effects.

### The ICT sector

The ICT-manufacturing sector is the most important sector in Taiwan. However, the experience of Japan and many European countries in the 1990s shows that having a strong ICT-producing sector does not guarantee rapid economic growth. To evaluate the importance of the ICT-manufacturing sector for

Taiwan's economic growth, we calculated the contributions of ICT production to manufacturing and aggregate output. We use the term 'ICT production' to refer to ICT manufacturing only, because Taiwan's software industry is still underdeveloped.

Two-digit manufacturing industry data include the following items in the electrical and electronic machinery industry: electrical machinery, household electrical appliances, data processing equipment, video and radio electronic products, communication apparatus, electronic components and parts. The category 'ICT manufacturing' is a suitable category for all these except electrical machinery and household electrical appliances. We have used the electrical and electronic machinery industry as the proxy of ICT manufacturing. The industrial production index has shown that, since the mid-1980s, the production of electrical machinery and household electrical appliance production has grown slowly. ICT manufacturing has therefore grown faster than the electrical and electronic machinery industry.

We use ICT value added to calculate the contribution of ICT production to the economy. Oliner and Sichel (1994, 2000) and Jorgenson and Stiroh (1999) have used the growth accounting method to calculate the ICT industry's contribution to growth. However, there are no Taiwanese data for the capital stock of computer hardware, software and communication equipment, so we could not use this method. Instead, we calculate the contribution of each industry to aggregate real GDP growth by multiplying the growth rate of the industry's real GDP by its share in aggregate real GDP.

*The ICT-manufacturing sector*

Table 6.4.3 shows several important statistics for the Taiwanese ICT industry over the period 1990–99. In 1999, the ICT-manufacturing industry accounted for 30 per cent of manufacturing GDP and 8.4 per cent of total GDP. These figures are double those of 1990, reflecting structural changes within the manufacturing sector. The value added of the communications industry was approximately 30 per cent of that of the electrical and electronic machinery in 1999. If we include the communications industry and the electrical and electronic industries in the ICT sector, the sector represents more than 10 per cent of the aggregate economy, the highest proportion in the world.[14]

We calculated the contribution of the electrical and electronic machinery to aggregate real GDP growth by multiplying the growth rate of the industry's real GDP by its share in aggregate real GDP. A similar method was used to calculate the contribution of the ICT manufacturing to investment. Our results show that ICT manufacturing contributed 1.52 percentage points to the economic growth rate in 1999, compared to less than 1 percentage point in the early 1990s. ICT manufacturing accounted for less than 10 per cent of economic growth in the early 1990s, 20 per cent over the period 1995–98 and 30 per cent in 1999. The electrical and electronic machinery accounted for only 8.4 per cent of aggregate real GDP but for 30 per cent of growth of

Table 6.4.3a Characteristics of the Taiwanese ICT-manufacturing industry (US$ million, per cent)[a]

| Year | Real GDP[b] ||| Fixed capital formation[b] |||| TFP growth rate[b] ||| Market value of listed stocks[c] || TSEC weighted stock price indices[c] ||
|---|---|---|---|---|---|---|---|---|---|---|---|---|---|
| | Growth rate of aggregate real GDP | As % of manufacturing GDP | As % of total GDP | Contribution to economic growth | Growth rate of real aggregate fixed capital | As % of manufacturing fixed capital formation | As % of total fixed capital formation | Electrical & electronic machinery's contribution to fixed capital formation | Growth rate of non-agricultural sector TFP growth | Contribution of non-agricultural sector | Growth rate of non-agricultural sector | As % of market value of listed stocks – electronic TFP | Growth rate of weighted stock index total | Growth rate of stock index – electronic (monthly average) |
| 1990 | 5.39 | 13.74 | 4.29 | 0.13 | 7.59 | 13.57 | 3.44 | -0.15 | 0.84 | 0.016 | 2.84 | -3.81 | — |
| 1991 | 7.55 | 14.43 | 4.48 | 0.54 | 9.23 | 15.58 | 3.83 | 0.82 | 1.40 | 0.135 | 3.82 | 0.47 | — |
| 1992 | 7.49 | 14.91 | 4.49 | 0.34 | 18.48 | 15.17 | 3.64 | 0.46 | 0.90 | 0.002 | 4.24 | -4.83 | — |
| 1993 | 7.01 | 16.41 | 5.05 | 0.60 | 11.97 | 16.19 | 3.52 | 0.28 | 0.85 | 0.082 | 5.46 | 3.31 | — |
| 1994 | 7.10 | 17.61 | 5.00 | 0.68 | 7.40 | 18.49 | 4.40 | 1.50 | 0.98 | 0.066 | 4.63 | 2.74 | — |
| 1995 | 6.42 | 20.38 | 5.76 | 1.30 | 7.31 | 22.33 | 5.78 | 2.38 | 1.32 | 0.178 | 13.39 | -1.76 | -1.96 |
| 1996 | 6.10 | 22.49 | 6.28 | 0.98 | 1.66 | 32.10 | 9.24 | 5.77 | 1.36 | 0.087 | 14.23 | 2.85 | 2.37 |
| 1997 | 6.68 | 24.79 | 6.92 | 1.22 | 10.65 | 38.75 | 12.93 | 7.09 | 1.78 | -0.013 | 25.60 | 1.64 | 7.60 |
| 1998 | 4.57 | 27.19 | 7.50 | 1.00 | 8.01 | 43.50 | 14.83 | 3.54 | 0.88 | -0.03 | 32.86 | -1.29 | 0.76 |
| 1999 | 5.42 | 30.11 | 8.40 | 1.52 | 1.83 | 46.58 | 21.83 | 2.16 | — | — | 54.48 | 1.43 | 4.86 |

Table 6.4.3b Characteristics of the Taiwanese ICT-manufacturing industry (US$ million, per cent)[a]

| Year | R&D in electrical and electronic machinery | Export value of technology-intensive product | Export value of information and electronic products | Trade surplus from information electronic products | | FDI of electronic and electrical machinery | Import and Export value of technology | Import value of technology-electrical and electronic machinery industry | Export value of technology-electrical and electronic machinery industry | Employees in electronic and electrical machinery industry | |
|---|---|---|---|---|---|---|---|---|---|---|---|
| | As % of total industry R&D | As % of total exports | As % of total exports | Amount of total trade surplus | As % of total trade surplus | As % of total FDI | Export/import value of technology | As % of total import value of technology | As % of total exports value of technology | Growth | As % of total employees in manufacturing/total employees |
| 1987 | 28.34 | 19.4 | 17.74 | 4,867 | 26.03 | 38.52 | 8.56 | 0.26 | 4.00 | 8.98 | 17.65/10.21 |
| 1988 | 38.48 | 22.6 | 19.30 | 5,439 | 49.47 | 18.05 | 4.54 | 0.17 | 0.01 | 1.89 | 17.93/10.06 |
| 1989 | 44.08 | 24.2 | 18.94 | 5,941 | 42.32 | 13.09 | 2.76 | 0.14 | 0.87 | -2.73 | 18.06/9.69 |
| 1990 | 51.82 | 26.7 | 18.97 | 5,251i | 42.01 | 16.38 | 6.39 | 0.20 | 0.95 | -2.79 | 18.75/9.55 |
| 1991[d] | 49.13 | 27.2 | 18.09 | 4,988 | 37.45 | 32.06 | n.a. | n.a. | n.a. | -1.04 | 18.91/9.35 |
| 1992 | 43.35 | 29.5 | 18.54 | 4,135 | 43.70 | 22.12 | 15.42 | 56.66 | 42.64 | 1.85 | 19.17/9.05 |
| 1993 | 46.89 | 31.4 | 19.49 | 4,532 | 56.44 | 18.68 | 9.28 | 46.04 | 30.39 | 0.78 | 19.40/8.79 |
| 1994 | 47.99 | 32.5 | 20.55 | 5,101 | 66.26 | 18.16 | 5.66 | 49.30 | 48.00 | 3.54 | 19.73/8.67 |
| 1995 | 56.76 | 36.5 | 23.43 | 7,066 | 87.15 | 42.42 | 3.99 | 62.00 | 32.75 | 3.67 | 20.75/8.68 |
| 1996[d] | 58.72 | 39.7 | 25.15 | 13,216 | 97.38 | 18.01 | n.a. | n.a. | n.a. | 1.80 | 21.52/8.95 |
| 1997 | 57.70 | 39.7 | 26.59 | 12,081 | 157.80 | 22.32 | 4.91 | 76.23 | 23.50 | 4.95 | 22.30/9.30 |
| 1998 | 62.09 | 41.1 | 27.73 | 7,308 | 123.53 | 32.25 | 2.37 | 75.62 | 78.46 | 2.55 | 22.82/9.52 |
| 1999 | 64.79 | 42.1 | 30.41 | 8,635 | 79.21 | 24.44 | n.a. | n.a. | n.a. | 1.14 | 22.92/9.63 |

Source: *National Income In Taiwan Area of the Republic of China*, *The Trends in Multifactor Productivity*, DGBAS, Executive Yuan, Republic of China. SFC Statistics, Securities and Futures Commission, Ministry of Finance, The Republic of China. *Monthly Statistics of Exports and Imports*, Taiwan Area, ROC, Ministry of Finance, ROC and *Industrial Statistics Survey Report*, MOEA.

TFP= total factor productivity; GDP = gross domestic product; FDI = foreign direct investment; n.a. = not available.

Notes

a Figures are monthly averages; b for the electrical and electronics machinery industry in manufacturing; c for the electronics industry, classified by the Taiwan Stock Exchange Corporation; d MOEA's Industrial Statistical Survey was not conducted in 1991 and 1996.

aggregate real GDP. The reason was that the ICT-manufacturing sector had a rapid growth rate, exceeded only by that of the securities and futures industry, and far exceeding growth in the traditional manufacturing industries. While the ICT sector has had a strong impact on Taiwan's economic growth, it is risky for Taiwan's economy to rely too heavily on it, because the fluctuations in the aggregate economy will be more closely associated with the ups and downs of the sector.

Not surprisingly, the electrical and electronic machinery's shares of real fixed capital formation in manufacturing and aggregate real fixed capital formation have increased steadily. In 1999, the electrical and electronic machinery accounted for almost half of fixed capital formation in the manufacturing sector and one-fifth of aggregate fixed capital formation. In the late 1990s, the electrical and electronic machinery contributed more than 2 per cent per annum to the growth rate of fixed capital formation, compared with less than 1 per cent before 1993, reflecting the rapid development of the ICT industry since 1993. Over 1996–99, the contribution of the electrical and electronic machinery to the aggregate fixed capital formation growth rate exceeded the aggregate fixed capital formation growth rate, reflecting the importance of the electrical and electronic machinery for Taiwan's industrial development.

Following Oliner and Sichel (2000), we used the value added of the electrical and electronic machinery as a share in total output as the weight to calculate the contribution of the electrical and electronic machinery to TFP growth in the non-agricultural sector. The electrical and electronic machinery output share is not high, so its contribution to TFP growth is quite low. Even in 1995, when the TFP growth rate of the electrical and electronic machinery reached its peak (3.1 per cent), the electrical and electronic machinery contributed about 0.2 percentage points to non-agricultural sector TFP growth. TFP growth of the electrical and electronic machinery has declined since 1995 and even became negative in 1997 and 1998. We observe both a rapid expansion of the electrical and electronic machinery and its declining TFP growth rate. One possible explanation is that the electrical and electronic machinery has become more efficient but not more productive, but this issue is beyond the scope of this chapter.

In 1999 and 2000, half of new listed companies and 40 per cent of new OTC-listed companies were electronic companies. The ratio of listed electronic companies to listed companies rose from 6.7 per cent in 1989 to 10.2 per cent in 1994 and 25.4 per cent in 2000. Also, affected by the US stock market's rising trend in the past few years, the stock prices of electronic companies have grown in the past four years and exceed the weighted stock index. For electronic companies, both the trading value as a share of total trading value and the market value as a share of market value in the Taiwan Stock Exchange Corporation have increased by a factor of four since 1995. The leading ICT manufacturing companies can therefore dominate the weighted daily stock

*The Taiwanese experience* 153

price index. These phenomena have encouraged electronic companies to become listed companies, to make it easier to finance their investment in the capital market.

R&D in the electrical and electronic machinery industry has accounted for more than half of total industry R&D. The definitions of the ICT sector are different in different countries, but Taiwan's share of R&D in the electronic machinery industry to total industry R&D (57.7 per cent) was higher than Finland's share of ICT R&D in total business R&D (51 per cent), which was the highest in OECD member countries (OECD 2000a).

The share of ICT exports in total exports has increased steadily over the past decade – by 15 percentage points (from 19.5 per cent to 34.5 per cent) from 1993 to 2000. ICT products have been the most important exports since the 1990s and the ICT sector has generated a large trade surplus despite the fact that ICT manufacturers have to import materials and components in order to manufacture ICT products.

ICT products account for a significant share of Taiwan's trade surplus. The share has grown to more than 50 per cent since the ICT-manufacturing sector developed in the late 1980s. In 1997, 1998 and 2000, the trade surplus from information and electronic products trade exceeded the total trade surplus of the country. ICT trade contributed significantly to the accumulation of foreign reserves. Without the ICT-producing sector, Taiwan would have a trade deficit.

The Taiwanese government has supported and facilitated inward investment and technology transfer, by helping local companies to develop specialised capabilities, identify export opportunities, and exploit those opportunities. FDI in the electrical and electronic machinery industry accounted for 20–30 per cent of total FDI in the past few years. Overall, Taiwan imports more technology than it exports. The electrical and electronic machinery industry accounts for three-quarters of total technology import and export values, respectively. These statistics indicate that FDI and technology imports are important channels for transmitting technology for the electrical and electronic machinery industry.

In 1999, the electrical and electronic machinery industry employed 9.63 per cent of all employees in Taiwan, the highest proportion in the world.[15] Compared with the share of the electrical and electronic machinery industry value added in manufacturing real GDP, the share of the electrical and electronic machinery industry employees in manufacturing employees was larger before 1995 than after. This suggests that ICT manufacturing has moved from being a labour-intensive industry to being a technology-intensive industry as technology has been upgraded and production moved offshore.

We next discuss the contributions of different sectors to production from 1982 to 1999. Tables 6.4.4 and 6.4.5 report real GDP growth rates and their contributions to total real GDP for different sectors of the economy and different manufacturing industries, respectively. The Taiwanese economy's structural change started in 1987. ICT manufacturing took off in 1993 due to the rapid growth of the world ICT market and the domestic development of

Table 6.4.4 Taiwanese GDP growth rates and contributions to GDP by sector, 1982–99 (per cent)

| Sector | 1982–99 | 1982–86 | 1987–92 | 1993–99 |
|---|---|---|---|---|
| Growth rate of aggregate real GDP | 7.32 | 7.84 | 8.21 | 6.19 |
| GDP growth rate (contribution to aggregate real GDP by sector) | | | | |
| *Agriculture, forestry, fishing and animal husbandry* | 0.86(0.06) | 1.78(0.13) | 1.42(0.08) | –0.28(–0.001) |
| *Industry* | 6.11(2.46) | 8.35(3.60) | 5.88(2.41) | 4.71(1.68) |
| Mining and quarrying | –0.21(0.0002) | –1.89(–0.02) | 3.74(0.02) | –2.39(–0.004) |
| Manufacturing | 6.36(2.07) | 9.72(3.39) | 5.05(1.70) | 5.09(1.43) |
| Electricity, gas and water | 7.96(0.20) | 10.5(0.26) | 8.31(0.21) | 5.84(0.15) |
| Construction | 4.87(0.34) | 0.81(0.02) | 10.90(0.52) | 2.59(0.16) |
| *Services* | 8.76(4.85) | 8.28(4.17) | 10.67(5.81) | 7.44(4.52) |
| Wholesale, retail trade, eating and drinking places | 9.05(1.35) | 8.06(1.08) | 11.24(1.61) | 7.87(1.31) |
| (1) Wholesale and retail trade | 8.04(0.91) | 7.67(0.84) | 9.30(1.01) | 7.23(0.87) |
| (2) International trade | 13.15(0.31) | 5.93(0.09) | 22.35(0.47) | 10.41(0.34) |
| (3) Eating and drinking places | 12.28(0.14) | 17.41(0.16) | 13.14(0.16) | 7.89(0.11) |
| Transport, storage and communication | 8.11(0.54) | 8.39(0.57) | 7.93(0.52) | 8.05(0.54) |
| (1) Transport and storage | 6.13(0.31) | 7.74(0.42) | 6.84(0.34) | 4.36(0.20) |
| (2) Communication | 13.19(0.25) | 11.15(0.15) | 11.62(0.18) | 16.00(0.38) |
| Finance, insurance, real estate | 9.88(1.63) | 7.32(0.9) | 14.11(2.27) | 7.92(1.54) |
| (1) Finance and insurance | 9.49(0.71) | 6.71(0.41) | 14.49(1.07) | 7.18(0.62) |
| (2) Securities and futures | 60.06(0.44) | 87.23(0.07) | 68.9(0.79) | 33.08(0.40) |
| (3) Real estate | 10.02(0.07) | 5.70(0.03) | 21.77(0.16) | 3.02(0.02) |
| (4) Dwellings services | 9.56(0.74) | 8.27(0.54) | 12.37(0.90) | 3.58(0.73) |
| Business services | 9.40(0.17) | 5.76(0.13) | 10.99(0.21) | 10.79(0.17) |
| Social, personal and related community services | 10.10(0.68) | 9.69(0.55) | 12.19(0.79) | 8.60(0.67) |
| Government services | 5.55(0.62) | 5.23(0.65) | 7.37(0.81) | 4.24(0.43) |
| Other services | 9.08(0.09) | 12.72(0.12) | 5.52(0.06) | 9.55(0.10) |

Source: *National Income In Taiwan Area of the Republic of China*, DGBAS, Executive Yuan, Republic of China.

the IC industry. We have therefore divided the sample period 1982–99 into three subsample periods: 1982–86, 1987–92 and 1993–99.

As shown in Table 6.4.4, the service sector grew more quickly than the industry sector over each subsample period. The difference between the two growth rates rose over time, reflecting the increasing importance of the service sector. Over the period 1993–99, the growth rate of aggregate real GDP averaged 6.19 percentage points per annum, the service and manufacturing sectors accounting for 4.52 and 1.43 percentage points respectively. Manufacturing was the fastest growing sector in the industry sector. The

*Table 6.4.5* Taiwanese manufacturing industries' growth rate and their contribution to GDP growth by division, 1982–99 (per cent)

| Manufacturing industry | 1982–99 | 1982–86 | 1987–92 | 1993–99 |
|---|---|---|---|---|
| Food products | 3.49(0.12) | 9.88(0.32) | 3.06(0.09) | –0.71(–0.01) |
| Tobacco | 2.26(0.01) | 1.44(0.01) | 5.98(0.03) | –0.34(0.0004) |
| Textile mill products | 1.68(0.07) | 7.61(0.27) | –0.25(–0.003) | –0.90(–0.02) |
| Wearing apparel and accessories | –2.64(0.01) | 6.80(0.19) | –5.28(–0.07) | –7.13(–0.04) |
| Leather, fur and products | –0.21(0.02) | 16.96(0.12) | –6.97(–0.03) | –6.69(–0.01) |
| Wood and bamboo products | –0.76(0.02) | 13.26(0.09) | –2.84(–0.01) | –9.00(–0.02) |
| Furniture and fixtures | 6.84(0.03) | 11.72(0.05) | 7.72(0.03) | 2.60(0.01) |
| Pulp, paper and paper products | 1.96(0.03) | 7.33(0.10) | –2.87(–0.02) | 2.25(0.01) |
| Printing processing | 4.51(0.02) | 6.69(0.03) | 5.80(0.03) | 1.85(0.01) |
| Chemical material | 8.91(0.17) | 14.92(0.28) | 6.12(0.11) | 7.00(0.14) |
| Chemical products | 11.79(0.06) | 18.82(0.08) | 8.92(0.05) | 9.22(0.06) |
| Petroleum and coal products | 6.24(0.11) | –0.70(0.005) | 7.11(0.13) | 10.45(0.18) |
| Rubber products | 3.21(0.02) | 8.83(0.05) | 2.92(0.01) | –0.55(–0.002) |
| Plastics products | 6.42(0.16) | 18.91(0.45) | 2.26(0.07) | 1.07(0.02) |
| Non-metallic mineral products | 5.02(0.06) | 4.85(0.06) | 8.60(0.10) | 2.08(0.03) |
| Basic metal industries | 8.60(0.16) | 12.75(0.24) | 7.38(0.14) | 6.69(0.13) |
| Fabricated metal industries | 8.22(0.17) | 12.54(0.23) | 11.51(0.25) | 2.31(0.05) |
| Machinery and equipment | 8.68(0.11) | 9.94(0.11) | 12.08(0.16) | 4.86(0.07) |
| Electrical and electronic machinery | 14.53(0.73) | 15.93(0.56) | 11.38(0.50) | 16.22(1.04) |
| Transport equipment | 4.36(0.12) | 4.53(0.12) | 10.10(0.27) | –0.67(–0.01) |
| Precison instruments and other industrial products | 1.96(0.06) | 11.70(0.25) | –0.55(0.01) | –2.85(–0.03) |
| Traditional manufacturing | 1.71(0.27) | 8.27(1.10) | –0.15(–0.01) | –1.38(–0.09) |
| Basic manufacturing | 7.15(0.87) | 10.71(1.32) | 6.75(0.84) | 4.96(–0.57) |
| Technology-intensive manufacturing | 9.53(0.97) | 10.69(1.10) | 8.85(0.91) | 9.28(1.0) |

Source: Same as Table 6.4.2.

communication industry and the securities and futures industry were the fastest growing industries in the service sector. The communication industry's contribution to the growth rate of aggregate real GDP rose sharply from 0.18–0.25 percentage points over 1992–97 to 0.57 and 1.05 percentage points in 1998 and 1999, respectively. Its contribution to the growth rate of aggregate real GDP doubled every year over 1997–99. The communication industry has only recently shown a significant contribution to economic growth, reflecting the recent liberalisation of communication policy.

Table 6.4.5 shows that, since 1987, traditional manufacturing industries such as food, tobacco, textiles, apparel, leather and wood industries have had negative growth rates, on average. The electrical and electronic machinery is one of few exceptions, with a two-digit growth rate and higher growth

Table 6.4.6 Rate of growth of production and contributions by total manufacturing and industrial production subgroups, 1987–99 (per cent)

| Subgroup | Rate of production growth | | | Contribution to growth rate of manufacturing sector | | | Contribution to growth rate of total industrial production | | |
|---|---|---|---|---|---|---|---|---|---|
| | 1987–99 | 1987–92 | 1993–99 | 1987–99 | 1987–92 | 1993–99 | 1987–99 | 1987–92 | 1993–99 |
| Industry | 5.02 | 5.04 | 5.00 | | | | | | |
| Manufacturing | 4.96 | 4.82 | 5.07 | | | | 4.33 | 4.50 | 4.17 |
| Electrical and electronic machinery | 11.63 | 9.25 | 13.68 | 2.38 | 1.57 | 3.19 | 2.19 | 1.47 | 2.92 |
| *Data storage equipment* | | | | | | | | | |
| Data storage equipment | 24.43 | 17.85 | 30.07 | 0.51 | 0.18 | 0.83 | 0.47 | 0.17 | 0.76 |
| Data storage media units | 37.69 | 16.60 | 55.78 | 0.03 | 0.02 | 0.05 | 0.03 | 0.02 | 0.04 |
| Data terminal equipment | 5.92 | 12.62 | 0.18 | 0.13 | 0.14 | 0.11 | 0.12 | 0.13 | 0.10 |
| Data I/O peripheral equipment | 22.43 | 17.93 | 26.27 | 0.09 | 0.04 | 0.14 | 0.08 | 0.04 | 0.13 |
| Computer components | 27.38 | 23.70 | 30.55 | 0.32 | 0.17 | 0.47 | 0.30 | 0.16 | 0.43 |
| Other computer equipment | 23.07 | 21.70 | 24.24 | 0.21 | 0.09 | 0.34 | 0.20 | 0.09 | 0.31 |
| *Electronic parts and components* | | | | | | | | | |
| Electronic tubes | 9.40 | 9.08 | 9.67 | 0.03 | 0.04 | 0.03 | 0.05 | 0.04 | 0.03 |
| Semiconductors | 14.21 | 21.15 | 8.26 | 0.30 | 0.31 | 0.29 | 0.64 | 0.29 | 0.27 |
| Photonics materials and components | 16.35 | 15.57 | 17.01 | 0.02 | 0.02 | 0.02 | 0.14 | 0.02 | 0.02 |
| Electronic parts and components | 6.67 | 4.68 | 8.37 | 0.05 | 0.04 | 0.07 | 0.03 | 0.04 | 0.07 |
| Other electronic parts and components | 20.97 | 16.96 | 24.41 | 0.70 | 0.39 | 1.0 | 0.04 | 0.36 | 0.92 |
| *Communication equipment* | | | | | | | | | |
| Wired communication equipment and apparatus | 18.04 | 9.00 | 25.79 | 0.15 | 0.05 | 0.25 | 0.28 | 0.05 | 0.23 |
| Telecommunication equipment and apparatus | 10.03 | 12.10 | 8.28 | 0.30 | 0.03 | 0.03 | 0.02 | 0.03 | 0.02 |

Source: *Industrial Production Statistics Monthly, Taiwan Area, The Republic of China*, Department of Statistics, Ministry of Economic Affairs, Taiwan, ROC.

rates in the second half of the 1990s. Over 1993–99, the electrical and electronic machinery attributed, on average, about 1 percentage point per annum to aggregate GDP growth. As shown in Table 6.4.3, this contribution is rising. In 1999, the electrical and electronic machinery alone already explained nearly 30 per cent of economic growth. If we add the contribution of the communication sector, the ICT sector contributed 1.57 and 2.57 percentage points (about 30 per cent and 45 per cent) of aggregate economic growth in

1998 and 1999, respectively. We can conclude that the growth in the electrical and electronic and communication industries explains a growing and significant proportion of economic growth in the late 1990s. The two industries had a significant impact on Taiwan's economic growth over 1998–99, accounting for 1 percentage point (100 per cent) of the change. This development in the electrical and electronic and communication services industries suggests that a shift to the new economy took place in Taiwan in the late 1990s.

Tables 6.4.4 and 6.4.5 are based on national income data. Adding data from industrial production statistics would add more information to the results, for two reasons. First, electric machinery is included in the electrical and electronic machinery industry and cannot be separated using national income two-digit industry data. If the electrical machinery industry had a positive growth rate, we would overestimate the contribution of the electrical and electronic machinery industry to economic growth. Otherwise, we would underestimate it. Second, industrial production statistics could provide production value and industrial production indices for subgroups in the electrical and electronic machinery industry. This would allow us to understand more about the industry distribution within the ICT-manufacturing industry.

Table 6.4.6 is based on industrial production data. It reports the growth rates of electrical and electronic machinery subgroups and their contributions to growth of the manufacturing sector and total industrial production. We choose not to report data for sound and video appliances or electric machinery, and report only three major products of Taiwan's ICT manufacturing: data storage equipment, electronic parts and components and communication equipment.[16]

The difference between the growth rates of real GDP and industrial production represents the change in the value-added ratio (original inputs:domestic gross output). If the value-added ratio increases, the real GDP growth rate is higher than the industrial production growth rate. Over all subsample periods, the growth rates of real GDP were larger than those of total industrial production and manufacturing sector production, reflecting the fact that over time the value-added ratio rose, on average, for industry and the manufacturing sector. According to the 1996 input–output tables, the value-added rates of information products, communication products and electronic parts and components were 18.8 per cent, 23.8 per cent and 31.8 per cent, respectively. The electronic parts and components industry has a higher value-added rate. From 1991 to 1996, the value-added rate decreased by 0.5 per cent in information and communication, but increased by 3 per cent in electronic components.

After excluding video and radio products and electrical machinery, the electrical and electronic machinery group accounted for 80 per cent of the growth rate of manufacturing industrial production. If video and radio products and electrical machinery are included, the figure is lower (70 per cent) because the growth rate of production of these products has been negative, on average, since the mid-1980s. If it were not for these products, electrical and electronic machinery would account for more economic growth than indicated in Tables 6.4.3 and 6.4.5.

From 1987–92 to 1993–99, the contribution of the electrical and electronic machinery industry to growth in total industrial production doubled, from 1.47 per cent to 2.92 per cent; the contribution to the growth rate of the manufacturing sector also doubled, from 1.57 per cent to 3.19 per cent respectively (Table 6.4.6). Over 1987–92, both data storage equipment and electronic parts and components can explain about half of the contribution of electrical and electronic machinery to manufacturing sector growth. Over 1993–99, subgroup industries in the data storage equipment industry grew much faster than in the electronic parts and components industry. Sixty-one per cent of the contribution of electrical and electronic machinery to manufacturing sector growth came from the data storage equipment industry, 30 per cent from electronic parts and components and 9 per cent from communication equipment. Data storage equipment and data storage media units are the fastest growing industries in the ICT industry.

The US Department of Commerce (2000) reported that ICT industries – hardware, software and communication – accounted for 1.6 percentage points of the US real GDP growth rate (5.0 per cent) in 1999. This was about one-third of the economic growth rate. Even excluding the software industry, Taiwan's new-economy phenomenon was already as significant as in the United States, at least in 1999. These results suggest that Taiwan relies too much on computer hardware and communication. In 2001, the world computer hardware market was in severe depression. The production and export value of the computer hardware industry, and investment in the industry, fell in the first half of 2001, striking a heavy blow to Taiwan's economy.

*ICT investment*

The above discussion has emphasised the ICT-manufacturing sector itself. Infrastructure and access are other important channels by which ICT can affect the aggregate economy, because they are the basis for ICT diffusion, e-commerce and e-business. Johnson (1997) has shown that ICT raises inequality in the short and medium run and that the greater use of ICT can explain increasing wage inequality and productivity differences in the United States, other developed countries and many developing countries.[17] Tables 6.4.7–6.4.9 report the first indicator – ICT expenditure or investment from different data sources. Table 6.4.10 reports other indicators commonly used, such as the numbers or penetration rates of mobile phones, computers, internet users and internet hosts.

Table 6.4.7 reports consumption, investment and export demand for ICT as a percentage of ICT final demand, using data from five input–output tables over 1986–96. The 1986 and 1989 input–output tables have industry data only for electronic products.[18]

Table 6.4.7 Taiwanese electronic products: ratios of consumption, investment and exports to final demand (per cent)

| Electronic products | Consumption (domestic/imports) | | | | Investment (domestic/imports) | | | | Custom exports | | | |
|---|---|---|---|---|---|---|---|---|---|---|---|---|
|  | 1986 | 1989 | 1991 | 1994 | 1996 | 1986 | 1989 | 1991 | 1994 | 1996 | 1986 | 1989 | 1991 | 1994 | 1996 |
| Electronic products | 6.4/1.3 | 5.6/2.5 | 5.2/3.2 | 4.6/2.8 | 5.8/1.8 | 2.3/4.7 | 3.2/5.4 | 5.2/4.5 | 5.4/3.5 | 7.1/4.2 | 81.9 | 79.4 | 80.8 | 81.8 | 79.5 |
| Data processing equipment |  |  | 4.3/1.6 | 4.2/1.7 | 5.6/0.7 |  |  | 6.7/7.1 | 7.3/5.9 | 10.6/7.5 |  |  | 79.4 | 79.9 | 75.2 |
| Computer products |  |  |  |  | 9.2/0.1 |  |  |  |  | 16.3/2.8 |  |  |  |  | 70.1 |
| Computer peripheral equipment |  |  |  |  | 4.0/1.2 |  |  |  |  | 10.3/14.7 |  |  |  |  | 69.4 |
| Data storage media units |  |  |  |  | 3.5/4.7 |  |  |  |  | 0.5/0.0 |  |  |  |  | 89.7 |
| Computer components |  |  |  |  | 2.3/0.1 |  |  |  |  | 0.4/0.0 |  |  |  |  | 98.8 |
| Video and radio electronic products |  |  | 11.3/12.8 | 14.4/14.6 | 30.8/15.7 |  |  | 4.7/2.1 | 6.2/2.7 | 7.7/2.0 |  |  | 68.2 | 59.3 | 43.4 |
| Communication apparatus |  |  | 14.4/0.6 | 10.7/1.1 | 8.4/2.2 |  |  | 20.1/15.3 | 23.7/11.8 | 25.7/10.4 |  |  | 49.0 | 53.8 | 51.3 |
| Electronic components and parts |  |  | 0.1/0.0 | 0.3/0.0 | 1.0/0.6 |  |  | 0.0/0.0 | 0.0/0.0 | 0.0/0.0 |  |  | 98.5 | 96.9 | 95.7 |
| Electronic tube |  |  |  |  | 1.7/1.3 |  |  |  |  | 0.0/0.0 |  |  |  |  | 97.6 |
| Semi-conductors |  |  |  |  | 0.0/0.0 |  |  |  |  | 0.0/0.0 |  |  |  |  | 92.7 |
| Opto-electronic components and materials |  |  |  |  | 0.0/0.0 |  |  |  |  | 0.0/0.0 |  |  |  |  | 94.1 |
| Electronic components and parts |  |  |  |  | 0.3/0.0 |  |  |  |  | 0.0/0.0 |  |  |  |  | 99.0 |

Source: Input–output Tables, Taiwan Area, Republic of China. DGBAS, Executive Yuan, Republic of China.

160   *The New Economy in East Asia and the Pacific*

*Table 6.4.8*  ICT expenditure as a proportion of GDP and investment, 1996 (per cent)

| Country | ICT investment as a percentage of GDP | ICT expenditure as a percentage of gross fixed capital formation |
|---|---|---|
| Taiwan | 3.3 | 6.5 |
| Korea | 6.1 | 7.7 |
| Singapore | 6.1 | 6.7 |
| Hong Kong | 6.1 | 7.6 |
| India | 1.9 | 2.9 |
| Indonesia | 1.9 | 11.3 |
| Thailand | 2.4 | 2.9 |
| Malaysia | 4.7 | 5.1 |
| Japan | 6.4 | 6.5 |
| Belgium | 5.6 | 12.1 |
| Denmark | 6.3 | 15.3[a] |
| France | 5.9 | 10.9 |
| Germany | 5.2 | 11.2 |
| Netherlands | 6.6 | 13.5 |
| Sweden | 7.6 | 17.4 |
| United Kingdom | 7.6 | 18.3 |
| United States | 7.7 | 20.6 |

*Source*: Ark and Timmer (2000), Table 13.

*Note*
a  ICT expenditure concerns IT hardware, IT services and software and telecommunication.

On average, consumption ratio to final demand was less than 10 per cent and quite stable. For video and radio electronic products, the proportion rose rapidly, from 24.1 per cent in 1991 to 46.5 per cent in 1996; most of the increase in demand came from domestic consumption demand. Investment as a proportion of total demand increased by 50 per cent over ten years, mainly as a result of increasing domestic demand. However, the ratios for consumption and investment were quite low. More than 70 per cent of ICT products (other than video and radio electronic products and communication apparatus) are exported. The main reason is that many Taiwanese manufacturers of ICT products serve as original equipment manufacturers and original design manufacturers to US and Japanese MNCs.

Table 6.4.8 is taken directly from Ark and Timmer (2000). It reports ICT expenditure as a percentage of GDP and ICT investment as a percentage of fixed capital formation in 1996 for seventeen countries. Ark and Timmer lacked ICT investment data and simply assumed that ICT investment was about 30 per cent of ICT expenditure. In Taiwan, the proportion of investment

Table 6.4.9 Investment in ICT as a proportion of investment by industry division (per cent)

| Division | 1986[a] | 1991[b] | 1996[b] |
|---|---|---|---|
| Agriculture and animal husbandry | 0.19 | 0.29 | 0.12 |
| Forestry | 0.05 | 0.54 | 0.00 |
| Fishing | 0.54 | 0.38 | 0.69 |
| Mining | 0.19 | 0.11 | 3.51 |
| Food products | 0.05 | 0.03 | 12.30 |
| Beverages and tobacco | 0.22 | 4.23 | 17.75 |
| Textile mill products | 0.90 | 1.56 | 5.70 |
| Wearing apparel and accessories | 1.57 | 2.40 | 1.08 |
| Wood and related products | 0.53 | 0.55 | 0.00 |
| Pulp, paper and printing | 1.95 | 0.87 | 9.98 |
| Chemical materials | 0.92 | 0.72 | 6.67 |
| Artificial fibre and plastic products | 0.77 | 0.80 | 20.78 |
| Other chemical products | 0.91 | 3.28 | 8.50 |
| Petroleum refinery | 0.13 | 0.44 | 1.09 |
| Non-metallic mineral products | 2.33 | 0.93 | 4.04 |
| Steel | 0.04 | 0.04 | 5.03 |
| Other metals and metallic products | 0.40 | 1.64 | 6.01 |
| Machinery | 1.73 | 3.05 | 0.87 |
| Household electrical appliances | 6.41 | 4.80 | 0.00 |
| Electronic products | 22.67 | 5.90 | 16.32 |
| Electrical machinery and other materials | 11.69 | 4.44 | 0.00 |
| Transport equipment | 3.17 | 4.48 | 4.65 |
| Other manufactures | 0.99 | 0.42 | 6.11 |
| Construction | 1.68 | 0.70 | 0.00 |
| Electricity | 2.16 | 1.83 | 18.27 |
| Gas and water | 0.19 | 1.55 | 16.20 |
| Transport, storage and communication | 3.04 | 3.35 | 4.11 |
| Commodity trade | 6.15 | 8.13 | 12.60 |
| Other services | 7.38 | 8.67 | 5.16 |
| *Average* | *4.56* | *5.30* | *7.48* |

Source: Investment matrix provided by DGBAS, Executive Yuan, Republic of China.

Notes
a  The electronic products industry includes video and radio products.
b  The electronic products industry does not include video and radio products.

as a percentage of fixed capital formation was lower than that in India, Thailand and Malaysia, but was roughly the same as for other NIEs. It was far behind that of advanced economies except Japan. Emerging Asian countries except Indonesia generally have low ICT expenditures and low ICT investment ratios. Taiwan has room to increase its ICT expenditure and investment.

Table 6.4.9 reports Taiwan's investment in ICT as a percentage of investment by industry division from 1986 to 1996.[19] The average ICT investment share

*Table 6.4.10* Indicators of ICT infrastructure and access to ICT in Taiwan

| Type | Year | Statistics | Source |
|---|---|---|---|
| Computer | June 1998 | 75 per cent enterprises | Report of the Use of Computer in ROC, 1998[a] |
| Computer | 1999 | 38.92 per cent households | The Survey of Family Income and Expenditure[a] |
| Cellular mobile phone | February 2001 | 1.8 million and a penetration rate of 88.4 per cent (1st in Asia) | Monthly statistics of ROC[a] |
| Phone lines | February 2001 | 566 per 1000 | Monthly statistics of ROC[a] |
|  | February 1999 | 98 per cent households | The Survey of Family Income and Expenditure[a] |
| Cable TV | 1999 | 67.87 per cent households | The Survey of Family Income and Expenditure[a] |
| Internet users | End of June 2001 | 7.21 million and a penetration rate of 41.7.0 per cent (15+) | Institute of Information Industry[c] |
| Home internet penetration | May 2001 | 40.8 per cent, 3rd highest in Asia region after Singapore and Korea | Net value[c] |
| Internet hosts in Taiwan | January 2001 | 15th in the survey. 1.10 million in the Asia-Pacific region after Japan and Australia or 6.0 per cent penetration rate (15+) | 'Internet domain survey' by Network Wizard[b] |
| Local broadband users | 2000 | 20,000 | The institute for Information Industry[c] |
|  | 2001f/cast | 0.21 million |  |
|  | 2002 f/cast | 0.83 million |  |
|  | 2003 f/cast | 2.87 million |  |
| E-commerce potential | 2000 | 10th among 47 countries. 3rd in the Asia Pacific region after Japan and Australia | META Group[d] |

*Sources*: a DGBAS, Executive Yuan, Republic of China. b http://www.isc.org/ds/WWW-200101/. c Business Weekly. d *Industry of Free China*, (Council for Economic Planning and Development, Republic of China).

*Note*: 15+ means age 15 or older.

rose by three percentage points (or 50 per cent) in the ten years from 1986 to 1996. In 1996, the share was 7.48 per cent, one percentage point higher than that reported in Ark and Timmer (2000). The 'Electronic products' division had a high ICT investment share, as expected. Also, not surprisingly, the ICT investment share in commodity trade doubled in ten years, and was above average in all three sample years (1986, 1991 and 1996). The ICT investment share in the 'Other services' division was above average only in 1986 and 1991, perhaps because the finance, insurance and real estate sectors were the earliest investors in ICT.

Table 6.4.10 reports selected indicators of ICT infrastructure and access to ICT. Enterprises' computer penetration rates rose from 67.5 per cent in 1997 to 75 per cent in June 1998. The insurance, finance and real estate sector had the highest computer penetration rate, as expected (95.4 per cent); the mining sector had the lowest (8.2 per cent). Moreover, the size of companies matters. In June 1998, the computer server penetration rate was 90.8 per cent for big enterprises (over 500 employees), 83.2 per cent for medium enterprises (100–499 employees), 60 per cent for smaller enterprises (50–99 employees) and 25 per cent for small enterprises (3–9 employees).

The household computer penetration rate doubled in five years, from 18.45 per cent in 1995 to 38.92 per cent in 1999. The ratio is 10–25 percentage points behind that for Australia (45.9 per cent in August 1998), the United States (58.5 per cent in 1999) and many advanced countries, but is higher than that for Japan (25.9 per cent in 1999), France (35 per cent in 1999) and Italy (17.5 per cent in 1997) (OECD 2000a).

Since Taiwan liberated its wireless telecommunications policy in 1996, new technology (such as the global system for mobile communications technology) and competition among mobile phone companies and operations have forced the prices and bills of mobile phones to fall dramatically. In 2000, Taiwan's mobile phone penetration rate was the highest in Asia. Household telephone and cable television penetration rates were also quite high. Taiwan ranked in the top three in the Asia Pacific region for home internet penetration, the number of internet users and the number of internet hosts. Rates were roughly equal to those of the Netherlands, Germany, the United Kingdom and France, and higher than those of Italy and Japan (OECD 2000a). But the gap between Taiwan and some industrialised countries is still wide. For example, in 1999, the Taiwan household internet facility penetration rate was 19.58 per cent,[20] a gap of about 10–20 percentage points.[21]

The 1999 Survey of Income and Expenditure also showed that there is digital divide within Taiwan. Income and degree of urbanisation are strong determining factors for computer and internet facility penetration (OECD 2001).[22] In the top 20 per cent of Taiwanese households as measured by disposable income, computer and internet facility penetration rates were

Table 6.4.11 Growth rate of total factor productivity by industry, 1979–98 (per cent)

| Industry | 1979–98 | 1979–86 | 1987–92 | 1993–98 |
|---|---|---|---|---|
| Total (industry and services) | 1.11 | 0.95 | 1.26 | 1.19 |
| Industry | 0.65 | 0.86 | 0.45 | 0.57 |
| Mining and quarrying | 2.34 | 2.22 | 3.76 | 1.08 |
| Manufacturing | 0.70 | 1.13 | 0.27 | 0.56 |
| Food products | 0.11 | 0.43 | –0.15 | –0.04 |
| Tobacco | 1.65 | –0.62 | 3.19 | 3.13 |
| Textile mill products | 0.55 | 1.47 | 0.09 | –0.22 |
| Wearing apparel and accessories | –0.55 | 0.39 | –0.80 | –1.56 |
| Leather, fur and products | –0.12 | 1.42 | –1.59 | –0.71 |
| Wood and bamboo products and furniture | 0.43 | 1.57 | 0.68 | –1.33 |
| Paper | –0.51 | 1.04 | –2.34 | –0.74 |
| Chemical products | 1.04 | 2.06 | 0.17 | 0.54 |
| Petroleum and coal products | –0.21 | 0.94 | –4.55 | 2.58 |
| Rubber products | 0.38 | 1.55 | 0.40 | –1.18 |
| Non-metallic mineral products | 0.54 | 0.10 | 1.60 | 0.07 |
| Basic metal industries | 0.61 | 0.74 | 0.42 | 0.64 |
| Fabricated metal industries | 0.23 | 0.32 | 1.11 | –0.76 |
| Machinery equipment | 1.01 | 1.68 | 1.25 | –0.12 |
| *Electronic and electrical equipment* | 0.89 | 0.56 | 1.12 | 1.11 |
| Transport equipment | 0.12 | –0.07 | 1.29 | –0.81 |
| Precision instruments and other industrial products | –0.24 | 0.57 | –0.41 | –1.16 |
| Miscellaneous manufacturing industries | –0.83 | –0.78 | –0.72 | –1.00 |
| Electricity, gas and water | 0.96 | –0.87 | 2.33 | 2.04 |
| Construction | 0.01 | –0.57 | 0.87 | –0.10 |
| Services | 2.05 | 1.09 | 3.10 | 2.27 |
| Commerce | 2.27 | 1.35 | 3.87 | 1.89 |
| Transport, storage and communication | 1.74 | 1.07 | 1.73 | 2.66 |
| Finance, insurance, real estate and business services | 0.61 | –0.22 | 1.76 | 0.56 |
| Social and personal services | 2.39 | 1.49 | 3.38 | 2.59 |

Source: *The Trends in Multifactor Productivity, Taiwan Area, Republic of China*, DGBAS, Executive Yuan, Republic of China.

69.62 per cent and 42.61 per cent, respectively. The comparable figures for the bottom 20 per cent of households were only 8.34 per cent and 2.66 per cent, respectively. There was also a divide between cities and villages: computer and internet facility penetration rates were, respectively, 46.35 per cent and 24.46 per cent for households in cities, 32.8 per cent and 14.55 per cent for households in towns, and 15.57 per cent and 6.35 per cent for households in villages. The digital divide is quite serious in Taiwan. It will affect the diffusion of new technology and thus increase income distribution inequality between cities and villages. There is also a divide between north and south. Let us use two of the largest cities – Taipei (north) and Kaohsiung (south) – as an example. In 1999, Taipei households had the highest computer and internet facility penetration in Taiwan (60.17 per cent and 39.24 per cent, respectively). In Kaohsiung, the comparable figures were only 43.34 per cent and 28.51 per cent. These figures support the idea that we must deal with the digital divide in Taiwan.

## Spillover effects

In order to cope with the fast drop of ICT product prices and maintain its competitiveness in the world market, the ICT sector must become more productive and more efficient. This will lead to faster TFP growth in the ICT sector. The 'new economy' may also lead to the diffusion of ICT, resulting in technological change in other sectors through spillover effects. This is also expected to lead to faster TFP growth, especially in the finance, insurance, real estate and business service sector, because they use more ICT than most sectors. To check whether there is a link between ICT and productivity growth, we examined industry TFP growth over time.

Table 6.4.11 reports TFP growth across industries from 1979 to 1998. The table is based on the official statistics for the sectoral breakdown of the economy. The sample period was divided into three subsample periods. In the industry and services sectors, TFP growth rose from 0.85 per cent over 1979–86 to 1.26 per cent over 1987–92 and fell to 1.19 per cent over 1993–98. The 'Services' sector had above-average TFP growth rates over all subsample periods, but its productivity growth was lower in 1993–98 than that in 1987–92. There is no evidence of acceleration of productivity growth in Taiwan's economy since the mid-1990s. Moreover, productivity growth has varied considerably across sectors and industries. Only two out of eight sectors – manufacturing and transport, storage and communication – displayed higher productivity growth in 1993–98 than in 1987–92. And within the manufacturing sector, many traditional and basic manufacturing industries had negative TFP growth rates in 1993–98. Average TFP growth rates in the electronic and electrical equipment industry remained roughly the same over two subsample periods (1987–92 and 1993–98). As shown in Table 6.4.3, the contribution of TFP growth in the electronic and electrical equipment industry to TFP growth in the non-agricultural economy was negative in 1997 and 1998. The production and use of ICT did not raise productivity growth in the ICT sector itself.

To examine the spillover effects of ICT, we focus our discussion on the finance, insurance, real estate and business services sector, which is the major user of ICT. This sector has the highest PC penetration rate, but its TFP

*Table 6.4.12* Input coefficient A of the Taiwanese electronic products sector, 1986–96

| Sector | Imported goods | | | Domestic goods and services | | |
|---|---|---|---|---|---|---|
| | 1986 | 1991 | 1996 | 1986 | 1991 | 1996 |
| Agriculture, animal husbandry | 0.00 | 0.00 | 0.00 | 0.00 | 0.00 | 0.00 |
| Forestry | 0.00 | 0.00 | 0.00 | 0.00 | 0.00 | 0.00 |
| Fishing | 0.00 | 0.27 | 0.14 | 0.00 | 0.00 | 0.00 |
| Mining | 0.00 | 0.02 | 0.01 | 0.00 | 0.01 | 0.00 |
| Manufacturing | | | | | | |
|   Food products | 0.00 | 0.00 | 0.00 | 0.00 | 0.00 | 0.00 |
|   Tobacco | 0.00 | 0.00 | 0.00 | 0.00 | 0.00 | 0.00 |
|   Textile mill products | 0.00 | 0.00 | 0.00 | 0.00 | 0.00 | 0.00 |
|   Wearing apparel and accessories | 0.00 | 0.00 | 0.00 | 0.00 | 0.00 | 0.00 |
|   Leather, fur and related products | 0.00 | 0.00 | 0.01 | 0.00 | 0.00 | 0.00 |
|   Wood and bamboo products and furniture | 0.05 | 0.09 | 0.10 | 0.00 | 0.00 | 0.00 |
|   Pulp, paper and paper products | 0.00 | 0.03 | 0.05 | 0.00 | 0.00 | 0.00 |
|   Printing | 0.01 | 0.12 | 0.02 | 0.00 | 0.02 | 0.00 |
|   Chemical materials | 0.00 | 0.01 | 0.02 | 0.00 | 0.00 | 0.00 |
|   Chemical products | 0.00 | 0.00 | 0.00 | 0.00 | 0.00 | 0.00 |
|   Coal and petroleum refinery | 0.01 | 0.02 | 0.01 | 0.00 | 0.01 | 0.01 |
|   Rubber products | 0.00 | 0.01 | 0.01 | 0.00 | 0.00 | 0.00 |
|   Plastic products | 0.01 | 0.02 | 0.15 | 0.00 | 0.01 | 0.02 |
|   Non-metallic mineral products | 0.00 | 0.00 | 0.00 | 0.00 | 0.00 | 0.00 |
|   Basic metal products | 0.00 | 0.01 | 0.00 | 0.00 | 0.00 | 0.00 |
|   Metal products | 0.15 | 0.10 | 0.09 | 0.00 | 0.02 | 0.02 |
|   Machinery equipment | 0.66 | 1.46 | 2.70 | 0.48 | 1.32 | 1.30 |
|   Electrical appliances and houseware | 2.22 | 1.26 | 1.99 | 0.58 | 0.82 | 3.16 |
|   Electrical machinery and other materials | 1.98 | 0.96 | 1.12 | 0.76 | 1.03 | 1.60 |
|   *Electronic equipment* | 18.60 | 13.47 | 12.17 | 26.08 | 29.52 | 31.09 |
|   Transport equipment | 0.49 | 0.40 | 0.46 | 0.07 | 0.04 | 0.14 |
|   Precision instruments and apparatus | 4.85 | 5.74 | 10.07 | 9.77 | 7.96 | 12.49 |
|   Other manufactures | 0.89 | 1.70 | 2.79 | 0.64 | 1.10 | 3.66 |
| Water, electricity and gas | 0.17 | 0.17 | 0.02 | 0.00 | 0.00 | 0.00 |
| Construction | 0.16 | 0.25 | 0.24 | 0.00 | 0.01 | 0.01 |
| Commerce | 0.01 | 0.01 | 0.01 | 0.00 | 0.00 | 0.00 |
| Transport, storage and communication | 0.17 | 0.30 | 0.30 | 0.00 | 0.07 | 0.04 |
| Finance and insurance | 0.00 | 0.01 | 0.17 | 0.00 | 0.00 | 0.00 |
| Real estate services | 0.00 | 0.00 | 0.00 | 0.00 | 0.00 | 0.00 |
| Business services | 0.51 | 0.44 | 0.66 | 0.15 | 0.45 | 0.87 |
| Social and personal services | 0.85 | 0.84 | 0.94 | 0.25 | 0.11 | 0.24 |
| *Information services* | 2.20 | 5.22 | 2.20 | 1.84 | 3.15 | 3.22 |

*Source*: Input–output tables, Taiwan Area, Republic of China, DGBAS, Executive Yuan, Republic of China.

*Table 6.4.13* The inversion of the coefficient matrix $(I-A)^{-1}$ or $[I-(I-M)A]^{-1}$ of the Taiwanese electronic products sector

| Sector | Imported goods and services | | | Domestic goods and services | | |
|---|---|---|---|---|---|---|
| | 1986 | 1991 | 1996 | 1986 | 1991 | 1996 |
| Agriculture, animal husbandry | 0.00 | 0.00 | 0.01 | 0.05 | 0.06 | 0.09 |
| Forestry | 0.01 | 0.00 | 0.00 | 0.14 | 0.02 | 0.03 |
| Fishing | 0.02 | 0.01 | 0.08 | 0.05 | 0.38 | 0.23 |
| Mining | 0.01 | 0.06 | 0.07 | 0.11 | 0.17 | 0.16 |
| Manufacturing | | | | | | |
| Food products | 0.00 | 0.01 | 0.01 | 0.07 | 0.10 | 0.12 |
| Tobacco | 0.00 | 0.01 | 0.01 | 0.05 | 0.07 | 0.10 |
| Textile mill products | 0.01 | 0.02 | 0.03 | 0.10 | 0.09 | 0.12 |
| Wearing apparel and accessories | 0.00 | 0.02 | 0.05 | 0.11 | 0.11 | 0.16 |
| Leather, fur and related products | 0.00 | 0.01 | 0.03 | 0.07 | 0.07 | 0.10 |
| Wood and bamboo products and furniture | 0.00 | 0.01 | 0.05 | 0.15 | 0.20 | 0.23 |
| Pulp, paper and paper products | 0.00 | 0.01 | 0.03 | 0.16 | 0.20 | 0.23 |
| Printing | 0.12 | 0.38 | 0.52 | 0.15 | 0.28 | 0.17 |
| Chemical materials | 0.01 | 0.02 | 0.03 | 0.08 | 0.10 | 0.12 |
| Chemical products | 0.01 | 0.02 | 0.02 | 0.09 | 0.10 | 0.12 |
| Coal and petroleum refinery | 0.02 | 0.06 | 0.06 | 0.04 | 0.06 | 0.05 |
| Rubber products | 0.00 | 0.01 | 0.02 | 0.10 | 0.14 | 0.17 |
| Plastic products | 0.01 | 0.02 | 0.04 | 0.13 | 0.14 | 0.33 |
| Non-metallic mineral products | 0.01 | 0.03 | 0.04 | 0.13 | 0.14 | 0.17 |
| Basic metal industry | 0.01 | 0.02 | 0.02 | 0.18 | 0.21 | 0.20 |
| Metallic products | 0.00 | 0.04 | 0.06 | 0.33 | 0.27 | 0.27 |
| Machinery and equipment | 0.71 | 2.02 | 2.14 | 1.08 | 1.98 | 3.39 |
| Household electrical appliances | 0.85 | 1.36 | 4.91 | 3.48 | 1.78 | 2.67 |
| Electrical machinery and other materials | 1.08 | 1.62 | 2.58 | 3.14 | 1.45 | 1.59 |
| *Electronic products* | 135.33 | 141.94 | 145.20 | 123.00 | 115.69 | 113.98 |
| Transport equipment | 0.15 | 0.10 | 0.34 | 0.89 | 0.71 | 0.83 |
| Precision instruments and apparatus | 15.32 | 13.45 | 19.53 | 6.31 | 6.96 | 11.89 |
| Other manufactures | 0.89 | 1.65 | 5.52 | 1.34 | 2.19 | 3.48 |
| Water, electricity and gas | 0.00 | 0.02 | 0.07 | 0.34 | 0.30 | 0.13 |
| Construction | 0.01 | 0.06 | 0.09 | 0.48 | 0.50 | 0.55 |
| Commerce | 0.01 | 0.02 | 0.04 | 0.12 | 0.14 | 0.15 |
| Transport, storage and communication | 0.00 | 0.14 | 0.13 | 0.33 | 0.46 | 0.48 |
| Finance and insurance | 0.00 | 0.04 | 0.05 | 0.11 | 0.13 | 0.32 |
| Real estate services | 0.00 | 0.01 | 0.01 | 0.11 | 0.09 | 0.10 |
| Business services | 0.22 | 0.89 | 1.58 | 0.81 | 0.74 | 1.01 |
| Social and personal services | 0.45 | 0.23 | 0.50 | 1.20 | 1.10 | 1.21 |
| *Information services* | 2.49 | 4.50 | 4.71 | 2.77 | 6.12 | 2.61 |

Source: Same as Table 6.4.12.

growth rate over 1993–98 was only one-third of that over 1987–92, suggesting that the wide use of ICT did not help to promote productivity growth in the sector. For a small, open economy with a less developed financial sector, like Taiwan, productivity growth in the financial sector is not expected to come from trade and innovations. Human capital and FDI may play some role, and TFP growth in the financial sector is expected to accelerate due to financial liberalisation, telecommunication liberalisation and uses of ICT. Instead, we observe a rising and then falling trend in productivity growth in this sector. Empirical evidence has shown that the use of ICT raised US TFP growth in the long run (finally solving Solow's productivity puzzle). We thus would not explain the rising TFP growth over 1987–92 as a spillover effect of ICT, because Taiwan's ICT use is still less developed than that of the United States. The rapidly falling TFP growth over 1993–98, with high computer and internet penetration rates, leads us to doubt that the use of ICT has stimulated TFP growth in Taiwan's finance, insurance, real estate and business services sector.

We used input–output tables to examine the possible ICT sector spillover effects through using ICT products as intermediate goods. Tables 6.4.12 and 6.4.13 report input coefficient A and the inversion of the coefficient matrix $(I-A)^{-1}$ of product of the electronic products sector into thirty-six sectors. We separated the imported electronic products sector and electronic products produced in domestic electronic products sectors while we calculated A and $(I-A)^{-1}$. To read the results easily, we report rotated tables that are part of the original input–output tables. The electronic product sector had the highest share of ICT goods as intermediate goods. This reflects the fact that the electronic product sector has to import a lot of electronic components and parts to produce final goods. The input coefficient and the inversion of the coefficient of domestic goods produced in the electronic sector into the electronic sector rose over time; for imported goods, the coefficients fell. All other sectors use a rather low percentage of electronic goods as intermediate goods. Tables 6.4.12 and 6.4.13 show that there are very limited spillover effects arising from the use of ICT goods and services as intermediate goods.

## CONCLUSION

This chapter examines the ICT sector in Taiwan. We focus our discussion on the impact of the ICT sector on production and trade. The ICT-producing sector has accounted for a significant proportion of economic growth, especially in 1999 (45 per cent, or 2.57 percentage points of the economic growth rate). The ICT sector is the most important sector in Taiwan as measured by investment, exports, trade surplus, R&D, FDI and capital market share.

In Taiwan, the new economy is manifested through the effects of ICT production but not through ICT use and its spillover effects in other sectors. As ICT manufacturers shift production to low-cost countries, they must upgrade

*The Taiwanese experience* 169

to more advanced products or process technologies if Taiwan's ICT-producing sector is to develop in the twenty-first century. On the other hand, if ICT infrastructure is built at an increasing rate and there is greater access to ICT technology, Taiwan will move quickly to levels demonstrated in advanced economies. The government has decided to abandon the 'go slow' policy and open the 'three direct links' policy that has been in place for many years. We look forward to seeing how the new policies will affect development of Taiwan's ICT sector in the near future.

## NOTES

1. See survey by Triplett (1999).
2. For example, Whelan (1999), Jorgenson and Stiroh (2000) and Oliner and Sichel (2000). Empirical results of Jorgenson and Stiroh (1999), Kiley (1999) and Gordon (1999) do not support the presence of a new economy in the United States.
3. 'APROC newsletter', in *Industry of Free China* (August 2000, 90: 8). The rankings are based on five major assessment categories: knowledge-based jobs, technological innovation, degree of transformation to a digital economy, economic dynamism and globalisation.
4. *Quarterly Forecast of International Trade*, Board of Foreign Trade, Ministry of Economic Affairs, Republic of China.
5. The formal definition of the ICT sector was based on the April 1998 meeting of the Working Party on Indicators for the Information Society (WPIIS) and subsequently endorsed and declassified at the September 1998 meeting of the Committee for Information, Computer and Communications Policy (OECD 2000b).
6. See Dedrick and Kraemer (1998). The rank is based on the IT IS report. Other reports rank Taiwan differently. For example, Reed Electronics Research ranked Taiwan as the seventh largest supplier of ICT goods in 1997. The difference may result from the definition of ICT goods or data sources, but there is no doubt that Taiwan is an important ICT production country in the world (OECD 2000a).
7. The term 'thin client' is used for a system where the software, data and processing power reside on the network rather than on individual personal computers.
8. The Industrial Technology Research Institute (ITRI) helped the semiconductor industry, flat-panel display industry, information industry, communication industry, optoelectronics industry, and electronic materials industry to develop new industrial technologies. See http://www.irti.org.tw/.
9. *Indicators of Science of Technology*, National Science Council, Republic of China.
10. According to the OECD definition, a high-tech industry is one that devotes more than 10 per cent of its sales to R&D expenditures.
11. *Indicators of Science of Technology*, National Science Council, Republic of China.
12. The Global Logistics Development Plan, which is the third stage of the Asia Pacific Regional Operations Centre (APROC) Plan (started in 1995) with the National Information Infrastructure (NII) Plan (started at the end of 1997) and the industrial automation and digitisation program as a sub-plan of the Plan to Develop Knowledge-based Economy in Taiwan (started in 2000).
13. See Barro and Sala-i-Martin (1995, Chapter 8) for a survey of technology diffusion.
14. Korea has the highest proportion of value added in the ICT sector to value added in the total business sector (10.7 per cent) (OECD 2000c).
15. Sweden has the highest share (6.3%) of ICT sector employment in total business sector employment in the OECD member countries. For the United States, the figure is 3.9%, for Japan 3.0% and for Korea 2.5% (OECD 2000c).

16 The OECD (2000c) definition of the ICT sector includes 3220 and 3230 industrial classes. They are close to sound and video appliances. We do not report relevant statistics for the sound and video and radio products sector because its contribution to the industry industrial production is rather small: sound and video appliances are not major products for Taiwan's ICT-producing sector.
17 See survey by Johnson (1997).
18 The 1991 and 1994 input–output tables have data also for four sub-industries: data processing equipment, video and radio electronic products, communication apparatus and electronic components and parts. The 1996 tables have sub-industry data only for processing equipment and electronic components and parts. With the exception of electronic tubes, electronic components and parts are intermediate goods: their consumption demand is zero. The investment demand for all electronic components and parts is zero.
19 There are twenty-nine divisions in total. The coverage of ICT investment is different for different years due to data constraints of the input–output tables. Investment in video and radio products is included in the ICT investment data only for 1986.
20 *Survey of Family Income and Expenditure*, DGBAS, Executive Yuan, Republic of China.
21 Data are from OECD (2000a). Gaps are 20 percentage points for Australia, Iceland, the United States and Sweden, and 10 percentage points for Norway, Denmark and Finland.
22 The education of household head, family structure and age structure also matter (OECD 2001).

# REFERENCES

Ark, B. van and Timmer, Marcel (2000) 'Asia productivity performance and potential at the turn of the century; an international perspective' (mimeo).
Barro, Robert and Sala-i-Martin, Xavier (1995) *Economic Growth*, McGraw-Hill.
Dedrick, Jason and Kraemer, Kenneth L. (1998) *Asia's Computer Challenge*, Oxford: Oxford University Press.
Gordon, Robert J. (1999) 'Has the "new economy" rendered the productivity slowdown obsolete?' (mimeo).
Johnson, George E. (1997) 'Changes in earnings inequality: the role of demand shifts, *Journal of Economic Perspectives* 11, 2: 41–54.
Jones, Charles I. (1998) *Introduction to Economic Growth*, W. W. Norton.
Jorgenson, Dale W. and. Stiroh, Kevin J. (1995) 'Computers and growth', *Economics of Innovation and New Technology* 3: 295–316.
—— (1999) 'Information technology and growth', *AEA Papers and Proceedings* 89: 109–15.
—— (2000) 'Raising the speed limit: U.S. economic growth in the information age', *Brookings Papers on Economic Activity* 1: 125–235.
Kiley, Michael T. (1999) 'Computers and growth with costs of adjustment: will the future look like the past?', Federal Reserve Board, Finance and Economics Discussion Series Paper 1999–36.
OECD (Organisation for Economic Co-operation and Development) (2000a) *OECD Information Technology Outlook*, Paris: OECD.
OECD (2000b) *A New Economy? The Changing Role of Innovation and Information Technology in Growth*, Paris: OECD
OECD (2000c) *Measuring the ICT Sector,* Paris: OECD.
OECD (2001) *Understanding the Digital Divide,* Paris: OECD.
Oliner, Stephen D. and Sichel, Daniel E. (1994) 'Computers and output growth revisited: how big is the puzzle?' *Brookings Papers on Economic Activity* 2: 273–317.

—— (2000) 'The resurgence of growth in the late 1990s: is information technology the story?', in proceedings of 'Structural Change and Monetary Policy', Federal Reserve Bank of San Francisco.

Triplett, Jack E. (1999) 'The Solow productivity paradox: what do computers do to productivity?', *Canadian Journal of Economics* 32: 309–34.

US Department of Commerce (2000) *Digital Economy 2000*. Washington DC: US Department of Commerce.

Whelan, Karl (1999) 'Computers, obsolescence, and productivity', Federal Reserve Board (mimeo).

# 7 Corporate strategies in information technology firms

*Yasunori Baba and F. Ted Tschang*

## INTRODUCTION

In information technology (IT) industries, the managerial strategies and behaviour of firms have three main characteristics. The first is the substantial degree of uncertainty and risk stemming from the incredible change in the business model used in these industries. The corporate risks include those of acting as well as of not acting; this leads to tremendous shake-ups in the industrial structure over time, with winners in one year becoming losers one or two years afterwards. Firms that hope to survive the transitions have had to reinvent themselves and create their core competencies anew.

The second characteristic is the rapidity of technological change and its critical impact on the firm's business revenue. Change occurs with the advent of a new technology (such as the internet) or a new mode of using technology (such as modular production), either of which can create a new market for a new entrant. Even if such change is expected, it may be disruptive.

The third characteristic follows from the changing nature of product life-cycles. In developed countries, there is a trend toward market saturation and post mass-production. We are witnessing a new product cycle in IT industries: standardised consumer electronics, audiovisual and even advanced IT products that were produced in Japan are now being produced in other Asian countries, particularly China. China is now the world's largest producer of digital video disk (DVD) players (38.3 per cent of the world market), colour televisions (CTVs) (24.6 per cent), video tape recorders (VTRs) (23.2 per cent), mobile telephones (12.9 per cent) and desktop personal computers (11.9 per cent) (Nikkei 27 July 2001).

In this paper, we will discuss how Asian IT firms cope with the emerging changes. On one hand, they must face new market opportunities developed by ITs; on the other, they must face imminent industrial uncertainty and possible decreasing returns. We seek to identify the experience of corporate transformation in markets characterised by the introduction of new ITs, illustrating the impact on business organisation and strategies in particular firms and industries. We examine this experience within borders and also across borders, attempting to show how these technologies have affected the boundaries of the firm and corporate behaviour. In particular, we ask how, in the global setting, firms can combine their technological assets with new markets to succeed in new business environments.

To put it another way, we highlight the diverse strategies firms use to create responses to uncertainty. Most case studies of IT firms involve US companies; we present two case studies of non-US firms with different national settings and corporate status: Sony and i-flex. These two examples illustrate two strategic corporate behaviours seen in Japan (Sony) and India (i-flex). Both companies faced different external challenges; both met those challenges with different strategies at the corporate and product level. Both companies changed the nature of the products they provide and their managerial and organisational strategies. In the following sections, we examine the two companies using the lens provided by our theoretical discussion. Essentially, our theoretical review suggests that ITs and the ways companies see their potential for new markets can be key aspects of the new types of strategic behaviour employed by Sony and i-flex.

Sony, like other Japanese IT firms, faces declining margins in hardware manufacturing and the loss of its core competencies, as it shifts its manufacturing sites to lower wage countries. It is using synergies among its many diverse business fields to develop a new strategy that both embraces the broadband internet and develops new market segments. Survival is sought by combining its hardware strengths (design and product development) and its strength in entertainment content, and by managing appropriate production and development networks. In this paper, we focus on the firm's strategy in the business of television games, one of its most strategic products.

Our second case is the Indian software firm i-flex. In an Indian software industry that is primarily based on contracting, i-flex is unusual in that it has taken a risk and developed a strategy to move into the higher value-added product market. This is one way to circumvent the gradual reduction in the labour cost advantages currently enjoyed by Indian software firms and the pressures on their profit margins. I-flex has had to work with more limited resources than Sony and has had to develop strategies involving an initially less innovative but practical growth trajectory, followed by taking a major risk. This involved leveraging off its original human resource capabilities, domain knowledge in banking and existing operating cash flows to fund the development of an innovative product and using the name recognition of its affiliate, Citicorp, to enter emerging markets.

In the next section, we discuss some of the theoretical background underlying the main concepts we rely on. We then discuss Sony and i-flex, respectively, and provide a comparative analysis of the companies. By doing so, we hope to shed light on how much we can generalise our findings to the Asia Pacific region.

## THEORETICAL BACKGROUND

### Generic technology and systemic sequence

In recent times, generic technology, particularly in IT, has disrupted markets for existing products and services, and triggered clusters of systemic innovations. An original radical innovation often prompts innovations in complementary activities, generating a new combination of physical products, software, and human services (David 1987; Langlois 1988; Teece 1988; Imai and Baba 1991; Baba and Imai 1992, 1993). This systemic sequence can also be seen in traditional industries, but it is particularly far-reaching in IT industries because it contributes to a proliferation of innovative linkages between the manufacturing and service sectors, between households and new services, and between new services and software industries, and because of the proliferation of applications from the dawning of the internet.

The systemic sequence currently works in a reciprocal or self-reinforcing fashion. This is not surprising. First, software, which was initially viewed as complementary to hardware, now tends to create a new demand for advanced hardware. Second, although service sectors still tend to set the agenda for the application of software, software providers are now acting as boosters for new fields of service. Third, the combination of hardware and software creates ample opportunities for the development of completely new kinds of service. Thus, competition frontiers are moving away from the traditional, stand-alone innovation (providing novelty in a single product or service) toward the systemic supply of an optimal combination of hardware, software and services. With the increasing use of the internet, interactions among hardware, software and services have been steadily progressing.

The traditional product cycle approach does not apply to this systemic sequence (Vernon 1966). The traditional, one-way, mechanical view assumes product standardisation and hence the decline of market size. In contrast, our systemic view suggests that in places such as Japan a cluster of innovations will emerge, with increasing returns. An original product such as Sony's PlayStation game machine may follow the traditional course of a product cycle, but when the product is combined with complementary assets or technologies like newly developed game software or broadband internet communications, additional service or market opportunities are generated inside Japan.

Thus, a suite of products and their associated life-cycles could emerge, with possibly mutually reinforcing outcomes, such as effects that lock in advantages over time. In 2000, exports in the Japanese TV game industry amounted to ¥1,117.8 billion, of which ¥539.9 billion (48.3 per cent) were hardware and ¥587.9 billion (51.7 per cent) software (CESA 2001). When a product is coupled with a series of new, neighbouring products or technologies, a cluster is likely to evolve and grow in an open-ended fashion. Under these conditions, several technological trajectories and market evolution patterns are theoretically possible, so the emergent industrial milieu has a heightened degree of uncertainty in the form of unpredictable events and unforeseen timing (David 1987; Arthur 1985, 1996).

## Modular technology and open networks

Recently, IT firms have been forced to focus on emerging industrial practices such as modularity (building a complex product or process from smaller subsystems that can be designed independently but that function together as a whole). By breaking up a product into subsystems or modules, designers, producers and users have gained enormous flexibility in both their design philosophy and their market strategy. Different companies can take responsibility for separate modules and be confident that a reliable product will arise from their collective efforts (Baldwin and Clark 1997; Oniki 1999). Module designers rapidly move in and out of joint ventures, technology alliances and subcontracts, and established industrial networks disperse the ownership and control over the sources of production.

The traditional type of network is the production network. Multinational corporations (MNCs) outsource many production activities to specialised firms in the Asia Pacific region. For instance, manufacturing and design are now outsourced to specialised contract manufacturers in Taiwan and Singapore.

Another type of network, created by advanced IT manufacturers like Sony, seeks to provide its electronic and software products as modular, user-configurable systems; in other words, users can select from different devices and be confident that they will be compatible for data exchange and other activities. Such networks include multiple developers or manufacturers who are allied through complementary product offerings. From economies of scope, these product-based networks can create increasing returns to developers, manufacturers and consumers, by way of larger product offerings and mass customisation opportunities.

## Prototyping culture

In this section, we discuss the theoretical framework of 'prototyping'. This will help us understand how prototyping culture is reflected in the development of hardware, software and new services, and how prototyping is used in the corporate strategy of modern IT firms (Ulrich and Eppinger 1995).

In conventional product development, prototypes were constructed at the final stage, when the design was already determined and fulfilled the stated requirements; prototyping was part of a trial-and-error process of shaping parameters and confirming the workability of the design in the real world. In modern IT firms, there is an increasing emphasis on building prototypes at the design stage, to assist in conceptualisation (Winograd 1996).

We argue that conventional prototyping is more beneficial for firms facing stable market environments and mildly complicated tasks which lend themselves to centralised, but team-based, organisation, where objectives are clearly defined and decision-making is a rational process (Baba and Nobeoka 1998; Baba 1998). In contrast, the latest IT products and related services like e-commerce face unstable markets. These new forms of prototyping practices are increasingly seen as marketplace success depends more and more on unexpected value derived from new experiences. The concept of a 'prototyping culture' has emerged in the responses to this new trend. To quote Michael Schrage (1996, p.193):

> ... a prototyping culture, like all cultures, is a mixture of the explicit organisational structures and the tacit understanding and practices of the participants. Within some innovative environments, prototypes effectively become the media franca of the organisation: the essential medium for information, interaction, integration, and collaboration.

Thus, prototyping acts as a medium of interaction for shaping and determining the product concept itself.

## SONY: EVOLVING BUSINESS IN THE TELEVISION GAME INDUSTRY

### Corporate behaviour

Sony's competitive advantage has traditionally been in electronics manufacturing; it has distinguished itself by its product quality and, especially recently, its design and product development. In the last decade, it has branched into activities such as entertainment content and financial services. This was partly due to the 'commodification' of its core manufacturing business, in which parts and components could be easily made by multiple manufacturers, and so pricing became highly competitive. We will show how, in order to transcend competition, Sony has taken major risks based on its 'hypothesis' of what a new market and its customer base could look like, and also engaged in overall organisational and managerial changes according to corporate strategies driven by this vision of combining technology with new markets.

### *PlayStation 1994–99: combining hardware with software*

The Japanese firm Nintendo released the first successful TV game machine only eighteen years ago. Since then, hardware platforms have improved rapidly and game software has steadily increased in quality. Accordingly, the TV game industry has grown substantially, as evidenced by the increasing population of players and the variety of game titles.

Since its launch in December 1994, Sony's PlayStation has become the world's leading home game machine, enjoying the top market share in Japan, North America, and Europe. As of March 2001, cumulative PlayStation production shipments worldwide had reached approximately 82.23 million units: 18.46 million units in Japan, 31.46 million units in North America, and 32.31 million units in Europe. Various software developers and publishers have continued to release a steady stream of software titles (3,542 titles in Japan, 1,104 titles in North America and 1,137 titles in Europe), lifting worldwide cumulative sales of packages to over 465 million units.

In order to illustrate the evolution of Sony's television game business, we shall begin by focusing on the market entry phase. Sony Computer Entertainment Inc. (SCE), a subsidiary company of Sony, released PlayStation in 1994. Technically, PlayStation outperformed other game platforms because it had a 32-bit central processing unit (CPU) developed in-house and it incorporated the newly introduced CD-ROM technology. However, it had to compete with the first movers in the industry, particularly Nintendo and Sega Enterprises. In a market where there are both scale and network economies, how did SCE successfully manage its market entry?

In searching for scale economies from a single-format software development, game software suppliers (software developers and publishers) have typically tried to back the winning horses in the hardware platform races. They have tended to weight the relative distribution of product format by estimating future game software sales based on the format/hardware platform that commanded a greater market share. Consequently, consumers also tried to purchase the better-selling format software, for fear of becoming 'angry orphans' (David 1987). Obviously, game machine manufacturers or platform holders were paying close commercial attention to the changing market share of different software formats due to emerging network externalities amongst consumers. This positive feedback surely contributed to the market structure of the industry and might have stabilised the position of the first movers.

When SCE entered the market, competent game software publishers were well established in the fields of action and role-playing games and supplied software to the first movers' formats. It was difficult for a newcomer to capture competent game developers or publishers in those fields, because the relation between format holders and publishers was already well established. SCE sourced software from few third parties like Namco, but took the strategic decision to develop a new game genre by entering into game software development.

To increase market share of its platform, SCE first opted to strategically lower the price of PlayStation hardware. The price dropped from ¥39,800 in 1994 to ¥29,800 in 1995, ¥24,800 and then ¥19,800 in 1996, ¥18,000 in 1997 and ¥15,000 in 1999. Increasing market share led to declining production costs through learning curves and scale economies, and contributed to SCE profits. SCE compensated for insufficient revenue from hardware sales by increasing sales of PlayStation format game software, because the publishers of PlayStation format games paid royalties under the licensing contract with SCE. Also, SCE attempted to extend the market beyond game enthusiasts to other members of the population, such as women, who were traditionally seen to be indifferent to TV games. As illustrated in the next section, the high development cost associated with new game genres could be justified if an

SCE were to succeed in entering a game software market with a high entry barrier.

SCE successfully increased its market share in the TV game industry. In 1997, 7.9 per cent of sales were of 8-bit machines, 3.2 per cent were 16-bit machines, 56.7 per cent were PlayStation, 27.8 per cent were Nintendo 64, and 4.4 per cent were Sega Saturn. The share of game software in different formats was as follows: 10.9 per cent for the 8-bit format, 6.3 per cent for the 16-bit format, 39.7 per cent for the PlayStation format, 24.6 per cent for the Nintendo 64 format, and 18.5 per cent for the Sega Saturn format (CESA 1998).

Figure 7.1 shows the market trends of PlayStation format products from 1995 to 2001. First, the growth rate of the Japanese hardware market started to decline as early as 1998. The trend resulted partly from possible market saturation and partly from declining market price and therefore revenue growth. Second, the Japanese software market and foreign hardware markets have been steadily increasing. With some additional marketing endeavours, these markets can be expected to grow for some time before maturing. Third, the North American and European software markets have expanded since the late 1990s, and the rate of growth has recently accelerated. Overall,

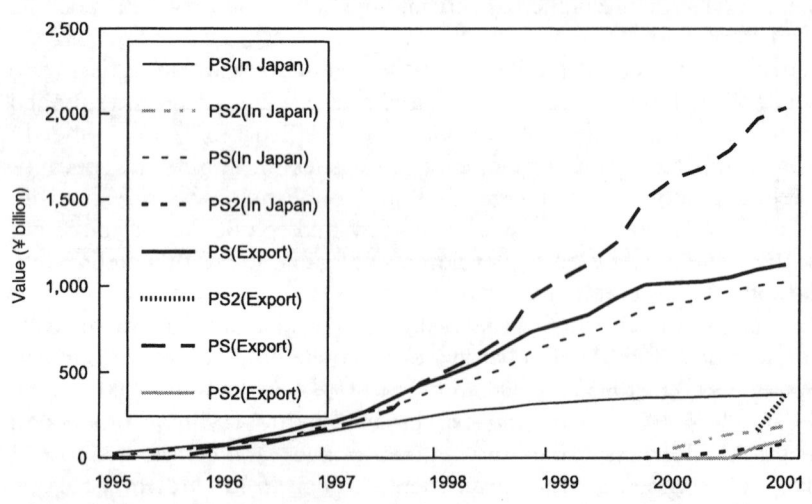

*Figure 7.1* Market trend of PlayStation format hardware and software

*Source*: Original data on PS2 hardware shipments and market price of each unit, and the amount of PS2 software shipments, comes from Sony Computer Entertainment Inc. US$1.00 = ¥123.65.

*Note*: The market price of software was assumed to be ¥4,000 per unit for PS format and ¥5,000 per unit for PS2 format.

these market trends clearly endorse the validity of Sony's strategy in the TV game business; a format holder often has to make an initial investment in hardware and to recoup the investment in the long run from the ensuing software business.

Acknowledging that Sony's decision to attack the first movers in the TV game industry was of great importance, we now want to shed some light on the firm's overall managerial and organisational arrangements. Top managers are expected to set out the future global image of the firm, with a rationale and a business trajectory of a certain type. For instance, in 1998 Sony's president Nobuyuki Idei proposed the term 'digital dream kids' to define the future direction of new product development. This business concept embodied the belief that Sony must continue to be a source of unique and enjoyable products that fulfil the dreams of its customers who are captivated by the potential of digital technology (Sony Corporation 1998).

As they branched out into risk-taking projects like the TV game business, the top managers of Sony strengthened the role of independent projects or 'internal companies' by having them take the form of an affiliated company. President Idei delegated the management of each project or business of an 'internal company' to its own executives. In selecting candidates for internal entrepreneurs, top managers welcomed the mavericks who did not fit comfortably into the traditional organisational management and culture. President Idei referred to SCE (with particular reference to Ken Kutaragi, the advocate of the concept of PlayStation) as an individualistic, artistic and creative team, emphasising the need for Sony to create internal companies that could provide a place for such business talent (Sony Corporation 1999).

In the late 1990s, the top Sony managers seemed to lack the farsightedness to prepare for the business contingencies of IT-induced innovation, but the very top management could be expected to predict future business trends and believed that newly emerging internal entrepreneurs could play critical roles in managing individual risk-taking projects.

*PlayStation 2: hardware, software and the internet*

In 2000, Sony upgraded its TV game business strategy by releasing a new game platform, PlayStation 2. The new system has highly advanced image-rendering capabilities, but is compatible with older PlayStation titles and can play new DVD-video software.

Sony had also been planning to use PlayStation 2 as a network terminal for digital distribution of content in 2001. PlayStation 2 is connected to a broadband network such as a cable television system, which allows it to offer wide-ranging forms of entertainment, to download directly PlayStation 2 software and to communicate interactively. Also in 2000, SCE established PlayStation.com (Japan) Inc. and began to use the internet for direct sales (but not downloading) of PlayStation 2, PlayStation 2 software, DVD-video software and related products. In 2000, SCE tied up with NTT DoCoMo Inc.

*Figure 7.2* Evolution of Sony's television game business development in the age of broadband communication

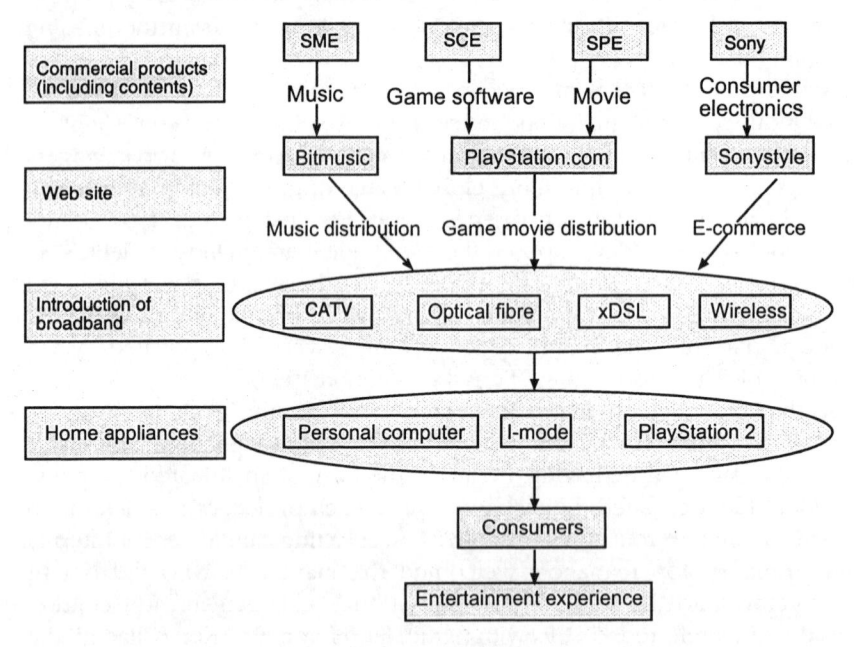

| Commercial products (including contents) | SME → Music | SCE → Game software | SPE → Movie | Sony → Consumer electronics |
|---|---|---|---|---|
| Web site | Bitmusic | PlayStation.com | | Sonystyle |
| | Music distribution | Game movie distribution | | E-commerce |
| Introduction of broadband | CATV | Optical fibre | xDSL | Wireless |
| Home appliances | Personal computer | I-mode | | PlayStation 2 |
| | | Consumers | | |
| | | Entertainment experience | | |

*Source*: Original information comes from Sony Corporation.

*Note*: SME = Sony Music Entertainment, Inc.; SCE = Sony Computer Entertainment Inc.; SPE = Sony Picture Entertainment, Inc.; DSL = digital subscriber line (various types).

to provide a new service that brought together PlayStation and the i-mode service of NTT DoCoMo. Already, several game publishers are selling titles compatible with a new networked service. Figure 7.2 summarises Sony's strategy for evolving its game business in the age of broadband communication.

Sony's current task is to progressively secure profits from the introduction of broadband communication by steadily incorporating services such as e-commerce and e-distribution. In the long run, Sony cannot secure its industrial position by simply relying on product innovation in electronics: there is an increasing tendency for manufacturing industry products to become commodified, resulting in price competition and decreasing returns. Faced with this situation, Sony has attempted to gain synergetic effects from its dispersed business activities. President Idei has used the term 'complex management' (the idea that 'the value of the group as a whole must be greater than the sum of the individual units') to refer to this need (Idei 1999).

In reforming its business, Sony has simultaneously adopted two different principles. First, it is trying to sustain its core competence in the traditional business by continuing its in-house development of hardware, particularly advanced semiconductors. In order to develop and produce a 128-bit CPU for PlayStation 2, SCE formed a joint venture with Toshiba, building a new production line with next-generation 0.18–0.15 micron process technology at Toshiba's plant. Sony's initial capital expenditure amounted to ¥130 billion. For PlayStation 2, SCE has developed a hard disk drive, adaptor, liquid crystal display, monitor, keyboard and mouse for connecting PlayStation 2 to networks. Sony is willing to sustain the technological competence and high brand image in electronics that it has inherited from its history in the hardware business.

The second principle concerns the fact that contemporary IT business environments are characterised by modular technology and open networks. If a firm can become the architect of a modular product system, creating the visible information or design rules, it can make use of complementary modules from other companies positioning themselves in each of the industrial activities comprising the product. If the firm is good at judging the quality of potential partners in each industrial segment and successfully builds the optimum partnership, a new combination of modular functions will allow the firm to derive faster and more efficient operations, resulting in better opportunities to gain the first-mover advantage (Idei 2001).

Sony had leveraged its strengths to secure the ownership of the PlayStation OS format, and had both the hardware and game content development capability. It is now exploring a series of 'soft alliances' with strategic partners. In doing so, it is fully taking into account the market conditions and other strategic factors arising from the coming broadband communication business. In order to build a new PlayStation 2 platform that could embrace future networked systems, Sony has formed strategic alliances with a number of key players. In particular, SCE has reached agreement with Cisco Systems to develop an IPv4/IPv6 dual protocol stack for PlayStation 2 for both the current internet and future broadband networks; with America Online Inc. (AOL) to create a framework for integrating interactive AOL features (such as instant messaging, chat and email) into game applications, including the development of a Netscape browser to optimise the PlayStation 2 environment; with Real Networks Inc. to embed its RealPlayer 8 into PlayStation 2; and with Macromedia Inc. to embed the company's Macromedia Flash Player web animation tool. Figure 7.3 shows Sony's strategic alliances for introducing networked services.

In the 1990s, Sony encouraged its autonomous business units and internal companies to form 'offspring', and it has used 'soft alliances' to decentralise its product platform management. This management style is flexible and speedy, but it has the disadvantage that it can be executed without proper monitoring and evaluation based on objective criteria. In 2000, Sony started to use the term 'integrated decentralised' management and undertook a series of corporate reforms. According to President Idei, integration provides a centripetal force to all its businesses; the top management holds the key to

*Figure 7.3* Sony's strategic alliances for introducing networked services

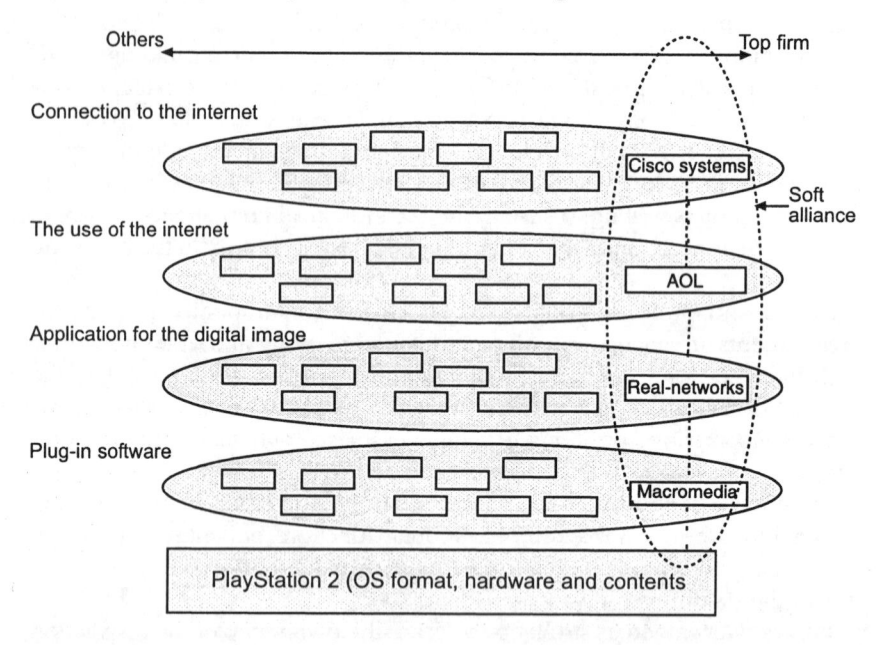

Others — Top firm

Connection to the internet — Cisco systems
The use of the internet — AOL
Application for the digital image — Real-networks
Plug-in software — Macromedia

Soft alliance

PlayStation 2 (OS format, hardware and contents)

*Source:* Original information comes from Sony Corporation.
*Note:* AOL = America Online Inc.

the optimal integration of the company's business endeavours (Sony Corporation 2001).

Under integrated decentralised management, the top management organisational structure has been redesigned and strengthened. The newly designed headquarters comprises three units: the 'global hub', which establishes a unified management policy for the whole group and sets individual strategies for groups; the new 'management platform', which serves the entire group by reorganising units into specialised supporting departments such as finance, legal services, personnel, intellectual property, information systems, environment and design; and the electronics headquarters, which oversees all electronics initiatives and aims to create valuable combinations of hardware and services. Sony also bought what became SCE and made it into one of the five basic business units of the whole company.

In order to cope with heightened uncertainty in the business, Sony also strengthened the financial control of each project by adopting objective criteria. In 1999, the company adopted a management metric – economic value added (EVA) – to evaluate performance and to increase returns on invested capital. Sony's electronics business was the first to apply this system. The result was a series of decisive actions, including the realignment of manufacturing facilities, the application of stringent standards to all new investments, and expanded outsourcing of production. Furthermore, by concentrating resources on the most attractive areas, the electronics business improved the efficiency of its invested capital and achieved a significant improvement in its EVA. In light of this success, Sony plans to use EVA to evaluate corporate management and investment decisions, and to examine operating results throughout the Sony Group (Sony Corporation 2001).

Let us briefly review the market performance of PlayStation 2. In 2000, SCE launched PlayStation 2 in Japan, North America and Europe; as of 31 March 2001, SCE shipped 10.61 million units of PlayStation 2 throughout the world. This is about three times the number of PlayStation products shipped in the corresponding period after its release. Various software developers and publishers released software titles for PlayStation 2. As of 31 March 2000, world-wide cumulative shipments of PlayStation 2 software were over 38.3 million units (17.2 million in Japan, 13.8 million in North America and 7.3 million in Europe). Software shipments have been about the same as PlayStation software during the corresponding sales period, but few titles have exceeded the million sales mark. Although Sony had proposed advanced game software as a candidate for the contents of PlayStation 2, the market may have partly settled on another candidate – the DVD movie – which has been strategically delivered at low prices. According to an SCE marketing survey conducted in September 2000, 70 per cent of consumers purchased PlayStation 2 for playing a new type of game, but 77 per cent used it for watching DVD movies (Sasaki 2001).

## Managing an innovative project

Traditional industrial products were mass produced and offered functionality, but a TV game offers value via an entertainment experience. This difference is vital to understanding the nature of the coming post-industrial or service society, because in the new era project managers or designers need to articulate what they think the public want of a new experience. Seen from the outside, good games, like good movies, are an almost serendipitous result, requiring knowledge of what might be an exciting sensory experience, gripping story, and so on. As is shown, the prototyping culture, the management culture, organisational arrangements and user relationships are important factors in choosing a successful innovative project.

Compared to other types of computer software, it is difficult to structure game software at the initial stage of the project. The task starts with a game concept, followed by a branching-off process, with project members adding contributions, eventually leading to a complete game package. The following summarises standard development processes in the Japanese TV game software industry. With modifications, this is applicable to both conventional (within a genre) series products and innovative products that are genre creating.

There are two main stages of game software development: planning and development. Planning consists of game inspiration, concept and design. At this stage, the game concept is turned from an initial inspired idea into a few elements of a simple prototype. At the end of planning, there is a playable set of game sequences, which is used to gain corporate approval for further development.

Specifications in the planning stage play an important role in the development stage. The specifications provide project members with a shared understanding of how the game is developed. Team members are expected to put in their best efforts on the game components for which they are responsible. In the process, prototyping is frequently adopted to adjust specifications so that the team can incorporate on the spot a series of new ideas conceived by team members.

TV games involve different types of components (for example, graphics, sound and character motion), so it is technically far from easy to evaluate the quality of individual components in advance. The second best approach is to build a prototype satisfying the original specifications, and to modify or redevelop the components to improve the quality of entertainment, while meeting the deadline for the release of the product. Finally, completed TV games are reviewed by the firms with the hardware licence before they are released to the market. The volume of the initial shipment is also determined. For example, in the games developed for the PlayStation platform, SCE determines the volume based on the number of orders derived from its retail outlets.

### *Parappa the Rapper*: an innovative project

We now want to discuss the development process for the TV game *Parappa the Rapper*, developed by SCE (based on Baba and Tschang 2002). We selected this case both because of the significant degree of innovation in the product and because it is representative of the new type of prototyping process seen in the industry. The game offered a new kind of entertainment in the field of TV games, and created a new game category and market segment (the rhythm–action game). More than a million copies have been sold since its release at the end of 1996, a landmark achievement for TV games. The game was acknowledged internationally to have set new standards for TV games. The game design was unique, with its new incorporation of rhythm and action, and its artistic quality was high. The target audience was also radically different: women, usually representing less than 10 per cent of purchasers for TV games, accounted for 40 per cent of the sales for *Parappa the Rapper*.

In the following discussion, we will compare specific features of the *Parappa the Rapper* project to those seen in traditional software development. The project can be described by three aspects: software development, organisational and management practice, and the way the user interaction process is managed.

## Software development

Project development followed an 'outward spiral' model – a term we use for a special variant of the spiral software development process (Boehm 1988; Cusumano and Selby 1995; Quintas 1995). In game software design, the spiral model is adapted to involve the stages of game inspiration, concept (fleshing out), design, development, and evaluation, as well as additional loops that repeat the same stages – redesign, revision of development and retesting/re-evaluation – as software prototype revision and refinement proceed. However, in an outward spiral, these repeated stages can lead to substantial restructuring or redesign of the game's basic concept or nature of play (Baba and Tschang 2002).

In a normal software project, components are designed to work logically with one another, but in a design-intensive activity where user experience is a key factor the parts must also reflect an aesthetic whole. The game project we evaluated used prototyping to allow the 'downstream' task (programming) to influence the 'upstream' tasks (game design, character design and so on). Further rounds of prototyping allowed the game to be refined. The game design continued to be highly changeable after moving to the development stage; any design changes to a component would reverberate back to the earlier structural design and throughout the basic structure of the product. Thus, we use the term 'outward spiral' for this variant of task flow where fundamental design structures and other components are continually revised. In contrast, in the traditional spiral, the design details are determined as the development proceeds. We posit that 'outward spirals' are more common in innovative products where wholesale revisions may be necessary in downstream development.

This potential for continuous revision makes the scrapping costs of unused component codes extremely expensive and the degree of reuse of the code very low. This continuous revision of the basic structure makes the 'outward spiral' prototyping approach in this case potentially more costly than the traditional spiral model.

## Organisation and management

The second notable feature of the development of *Parappa the Rapper* concerns organisation and management. As a general principle, the nature of the task at hand determines the most appropriate organisational form, ranging from 'organic' to 'mechanistic', to use the terms of Burns and Stalker (1961). SCE adopted project-based organisational forms, and organic rather than mechanistic organisation. For each project, this allowed the best teams to be constituted from a variety of labour pools, and for the teams to learn within the scope of the project. This became indispensable because TV game software requires the integration of complex components and a wide variety of knowledge and skills.

The people in the team to develop *Parappa the Rapper* were all drawn from the informal network of the project leader, Matsuura, who took a central role in game design, as well as music composition and performance. Matsuura was both a core member and a catalyst in this respect. Despite his central role, the project's defining characteristic can still be considered 'emergent', in the sense that there is an element of self-organisation from which aggregate properties arise somewhat unexpectedly.[1]

The 'emergent' nature of the work requires creative management abilities (Hobday and Brady 1998). A spiral development approach is not sufficient for the realities of game software development tasks, particularly creative ones. The *Parappa* model could have suffered from unstable work processes and therefore be more likely than other models to undermine successful project scheduling. The development of a highly innovative product in the TV game industry requires management to exercise a combination of freely exercised creativity and control over the project.

To ensure that the creative aspects survived, Matsuura played the role of a low-profile producer. This allowed him to maintain the egalitarian atmosphere needed for effective collaboration among specialists with different backgrounds. His low-profile manner enabled each project member to work autonomously towards the common goal, giving their utmost with their respective abilities. This sort of 'management of creative personnel' ensured that creativity flowed from the bottom up.

At the same time, the highly creative nature of the project could make for chaotic conditions, such as when strong decisions had to be made on choices that would have taken the project in opposing directions. Matsuura had to control this type of situation through his role as a benevolent dictator. He played this role by listening to all comments and opinions and then making an informed choice, if necessary killing off unauthorised or competing solutions. In short, the project leader must take some risk in structuring the project, to ensure that the product's integrity and vision remain true (Clark and Fujimoto 1991).

*User interaction*

The third feature of the *Parappa* project concerned user interaction. Game software provides users with a kind of entertainment that is difficult to structuralise (Newell and Simon 1972), so companies try to pre-empt user preferences by developing channels to users, establishing a series of the same type of game. The spiral model for game software products with well-defined markets has a relatively high degree of user–developer interaction in order to define new user requirements.[2]

However, the business fraternity has started to believe that tailoring to user needs or banking on user feedback does not automatically guarantee that an innovative game will branch out. It is hard for companies to get concrete information on what new features or arrangements users would like, perhaps because users themselves do not know what they like until they see it. In part because of this, the SCE tends to take a creator-driven approach in developing innovative games. When an innovative product like *Parappa the Rapper* is developed, the project does not have a user–developer interaction, and the user's preferences have to be 'guessed' or approximated by way of feedback from team members. The designer may even try to ignore the conventional user's requirements at the initial stage, in the hope of creating a more innovative trajectory for the game's design.

Marketing was another crucial element in the market success of *Parappa the Rapper*. The game represented a new game category, so the company could not rely on traditional users or media for advertising. The targeted user was not the manic player, but new and occasional users such as females and family members, especially those who are young and fashion-conscious. SCE developed a very innovative marketing scheme for the innovative product.

Overall, SCE's strategy for entering into the game software market despite the high entry barrier was to allow a one-year delay in the development and associated cost over-runs. In a sense, *Parappa the Rapper* benefited from SCE's strategy and from the overall support and farsighted leadership of the parent company, Sony.

## I-FLEX: AN INDIAN SOFTWARE COMPANY THAT CREATES PRODUCTS

### India's software industry

India's software industry is an oasis of high-tech development in a large developing country. Arora et al. (1999), Evans (1995) and Heeks (1996), among others, have pointed to its largely service nature, which has moved from a 'body shopping' phase of temporarily locating personnel directly in overseas client sites, to offshore development, where the software work is done in India for the overseas client. Most major software enterprises in India provide offshore development services to Fortune 1000 and other MNCs, either as contractors or as offshore in-house development centres.[3] However, recent work (Tschang 2001) has suggested that some companies are trying to move to higher value-added ground in the form of higher value-added services, and products.[4] Figure 7.4 is a highly stylised representation of this value addition.[5]

Many Indian start-up companies have tried to develop products but most have failed to penetrate the product market. I-flex is one of a few such companies to have survived in providing an own-brand product. Other companies have tried to provide higher value-added services. Some firms provide consulting services; others, such as many of the established offshore development firms, provide full end-to-end contract services to multinational clients, ranging from requirements analysis to design and implementation.

As Figure 7.4 illustrates, products have a higher financial and technological risk than higher-value-added services, offshore development services or body

*Figure 7.4* The changing nature of software development activities in India

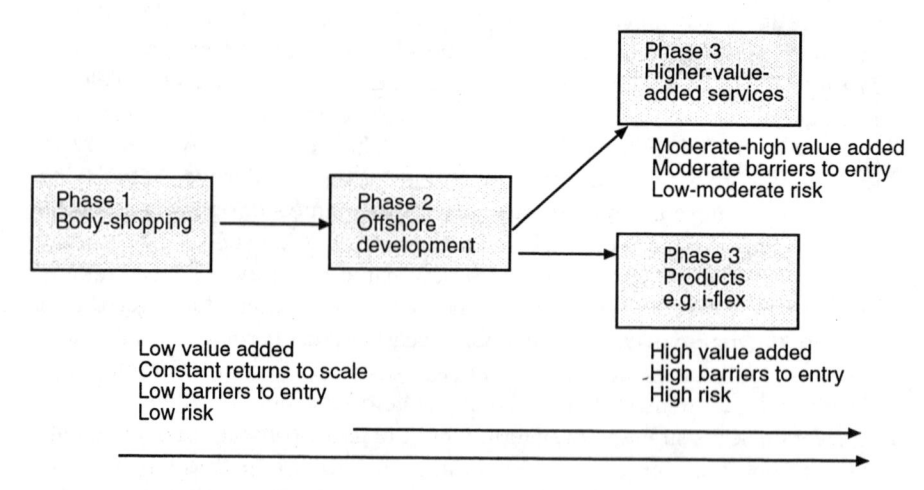

```
Phase 1               Phase 2                    Phase 3
Body-shopping   →     Offshore        →          Higher-value-
                      development                added services

                                                 Moderate-high value added
                                                 Moderate barriers to entry
                                                 Low-moderate risk

                                                 Phase 3
                                                 Products
                                                 e.g. i-flex

Low value added                                  High value added
Constant returns to scale                        High barriers to entry
Low barriers to entry                            High risk
Low risk
```

shopping. However, products have a higher degree of value added and potential profits, because of potentially increasing returns to scale (in other words, lower unit costs for each subsequent additional product sold). The barriers to entry in product markets are relatively high for local Indian firms, because product development requires quite large resources and takes a relatively long time. Nevertheless, the barriers to entry appear relatively low to established global firms. For example, a huge commitment of resources by firms like Microsoft provides them with a rapid inroad into a potentially lucrative new domain. Both types of higher-value-added activities – services and products – require relatively great domain knowledge: that is, software development firms need particular knowledge of the customer industry's business.

Software professionals' wages are rising rapidly in India, but the critical mass of talent continues to pull MNCs seeking software talent into India. Furthermore, firms like Infosys, Wipro and TCS that have commandeered the market are continually developing software engineering processes and domain expertise, which allows them to develop increasingly complex projects and take on more independent tasks. For example, they handle an increasing amount of overall information technology (IT) work for MNCs. This requires effective management of the organisation's knowledge, including how to disseminate the knowledge from each project that has been acquired by the software developers. This whole effect can serve to lock up their advantages. Given their continued advantage in providing services, these offshore development firms are loath to take on product development, and have had trouble making money from products.

Ultimately, this raises the issue of why product development should be carried out. As mentioned above, products offer a greater potential for increasing returns than do services, which have fairly constant returns to scale (that is, constant costs with increasing projects) and therefore a physical limit to profit growth. On the other hand, success in making products requires some control over the marketplace, and incumbents tend to have this locked up in a market for an established technology. This makes it very difficult, if not impossible, for smaller newcomers to compete with established MNCs such as Microsoft in their product space. New entrants have to devise a strategy that combines a market strategy and a product development strategy. A likely market for a new entrant to pursue is one where the incumbents are relatively uncompetitive – for example, a new market, a developing market or a market where incumbents are locked into older technologies. This is the case for i-flex's entry into the financial sector, where the incumbents tended to be specialised firms dealing with legacy systems. I-flex also chose to aim at more easily penetrable markets such as emerging markets or markets in developing countries. As a product development strategy, it chose initial accumulation of returns, followed by the use of those returns to fund a radical overhaul of its product platform in order to leapfrog its international competitors.

## The development of i-flex

### Beginnings

The Indian government liberalised foreign investments in 1986. At that time, Citicorp sought to develop software in India. It set up Citicorp Overseas Software Limited – an offshore development subsidiary – to work with Citibank. Several organisations that had relations with Citibank wanted to use the system. In 1992, Citicorp spun off 48.3 per cent of the company as Citicorp Information Technology Industries Ltd (Citil) and provided venture capital for it. It initially invested a million dollars, with a mandate to go after any customer, including Citibank competitors. Citicorp is still the largest minority shareholder.

Early on, the Citil management decided to stay focused on financial services, and not just become a 'boutique shop' serving various customers' service needs and be spread out over multiple sectors. In particular, its management chose to avoid the usual types of outsourcing work associated with offshore development firms, such as bidding for the Y2K market. Rather, i-flex focused on the client–server model (in 1995). This has taken it on a different path from that of most Indian firms.

Citicorp provided initial technological and domain expertise in the financial area. Some Indian software firms, such as offshore development centres, have focused on services and broadened their expertise: some companies are spread across 10–12 industries, including the financial sector. In contrast, i-flex (then Citil) decided early on to focus on only the financial sector, to increase its product offerings over time and to climb the value chain by making products.

The name of Citicorp helped Citil to establish some name recognition with customers, but only in the beginning. Later Citil's own project successes increasingly helped it to attract new customers, especially in developed country markets. The company grew much faster as an independent than as a Citicorp entity, attaining over 100 customers in six years. With its new flagship product, Flexcube, it began to work with relatively large Citicorp competitors. In 2000, Citil was renamed i-flex, partly to avoid being too closely associated with Citibank and partly to be more closely associated with the new flagship product Flexcube (I-flex 1999).

The i-flex company had two main characteristics: a focus on vertical financial services, and an early emphasis on products. These characteristics involved the creation of 'value-added relationships' with customers, meaning longer-term relationships that sometimes started with using customers as test sites, and that also eventually resulted in further additional components in i-flex's suite of products. The company's focus on long-term relationships differed from other banking product companies mainly because of its early exposure to Citicorp as an internal customer.

I-flex's main focus is on the typical bank's back office – that is, helping banks to manage functions other than dealing directly with customers. Its work is a highly applied kind of development; it is not scientifically driven and it is not fundamental systems software development like Microsoft. I-flex is at the 'higher end' of the food chain: it may integrate other base products into its system. For example, it has embedded Oracle's relational database management system (RDBMS) in its own applications.

I-flex's strategy has been rewarded by fairly impressive financial results. It has recorded a compounded annual growth rate of 70.17 per cent in revenues, putting it at or higher than the Indian industry average. I-flex's profit before taxes grew by an average of 69 per cent from 1995 to 2000; its revenue grew by an average of 55 per cent over the same period. In 2000, operating revenue was 50 per cent in outsourced software development services and 50 per cent in packaged products such as Flexcube, the latter moving to 54 per cent in the second half of 2001. I-flex currently has over 1,600 personnel, of whom about 1,200 are in the product line organisation, the others in consulting. I-flex aims to employ 2,000 personnel by the end of 2001.

I-flex's growth can be broken down into roughly three phases. The first phase involved a relatively simple product, Microbanker, which was targeted at emerging markets customers. The second phase involved a sophisticated product, Flexcube, which was modular and more advanced than its competitors' products. The third phase involved broadening its offerings in the financial sector to include consulting and other services. Figure 7.5 shows these phases, and briefly describes the characteristics of the products and the market.

*Corporate strategies in information technology firms* 191

*Figure 7.5* I-flex's products, product characteristics and markets over time

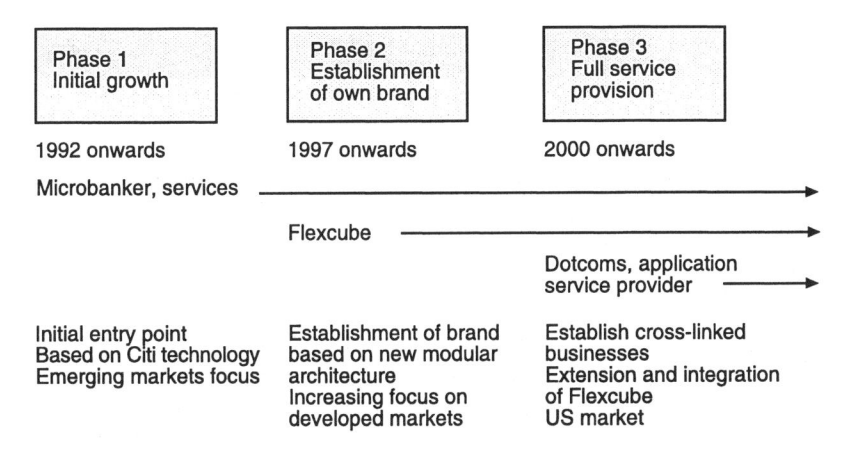

| Phase 1<br>Initial growth | Phase 2<br>Establishment<br>of own brand | Phase 3<br>Full service<br>provision |
|---|---|---|
| 1992 onwards | 1997 onwards | 2000 onwards |

Microbanker, services ──────────────────────────────────►

                               Flexcube ─────────────────────►

                                                        Dotcoms, application<br>                                                        service provider ──────►

| Initial entry point<br>Based on Citi technology<br>Emerging markets focus | Establishment of brand<br>based on new modular<br>architecture<br>Increasing focus on<br>developed markets | Establish cross-linked<br>businesses<br>Extension and integration<br>of Flexcube<br>US market |

*First growth phase*

In 1992, i-flex developed Microbanker, its first successful product, which was based in part on its internal work for Citicorp. To define it, i-flex relied on feedback from customers, including its own ex-bankers, and requests for proposals to help to define its features. Microbanker was targeted to emerging markets, and small to medium-sized banks. Microbanker was built on a relational RDBMS core, but the company did not focus on the graphical user interface (GUI), as it did for its follow-up product, Flexcube.

*Second growth phase*

The second phase of growth was defined by SCE's next main product, Flexcube, which is its current flagship product. Flexcube was released in 1997, quickly became a huge seller in its market, and gave i-flex its market brand. Flexcube was built from scratch, and developed in a modular fashion that allows i-flex to extend it fairly easily with almost any feature, and also allows customers a choice of modules and technology. Up to 2001, i-flex had managed to replace its customers' existing products (Microbanker) with Flexcube without losing a single customer. In 2000, it had sixty-four customers for Flexcube and another 110 using Microbanker, together putting i-flex in the top five in retail and corporate banking in terms of numbers of customers. In 2000, i-flex signed up eighteen countries, quickly moving into Asia and other countries.

Most of i-flex's competitors are in corporate or retail banking only, but i-flex chose to go into both. In 1995, it focused on the corporate banking business, but in 1998 it realised the potential in retail banking and integrated the products, serving the two markets together.

The total product rewrite resulting in Flexcube was a big risk, especially at a time when the company did not have a brand name to assure customers. It had to make a product that would suit customers it did not yet have. It chose an initial set of customers as beta sites. This involved tremendous financial and management risk. Finance was completely internally generated, but the services business and the earlier Microbanker product provided the necessary funding to develop Flexcube and get over its uncertain beginning.

*Third growth phase*

I-flex's third phase involved the development of new businesses and the broadening of its services. For example, it developed a consulting arm in 1999, and it can now provide a full contracted systems solution to customers. It has also linked different business arms: it has coupled Flexcube to other services and alternative means of delivery (for example, delivering Flexcube's services through its application services provider).

I-flex is focusing more and more on products, but it has also built up a development services group, consisting of consulting services, centres of excellence, development services and infrastructure facilities management. Its services business relies on its centres of excellence, which focus on specific areas such as internet banking, Java for financial services, customer relationship management (CRM) and knowledge management. All require tremendous domain knowledge, and local knowledge is also essential for the consulting arm. I-flex has set up entire electronic data processing departments for customer banks, and advises foreign banks on their India operations (IBS 1999).

I-flex has incubated and established new businesses. For example, it has set up business internet portals (dot.com companies), and in 2000 it set up an application service provider (ASP) business unit. One vertical business portal offers a full range of services and products to investors and other market participants in the stock exchange of India. This brings investors, brokers, banks and depositories into one portal. Another joint venture is a vertical business-to-consumer (B2C) and business-to-business (B2B) financial 'infomediary' portal, consisting of dotexplaza.com and TimesofMoney.com, the latter being a joint venture with the Times Group. The second business line is Flexcel International, an ASP which serves applications such as Flexcube applications on a subscription basis from a central data centre. As of 2001, i-flex had 280 customers in 74 countries, of which 195 are product customers.

Currently, the depth of i-flex staff's domain and technical knowledge allow it to take a customer over the full product life-cycle, from the conceptual stage to implementation and maintenance. Thus, it is able to consult on systems and also to develop the solution and implement it – an ability that customers value when they are seeking a full service solution for banking software needs. I-flex's centres of excellence are also important in supporting the development and application of new technologies. In terms of technological capability, i-flex has subscribed to international software engineering quality standards. It is now at SEI CMM (Software Engineering Institute capability maturity model) level 5, the highest level of software engineering competence.[6]

## Strategy

In the following sections, we discuss how i-flex has been able to record such strong and steady growth while developing its capabilities and product sophistication. In essence, it has four main strategies.

### 'Climbing the value chain'

An overt part of the i-flex strategy was to climb the value chain, both in terms of product functionality and in terms of broadening. The aim was to undertake the higher-value-added activities seen in the third phase of service provision. This required a conscious effort that involved putting the main resources into product development but also diversifying into services that made sense from the point of view of the product (for example, consulting).

As we have mentioned, it is much more risky to base a business on products than on services. However, the rewards are also immense. I-flex devoted a substantial amount of human resources to the development of Flexcube, involving a more radical overhaul of its design philosophy, even without any guaranteed customers at the time. This is now paying off substantially, with increased sales.

The i-flex management team differed from other Indian firms that were less successful in products by being committed to long-term product development and by being willing to fund products out of other stable revenue sources. The fact that the company was not listed on the stock market may have been important too, because it meant there was probably less pressure about long-term decisions.

### Marketing strategy

The marketing strategy involved slowly building a name brand, particularly as i-flex made a transition from emerging to developed markets. Although non-US markets offered lower revenues per customer, this was a lower-risk kind of business, and offered faster growth rates, so emerging markets offered a more attractive initial market entry point. I-flex's Citicorp affiliation helped to secure initial customers but, in moving to developed markets, its gradual accrual of project successes helped it to bring in bigger, more established customers. Thus, its successes in Singapore and Europe helped it to attract the interest of customers in the United States. In developed markets, the customer profile includes multiple banking locations (for each customer), bigger customers, and specialisation in either corporate or retail banking. The typical emerging markets customer has both retail and corporate activities. I-flex faced less competition in emerging markets, often only with local, ageing systems. Moreover, Microbanker was actually priced lower than the equivalent product of Kindle Systems, its main competitor (IBS 1999). In

1999–2000, 29 per cent of i-flex's revenue base was located in Europe, 29 per cent in the Middle East and Africa, 23 per cent in the United States and 14 per cent in the Asia Pacific region (including India), making it a fairly evenly spread global company (I-flex 1999). As i-flex moves away from India, it must learn to be a more global organisation – one that can deal with all kinds of customers with local needs, while maintaining its Indian development facilities. All i-flex's competitors are doing business this way.

## Building a customer base

In the financial sector, customers are won over one at a time. Unlike other types of products, banking products involve a relatively small set of large institutional customers, so, as long as i-flex could build a customer base out of relationships and recommendations or past successes, it could increase its market share. In contrast to most of its competitors, i-flex developed its market by building long-term relationships and markets based on product loyalty, trust and alliances.

In developing its strategy, the essential question i-flex faced was how it could couple growth in emerging markets with products in a scalable way (that is, in a way that allowed it to extend its product platform easily). While focusing on building relationships, i-flex found that it could scale up its product line, architecture and size even as it operated across more and more countries. One way of developing such a growth strategy is to use a modular platform that allows the company to integrate the new product requirements of new or pilot stage customers. The company learns from its customers' needs and develops prototype systems for a pilot set of customers. For instance, i-flex's customers in the United States wanted a system, which, by creating new opportunities to extend the product system, led to four new customers. Four banks in Thailand wanted a mutual fund operation; this needed a complete foreign exchange and money market subsystem, which i-flex had already built in. I-flex agreed to develop the systems for these banks, on the condition that it could resell the products to other banks. At the same time, it would provide free upgrades to those first customers. This mutual fund module became part of the Flexcube suite; today, sixteen mutual fund companies in South Africa use it. I-flex is involved in a number of arrangements like this. Every time a customer wants a new subset, i-flex can build it into its system.

I-flex also has strategic alliances with different technology suppliers and partners; this brings the company into contact with different markets and allows it to improve its technology – technology by being part of a kind of modular technology network. The companies include Oracle, Microsoft, Intel, Compaq, Sun, Hewlett Packard and IBM. I-flex's thirty-three business partners allow it to enter new markets. Oracle provided a seed licence for development tools because it (Oracle) saw the potential for success and was convinced of the importance of i-flex's expertise in banking (that is, its domain knowledge).

Finally, two new lines of business – the ASP and the vertical business portal – allow i-flex to work at a deeper level with certain partners or partners that could not be reached before, and to avoid the limitations of product-driven business or simple alliance with business partners. These ventures also bring in additional users, which further increases i-flex's customer base. Other benefits can accrue across multiple business lines; for example, the ASP has been coupled with Flexcube, which allows i-flex to generate revenues from Flexcube in new ways.

Some customers do not have domain knowledge, and this is where i-flex's services group comes in. This kind of cross-selling (that is, the coupling of products and services) is a powerful combination for business. I-flex's products business yields continuous revenue increases, while its service business brings a different type of work and revenue stream. For instance, for a mid-sized bank, every $1 paid in licence fees brings in an additional $2 in operations and services. I-flex wants partners and locals to get a portion of this. To accomplish this, it will continue to build relationships and to couple its services with products. This also needs a quality marketing support team.

*Product development (Flexcube)*

Product development includes the choice of technology – for Flexcube, this involved a very modular design, which was innovative within the industry – and an organisational structure that facilitates it. The key innovation that i-flex brought to the banking software market was the concept of building the Flexcube product with an extensible architecture – one done entirely from the ground up.

Flexcube is highly modular: new components can be added relatively effortlessly to the standardised framework. Around this extensible core, people can add more and more functions. The greater the number of functions, the easier it is to attract new customers, because of the potential for customising the system. This provided flexibility for banks to offer new banking products with no code changes. I-flex also borrowed ideas from other successful products to complement this design approach; for example, it used features from Microsoft Office (such as a task bar at the bottom of the screen that allowed different applications to be opened from the window).

The design allows i-flex to build skills rather than waste them in time-consuming code changes. I-flex systems allow a choice of parameters and products; even accounting problems can be 'parameterised' so that systems can be built by customising the products rather than revising the product code. This innovative product design and architecture allows for flexible and fast growth. New ideas can be easily brought into the product architecture. Today, i-flex implements design changes in nine months, compared with three to four years in the past.

## Organisational structure

I-flex measures and delivers results jointly as a team, and it rotates its people to ensure an even distribution and dissemination of knowledge. A typical team includes employees coming all the way from marketing to implementation. For example, when i-flex had to map out its latest technology strategy for the following two years, it included the marketing and sales teams, who had just as big a say in the process as others. Another example was the design idea for Flexcube, which involved a collective thought process, like all processes in the organisation. In this way, i-flex considers its team processes to be different even from those of MNCs, who typically assign a single technical leader to a project and require the rest of the team to do parts of the work. In technical product development, skills are not the barrier, but the company needs the ability to support a long-term continuous product cycle. It needs people to stay with it for the long term.

I-flex also brings in outside knowledge through consulting firms and subject matter experts for domain work. For example, its new module on derivatives and options required it to bring in an expert from the United States to completely transfer his knowledge to the organisation.

## CONCLUSION

By comparing two cases of Japanese and Indian firms known for their excellent business records, we have shown how modern IT firms in the Asia Pacific region have tried to combine new technological development and new markets. By describing the behaviour patterns involved, the paper provides a package of 'stylised facts' that can be adapted to competitive settings of emergent IT business. Admittedly, other firms have succeeded in these countries, but few, if any, have been as innovative and forward looking as Sony in Japan, which has made a transition from hardware manufacturing to software and services development, or as successful at moving from software contract services to software products as i-flex in India. Table 7.1 compares some strategies of the two firms.

Currently, hardware-based IT firms like Sony face a kind of predicament. On the one hand, firms are experiencing the increasing 'commodification' of products, in which production is suffering from the inescapable fate of decreasing returns and firms may move from one location to another searching for cheaper labour. On the other hand, by placing too great an emphasis on satisfying customers' current needs, firms may fail to adopt or adapt new technology that will meet customers' unstated or future needs (Christensen 1997). Consumer-oriented companies like Sony face a challenge in post-industrial society: firms have to provide consumers with new services or valuable experience for staying in the market but it is difficult to articulate these new services or experiences.

An equally difficult, if not greater, problem is faced by emerging companies or companies like i-flex in developing countries. Here, we have an Indian software company competing with established MNCs that have formidable resources and name brands. The traditional advantage for Indian software firms was low-cost, high-skilled labour. However, these firms face inevitable competition, which will compress their profit margins; their wage differentials

*Table 7.1* Comparison of combinative strategies of Sony and i-flex

|  | Sony | I-flex |
|---|---|---|
| Market mindset | Hypothesis (future needs) driven | Pragmatic (current needs) driven |
| Corporate culture | Prototyping for realising potential market | Building a relational customer base |
| Business strategy | Diversification from manufacturing to contents and services. Searching for synergistic effects from businesses | Moving to products, then to extensible products, and now to synergistic products and services |
| Technological strategy | Using generic and modular technologies | Introducing the extensible modular product concept |
| Risk | Very high strategic risk | Product risk higher than that of services |
| Risk management | Use of open network and newly strengthened top management with monitoring/evaluation tool | Gradual staged risk taking |

with other countries will decrease accordingly, as is already happening. Furthermore, these advantages are increasingly more easily usurped as would-be multinational clients decide to locate their own development centres in India.

As is explained in this paper, the strategies developed from a firm's previous experience or a rational analysis of the situation have clear limitations when it comes to dealing with unpredictable events. Such uncertainty in the business climate requires a series of changes in technology and market strategies and managerial and organisational arrangements.

Table 7.1 shows how Sony tried to solve its predicament by developing a hypothesis-based market strategy to take advantage of technological disruptions triggered by the introduction of the internet. Sony's corporate mindset addressed consumer needs in an unforeseeable future by forming a hypothesis of coming societal change. To base strategy on this was a major risk. In following its ambitious hypothesis, Sony used the potential of the information technologies to help it diversify from a manufacturing to a full-fledged IT firm for the coming era, one that covered both content and service development. The planned evolution of Sony's TV game business is assumed to contribute to the firm's long-run growth through the business's synergistic effects with other widely dispersed business fields.

By embracing a new prototyping tradition, Sony's corporate culture is largely characterised by a developer-driven approach. Our observations on Sony and the practices used in developing the *Parappa the Rapper* suggest that, in order to provide products or services whose selling point is originality, firms have to deliberately create disruptive changes within individual projects by realising business opportunities through prototyping practices.

Here, we should recapitulate that hypothesis-based management or prototyping practices in Sony or SCE exhibit several contradictory requirements: the idea of a risk-taking project or genre-creating game often stems from internal entrepreneurs or from the creativity of a single maverick. Hypothesis-based management and prototyping practice can also be theoretically assured by monitoring by the top management and scientific evaluation through objective criteria. Here, there is another kind of 'innovator's dilemma'. By putting too great an emphasis on new technology to meet a customer's future needs, a firm can also eventually fall behind, should it fail to manage its strategic risks. These risks are partially managed: products that are designed in a modular way in cohesion with networks of other firms will lessen the resources demanded of the firm that sets the format standard, and also allow it to rely on other firms' manufacturing or other competitive advantages.

As shown in Table 7.1, i-flex's characteristics and strategy aimed for a 'pragmatic' model that varies from the Sony model of 'hypothesis-based' development. I-flex aimed to cater to the immediate market's demands for software development by efficiently providing the market with reliable software meeting those needs. To some extent, this was an artifact of its customer base and its need to couple product development and piloting with longer-term customer relationships. Being corporate in nature, i-flex's customer base was much smaller than the customer base for a typical game software product, making relationship-based marketing important to the company. Its greater emphasis on relationship-building than even its competitors also helps to keep customers locked in to i-flex solutions, as customers become comfortable with staying with i-flex's products over time.

Although it was the offspring of a well-known parent, i-flex had to make a transition to its own name brand, and with its own innovative product. The domain knowledge, process engineering and team orientation that i-flex used to create its innovative product Flexcube represents a created competitive advantage, partly constituted out of existing advantages. However, as witnessed by the failure of other products developed from the Indian software industry, this alone was not sufficient; the innovation in i-flex's strategy was to combine these additional assets with an initial marketing focus on emerging or developing country markets to build a track record.

Because Flexcube's offerings were extensible, i-flex could develop more offerings as its customers increased, providing a type of increasing return of benefit to its customers. With Flexcube, i-flex made a strategic product decision that increased its risk but that also provided a long-run competitive advantage through its extensible nature.

I-flex had far fewer resources than Sony and was an originally smaller player in its market, so the i-flex model was tempered to some degree, making efficiency more important than risk-taking. However, this did not eliminate the need to manage risk or to take risks, because i-flex sought a radical departure from its competitors.

## PREDICTIONS

In focusing on PlayStation 2's market segment, Microsoft has been behaving as an aggressive competitor with a strategy to attack the incumbents. In March 2000, Microsoft pre-announced market entry of the X-box for November 2001. The specifications and price of X-box are strategically set to match those of PlayStation 2. Obviously, Microsoft plans to compete in emerging content and service industries that are driven by TV game products. It will do this through its holding of the Windows OS format. In response to the new attacker's strategy, Sony tried to achieve its planned global PlayStation 2 shipment of 20 million units in 2001 by all means, including by reducing the unit price from ¥39,800 to ¥35,000 within six months after its release in Japan.

Given these new developments, how can we evaluate the future market status of PlayStation 2, which was one of the most important Sony strategic products at the beginning of the twenty-first century. PlayStation 2 format game software shipments were about the same as PlayStation software during the corresponding sales period, but sales in general have remained lower than SCE's business expectations. The performance may be linked with sluggish sales of game software overall in the late 1990s. In the Japanese market, the value of shipments of game software reached a peak of nearly ¥600 billion in 1997, but subsequently decreased, to ¥293 billion in 2000. SCE plans to start a trial of the *Final Fantasy XI* from December 2001 by releasing an online version. Although *Ultima Online*, a world-famous online game, attracts 250,000 players over the world, SCE plans to attract as many as a million players to its online games in the near future.

The incorporation of generic technologies entails great uncertainties in the TV game business, but we suggest that the entire business will evolve through its interplay with evolving market selection criteria – that is, which characteristics of a particular technology are the ones that users will eventually come to value most highly. Accordingly, it is highly likely that the market could act against the business hypothesis articulated by Sony. In that case, Sony would have to face a difficult problem in trying to recoup its ¥700 billion investment in the development of the PlayStation 2 project alone.

Throughout its short history, i-flex has managed a difficult task of competing with market incumbents even as it developed new areas of expertise. However, like Sony, the strategies that i-flex is using could also be used by its competitors. In other words, there is nothing to stop its competitors or new entrants from using relationship-based marketing to attack emerging markets first, or to take a risk in developing a new generation product based on modular technology.

I-flex's competitors include international vendors to emerging markets of products like Temenos's Globus, Kindle Systems' Bankmaster and MKI's Opics. Most are based in the United Kingdom. In non-emerging markets i-flex also faces intense competition from established banking software companies like the Irish company Kindle Systems. Many of its international competitors are now in India undertaking development, but most are still upgrading old applications in their Indian facilities. The choice of technology is equally important. I-flex's competitors still appear to be locked into an inherited code that cannot be re-engineered into more modular systems.

One thing that i-flex may have going for it is its unique combination of intellectual assets, which cannot be accumulated suddenly. Another could be its ability to execute, which has so far been flawless and which is a valuable asset. Finally, if i-flex can somehow attract more larger customers, or a larger market share, and increase in size fast enough through its modular technology and extensible strategy, it will no longer have to formulate marketing strategy as a catch-up competitor.

## NOTES

1. It is quite likely that this worked better here because the attributes of the group members are so difficult to define in a creative project. For instance, in a typical organisation, the role of Ryu (the rap musician) would not be uniquely defined, and he would probably have been confined to musical score creation. But the emergent manner of the project was flexible enough to allow him to contribute to the bicultural attributes of the software as well, thereby taking on a larger role. Thus, the division of labour and progress of development can be seen as interactive, if not codependent.
2. In the case of *Final Fantasy*, a large series of games was developed, involving reliance on multiple channels for obtaining user feedback, such as user surveys and internet chat boards. Once a product achieves market success, companies tend to think about incorporating the user feedback into the next game in the series. This results in a series type game like *Final Fantasy*, and locks users in to purchasing from that series.
3. The attitudes of domestic companies towards their business contrasts with those of MNCs, which typically use Indian offices as localising platforms. However, a few MNCs (for example, Novell, Borsch and GE) act as resource back-ups in the sense that they actually do development work as an equal member of the international development group.
4. This case study is derived from a larger ongoing study of research and development and product development in the Indian software industry (Amsden et al, forthcoming). The authors express their appreciation to Professor Amsden for letting us use the material on i-flex.
5. This figure was developed on the basis of interviews as well as insights from industry experts, including a former chief technical officer of Wipro Technologies.

6  As measured by the capability maturity model (CMM) of the Software Engineering Institute (SEI).

## REFERENCES

Amsden, A., Tschang, T. and Sadagopan, S. (2003). 'Measuring technological upgrading in the Indian software industry: a framework of R&D capabilities and business models', working paper, Asian Development Bank Institute.

Arora, A., Arunachalam, V.S., Asundi, J. and Fernandes, R. (1999) 'The Indian software industry', Carnegie Mellon University. Report to the Alfred Sloan Foundation. http://www.heinz.cmu.edu/project/India.

Arthur, W.B. (1985) 'Competing technologies and lock-in by historical small events: the dynamics of allocation under increasing returns', Center for Economic Policy Research, 43, Stanford: Stanford University.

—— (1996) 'Increasing returns and the two worlds of business', *Harvard Business Review*, July–August: 100–109.

Baba, Y. (1998) *Digital Kachi Souzou* [Digital Value Creation], Tokyo: NTT Press.

Baba, Y. and Imai, K-I. (1992) 'Systemic innovation and economic growth', in Heerje, A. and Perlman, M. (eds) *Papers in the Schumpeterian Tradition*, Ann Arbor: University of Michigan Press.

—— (1993) 'A network view of innovation and entrepreneurship', *International Social Science Journal*, 135: 21–34.

Baba, Y. and Nobeoka, K (1998) 'Towards knowledge-based product development: the 3-D CAD model of knowledge creation', *Research Policy*, 26: 643–659.

Baba, Y. and Tschang, T. (2002) 'Product development in Japanese TV game software: the case of an innovative game', *International Journal of Innovation Management*, 5.

Baldwin, C.Y. and Clark, K.B. (1997) 'Managing in an age of modularity', *Harvard Business Review*, 75, 5: 84–93.

Boehm, B. (1988) 'A spiral model of software development and enhancement', *IEEE Computer*, 21, 2, May: 61–72.

Burns, T. and Stalker, G.M. (1961) *The Management of Innovation*, London: Tavistock.

CESA (Computer Entertainment Software Association) (1998) *White Book of the Japanese Game Software Industry, 1997*, Tokyo: CESA.

—— (2001) *White Book of the Japanese Game Software Industry, 2000*, Tokyo: CESA.

Christensen, C. (1997) *The Innovator's Dilemma: When New Technologies Cause Great Firms to Fall*, Boston MA: Harvard Business School Press.

Clark, K. and Fujimoto, T. (1991) *Product Development Performance: Strategy, Organization, and Management in the World Auto Industry*, Boston MA: Harvard Business School Press.

Cusumano, M. and Selby, R. (1995) *Microsoft Secrets*, New York: The Free Press, Simon and Schuster.

David, P.A. (1987) 'Some new standards for the economics of standardization in the information age', in Dasgupta, P. and Stoneman, P.L. (eds) *The Economics of Technology Policy*, Cambridge: Cambridge University Press.

Evans, P. (1995) *Embedded Autonomy: States and Industrial Transformation*. Princeton NJ: Princeton University Press.

Heeks, R. (1996) *India's Software Industry*, New Delhi: Sage.

Hobday, M. and Brady, T. (1998) 'Rational versus soft management in complex software: lessons from flight simulation', *International Journal of Innovation Management*, 2, 1: 1–43.

IBS (International Banking Systems) (1999) *IBS Report 1999*, Martin Whybrow, IBS Publishing (supplemented with 2000 Sales League Table).

Idei, N. (1999) 'Nihon Keizai Runesansu Sengen', *Bungei Syunjyu*, March: 94–102.
—— (2001) 'Beyond the narrowband era: the future of management', *Diamond Harvard Business Review*, June: 30–45.
I-flex (1999) *I-flex Annual Report 1999–2000*, I-flex Solutions Ltd.
Imai, K-I. and Baba, Y. (1991) 'Systemic innovation and cross-border networks: transcending markets and hierarchies to create a new techno-economic system', in Organisation for Economic Co-operation and Development (OECD) (ed.) *Technology and Productivity: The Challenge for Economic Policy*, Paris: OECD.
Langlois, R.N. (1988) 'Economic change and the boundaries of the firm', *Journal of Institutional and Theoretical Economics*, 144, 4: 635–57.
Newell, A. and Simon, H.A. (1972) *Human Problem Solving*, Englewood Cliffs NJ: Prentice Hall.
Oniki, H. (1999) 'Japanese/US comparative advantage: width and depth of co-ordination, in Macdonald, S. and Nightingale, J. (eds), *Information and Organization: A Tribute to the Work of Don Lambarton*, Amsterdam: Elsevier Science B.V.: 107–214.
Quintas, P. (1995)' Software innovation in the context of complex product systems', Working Paper, Science Policy Research Unit, University of Sussex, May 1995. www.sussex.ac.uk/spru/cops.
Sasaki, H. (2001) 'Sony no broadband senryaku', Tokyo: Nihonjitugyoshuppansha.
Schrage, M. (1996) 'Cultures of prototyping', in Winograd, T. (ed.), *Bringing Design to Software*, New York: ACM Press.
Sony Corporation (1998) *Annual Report*, Tokyo: Sony Corporation.
—— (1999) *Annual Report*, Tokyo: Sony Corporation.
—— (2000) *Annual Report*, Tokyo: Sony Corporation.
—— (2001) *Annual Report*, Tokyo: Sony Corporation.
Teece, D.J. (1988) 'Technical change and the nature of the firm', in Dosi, G., Freeman, C., Nelson, R., Silverberg, G. and Soete, L. (eds), *Technical Change and Economic Theory*, London: Pinter.
Tschang, T. (2001) 'The basic characteristics of skills and organizational capabilities in the Indian software industry', Working Paper #13, Asian Development Bank Institute, http://www.adbi.org/publications/wp.
Ulrich, K. and Eppinger, S. (1995) *Product Design and Development*, McGraw-Hill Inc..
Vernon, R. (1966) 'International investment and international trade in the product cycle', *Quarterly Journal of Economics*, 80, 2: 190–207.
Winograd T. (ed.) (1996) *Bringing Design to Software*, New York: ACM Press.

# 8 Intellectual property protection and capital markets in the new economy

*Keith E. Maskus*

## INTRODUCTION

The global economy continues to see considerable progress in the development and use of technologies and products characterising the so-called 'new economy'. In such areas as computer software, personal computers, electronic commerce, internet transmission of digital products, telecommunications, compilation of databases and biogenetics, innovation has at its core the development and use of information. Information is often costly to produce but is essentially a public good: non-rival and difficult to exclude without legal protection. Thus, intellectual property rights (IPRs) – patents, copyrights, trademarks, trade secrets, and related devices – have become a central issue in countries that wish to promote such innovation.

The existence or prospect of the exclusive rights awarded by IPRs can be important in determining the ability of entrepreneurs and firms to attract capital for their R&D and expansion programs. In advanced innovation systems there is an important complementarity between capital spending (including R&D) and the protection of new technology and market positions through IPRs. However, while IPRs promote innovation and encourage commercialisation of new information products, they also support market exclusivity that can restrict access to consumers and limit competition by rival enterprises. In that context, excessive protection for intellectual property can distort capital allocation and retard productivity growth.

As always in the IPR area, there is a tradeoff between encouraging invention through protected market positions and promoting access to new information and innovative competition by rival firms. For countries of East Asia and the Pacific region, it is not easy to strike the right balance between strong protection, favouring original innovators, and limited protection, favouring imitation and access. The interests of Japan, Malaysia and Vietnam are as different from one another as are those of the United States, Mexico and Guatemala. A 'one-size-fits all' approach in the Asia Pacific region might lead to some rationalisation gains from policy similarity, but is unlikely to meet the needs of all countries.

The proposition that a balance is difficult to strike does not excuse countries from failing to enact or enforce intellectual property protection of particular concern to information-intensive sectors. Often these industries are characterised by significant network economies that strongly encourage growth in the use of their services and products by businesses and consumers. However, such use may be effectively conditional upon adequate copyright protection or other intellectual property policies. Refusing to protect property rights in information risks increasing technological isolation from the global knowledge frontier. It also could limit the supply of investment capital, from both domestic and foreign sources.

It is an open question as to how intellectual property protection and capital investment are interrelated; I analyse this issue in this chapter, with particular reference to the new economy and the Asia Pacific region. If IPRs were perfectly calibrated and capital markets were efficient and flexible, as in textbook models, investment would be drawn into economically valuable innovation without unduly damaging dynamic competition. However, there are imperfections on both sides of this equation, making the interactions complex and raising policy challenges. This chapter should be read as largely exploratory, for there is virtually no economics literature on the joint properties of protection and capital markets. Indeed, typical analysis of the effects of IPRs ignores problems in financing research and development (R&D).

In this chapter I argue for a 'nuanced' approach to intellectual property protection for new-economy sectors in the Asia Pacific region, taking account of complications arising in capital markets. While there is an argument for relatively strong standards in the relevant technologies, countries still could vary their effective protection through competitive limitations on rights. For technologically advanced nations with strong innovation systems, the objectives are domestic development and acquisition of technologies and digital products. For poorer nations, the goals are to have adequate and affordable access to useful information and to encourage local innovation and technology adoption. In both cases, flexible capital markets are important for growth, as are an abundant supply of skilled labour and other factors.

Ultimately, upgrading the protection of IPRs will do little to improve prospects for innovation, technology transfer or investment if the underlying economic environment is not conducive to those processes. Thus, policies to encourage the development and use of new technologies and products should be complemented by programs to enhance technology and information infrastructure, entrepreneurship, and market competition.

In the following section I provide a brief comment on the meaning of the new economy and its importance. I then provide a theoretical perspective on interrelationships between capital markets and IPRs, with an application to information technologies. I overview the types of intellectual property protection that are employed in new-economy sectors and consider the likely development interests of Asia Pacific economies in that regard. I then review briefly the limited evidence available on how investment and intellectual property protection interact. Finally, I provide some concluding remarks.

## A BRIEF PERSPECTIVE ON 'THE NEW ECONOMY'

'The new economy' is one of those dissatisfying and vague phrases that can mean almost anything involving information generation and management. My preference is to set out a set of essential technological characteristics that many analysts seem to have in mind when they use the term. Those characteristics are as follows.

First, activities tend to involve the creation, organisation or use of extensive volumes of information, usually in an electronic and digital format. These tasks are 'new' in that software and electronic interoperability permits manipulation of far larger data sets than was previously the case. Second, dramatic improvements in communication technologies and infrastructure permit wider and cheaper distribution of information and content.

Third, there tend to be high investment and R&D costs in developing or organising information, leading to potentially large economies of scale. These economies may be enhanced by significant network effects, whereby users prefer to adopt a standardised form of technology use or dissemination. Fourth, many of the products and technologies involved have relatively short product life-cycles. In part this is because they may be easily reverse-engineered or copied and distributed widely at low cost. Because creators must have some means of recovering expected development costs, temporary exclusivity in production and distribution rights is essential for innovation.

These characteristics have profound implications for economic organisation and growth. While it is ridiculous to speak of the end of business cycles, enthusiasts ascribe to the new economy several positive macroeconomic impacts, including a significantly lower unemployment rate that is consistent with moderate inflation. However, the sustainability of these impacts is still unclear. Gains should be more evident at the microeconomic level, because improvements in information technologies and communication environments promise higher long-term productivity benefits to sectors that employ them. Lipsey[1] provides several examples of the application of computing power to process control, innovation and diagnostics where various sectors have achieved higher productivity and lower costs. The economies of scale inherent in performing research and producing and marketing digital products lie at the core of extensive merger activity in such sectors as pharmaceuticals, telecommunications, and entertainment services. Moreover, improvements in information-based product design and electronic distribution systems should promote entry of new goods and lower distribution costs, thereby expanding consumer choice. As suggested earlier, however, these trends offer both pro-competitive and anti-competitive characteristics.

For the purposes of this chapter, I refer to the new economy as the outcome of the systemic process of encouraging technical progress in information development and its incorporation into production systems. This process, if competitive and adopted effectively, should generate real gains in productivity and income. It stems directly from rapid evolution in software, digital memory expansion, the internet and electronic commerce, and telecommunications. These new technologies have made possible research and innovation in entirely new sectors, such as biogenetics, that could not exist readily without significant computing power, and that have expanded innovation prospects in biomedical research. Another example is computer-aided design in such areas as architecture and film production. Thus, a broad definition of the new economy would encompass all sectors in which innovation and design make intensive use of information technology. These are often referred to as 'knowledge industries', in recognition of the technical skills required to compete. Indeed, it is more informative to think of the new economy as really a 'knowledge economy' in which access to information is central for competitive success.

Numerous studies, both in this volume and published elsewhere, attest to the dynamism and growth of these sectors and their continuing penetration into broader production processes in the industrialised countries.[2] While these figures are difficult to measure and forecasts are at times grossly inflated, there is little doubt that consumer and business markets for e-commerce, electronic databases, software and telecommunications will continue to grow rapidly for some time. In many developing countries adoption of such technologies is only in its infancy, with substantial room for expansion.

While telecommunications equipment and computer hardware might have sufficient market lead times to fund their development, many of the inventions and creations supporting knowledge sectors depend on intellectual property protection. Digital products and software are easily downloaded and copied in the absence of such effective safeguards as copyright and anti-circumvention technologies. The growth of e-commerce depends on distribution software and secure encryption programs, themselves subject to copying. Biotechnology firms depend on patents to acquire capital for funding experimentation and trials, for without such patents their successful products could be readily imitated by second-coming firms. Accordingly, IPRs are integral to the processes by which these new technologies are developed and marketed.

A final comment is useful at this point. One fundamental feature of information and communication technologies is that they permit firms to extend their marketing reach far beyond local and national boundaries. For this process to succeed, however, firms must find ways of engendering in their customers, most of whom will be distant, confidence and trust in their ability to perform. Traditional business methods based on face-to-face contact and informal contracting mechanisms are unlikely to flourish in the new environment. In some contexts, IPRs help in substituting formal contracts (acceptable to a wider and more anonymous customer base) for informal contracts and more limited marketing methods. In that context, IPRs are a useful complement to the spread of the new economy and of particular relevance to developing economies in the Asia Pacific region.

## ANALYTICAL PERSPECTIVES ON INTELLECTUAL PROPERTY RIGHTS AND CAPITAL MARKETS

The interactions of intellectual property protection and investment capital are complex because of numerous complications in markets for both information and finance. The interactions inevitably vary with economic circumstances, including development levels and competition processes. A comprehensive treatment is beyond the scope of this chapter. Rather, I discuss fundamental concepts that are most relevant for development policy in knowledge-intensive sectors.

Capital resources are used to finance R&D expenditures, and IPRs help earn a return on that R&D, making them closely related. In the simplest conception, intellectual property protection generates temporary market power sufficient to appropriate enough rents to cover R&D costs, including financing costs. Implicitly this idea assumes that capital is readily available to inventors and creators at some market rate. Borrowers would assess the likelihood of research success and the potential for registering IPRs in deciding their credit needs in the development stage. However, patents and copyrights cannot be calibrated so finely that they achieve a fully optimal degree of innovation, suggesting that if IPRs are overly strong (or weak) the allocation of capital could be distorted. If, on the other hand, capital were constrained and borrowers faced credit limitations, not all socially valuable inventions would be financed or purchased. To the extent that credit limitations exist because of uncertainty about the ability of particular enterprises to safeguard rights to a new technology, stronger IPRs could help overcome this difficulty.

Thus, circumstances in either area affect efficiency in the other. Countries protect intellectual property in order to provide ex ante incentives to undertake innovation by defining ex post exclusive market rights. Failing to establish such rights invites second comers to copy and sell the products developed by original inventors while incurring far lower R&D costs. The problem is the extent to which inventors may appropriate sufficient surplus from users to recoup initial costs. In many industries, such natural market mechanisms as lead times, first-mover reputation advantages and the difficulty of reverse-engineering new products may accomplish this in the absence of intellectual property protection. However, these advantages are small in many new-economy sectors, including software, digital products, databases, and biotechnology. It may be costless to copy and distribute digital goods and software, and it is often straightforward to reverse-engineer pharmaceutical products and plant varieties. Appropriability problems are severe in these areas, so domestic innovative sectors may not develop if there is no intellectual property protection, and investment resources may not be made available except at sufficiently high premiums (or accelerated repayment terms) to reflect both the uncertainty of successful research outcomes and the risk of rapid imitation.

## Intellectual property rights

For countries interested in expanding domestic innovation and value added in knowledge sectors, therefore, appropriate policies for both IPRs and capital markets are important. On the intellectual property side, the mix between the costs and benefits of strong protection is complex. For example, anecdotal evidence suggests that software development has flourished in the absence of strong copyright enforcement in China and, earlier, in Taiwan (Maskus 2000). Beijing boasts its own 'Silicon Valley', called Zhongguancun, which generates substantial turnover in electronic products and software. However, this success seems due more to public support and the existence of universities than to original software development.[3] Relatively little innovation occurs in the district. Further, the programs written in both countries have been niche applications of business and personal software without wide application. This has avoided infringement. Firms tended to remain small unless they found technical means for protecting software from competitors, sacrificing scale economies. In both countries, software firms have become voices for stronger copyright enforcement.

At the same time, however, stronger copyright protection may raise costs to consumers and software users as they come to rely less on pirated copies, conceivably slowing down the incorporation and diffusion of information technologies into local economies. It is easy to exaggerate this problem, because the dynamic response of software sellers is typically to reduce considerably the prices of legitimate software products as markets become thicker, particularly if distribution systems are competitive. On balance, there are liable to be net gains from copyright enforcement in countries with a capacity to use and develop software.

Consider another example, the emerging biotechnology sector. China has ambitious plans to deploy bio-engineered seeds into agricultural production and to develop a bio-pharmaceutical sector with innovation potential. In part this development has occurred through research in government laboratories and universities, but there is an increasing recognition that the fruits of the research need to be commercialised through private entrepreneurial channels. Numerous biotechnological enterprises, with a mix of public and private ownership, now undertake applied research with a view to registering patents in China, the Asia Pacific region, and beyond (Maskus 2000). This represents an intriguing shift for China's development policy. Until recently the pharmaceutical sector was built on generic production of traditional medicines and imitated formulations, which helped restrain drug prices. With patents in place, the industry will shift more into innovation and licensed production at higher prices.

Consider also the internet, which many developing countries in the region are working to integrate into their economies. As several chapters in this volume discuss, effective and widespread access to the internet depends primarily on the existence of an efficient telecommunications infrastructure, appropriate pricing schemes, and competition among internet service providers. Intellectual property protection is a smaller, yet important, concern.[4] Firms in the Asia Pacific area, including small enterprises and start-ups, may use the internet to extend their marketing reach and to procure input supplies competitively. Protection of trademarks and domain names can spur competition, improve international marketing, and enhance client trust. Copyright protection increases the willingness of domestic and foreign entrepreneurs to place new content into electronic distribution streams, despite the difficulty of enforcing such rights. Accordingly, IPRs can increase the flow of products and information available. However, copyright protection that significantly curtails the fair use of protected material by researchers, students, libraries and other public users could be costly in development terms. Similarly, protection for electronic databases without much originality in creation is problematic for developing countries, for it could restrain access to foreign information compilations without inducing much domestic innovation.

A further complication is that some new-economy sectors are characterised by significant economies of scale, and IPRs can play an important role in achieving and sustaining those economies. There are high fixed costs in developing software, producing a movie, designing a logic semiconductor, and testing a new bio-engineered medicine, but the marginal cost of producing and distributing copies of those products is small. For developing countries, the gains from protecting intellectual property are that inventive firms are more likely to offer their goods and technologies on markets that have adequate IPRs. The costs arise from limiting the ability of local firms to imitate and reverse-engineer these goods, making it difficult to learn frontier technologies and overcome the entry barrier represented by the need to undertake original R&D. As a consequence, investment capital may shun attempts by local enterprises to compete with large international firms through developing new technologies.

Increasing returns stem also from network effects, which may characterise software, logic technologies, and telecommunications products. A network externality exists when the advantages of joining an information technology network, linking to a communications system or purchasing a piece of software rise with the number of users. The benefits may be as simple as enjoying more people to communicate with, but there are technological and pecuniary gains as well. Users prefer interoperable software and communication protocols in order to share work files. Larger networks attract complementary software that can improve connections and, being subject to economies of scale, can reduce computing costs. Developing countries can benefit from both the consumer externalities and the ability to develop new network applications.

However, network economies have potentially important impacts on competition. Significant advantages accrue to those suppliers that can expand their 'installed base' of users most rapidly, because of the associated declining costs and rising demand. Once a network or program achieves a critical size, it is possible for users in other systems to switch to it in rapid succession, potentially destroying the competitors. This 'tipping effect' and the tendency for users to get locked into use of a single network or software standard is thought by many to be a source of considerable market power and has been the basis of anti-trust actions in the United States against Microsoft. It is evident that competition in developing countries might be particularly affected if there are few available alternative technologies.[5] This potential for concentration in particular sectors may be more associated with technical standards than with IPRs per se. However, if firms that develop standards prefer to keep them proprietary for the purposes of licensing, intellectual property protection and trade secrets are the primary methods for doing so.

A related issue is that, in many new-economy sectors, patents on a specific technology can generate a dominant position if their owner chooses not to license it to horizontal competitors or firms that compete downstream with vertical licensees. More commonly, competing firms may register patents that mutually set out claims on products or technologies needed by each other. Such a situation supports patent pooling arrangements, which may be anti-competitive and could markedly raise costs of follow-on innovation in user countries. Finally, patent litigation is costly and its threat alone may deter rival firms from trying to compete in key technologies.

## Capital markets

This list of complex costs and benefits of intellectual property protection shows that it is not straightforward to calibrate IPRs efficiently. Yet even if a country could establish efficient, dynamically pro-competitive intellectual property regimes, it could be frustrated by weak capital markets. As discussed earlier, innovation and market development can be costly, requiring some form of start-up financing. Without access to capital, small firms and entrepreneurs would be heavily restrained from turning potential intellectual property into economic value. In contrast, there is little doubt that, in the deep and open capital markets in most developed economies, capital is readily allocated into new and potentially profitable inventive activities.

Most capital market failures distort investment choices across many areas, not just new information development and use. Thus, basic distortions in capital markets – directed lending, inadequate financial standards, non-transparent supervision and the like – should be removed independently of their impacts on technology development. Official restraints that limit the emergence of deeper capital markets, including equity finance, venture capital, and foreign direct investment (FDI), may be directly relevant in discouraging the development and use of intellectual property. In countries where financial markets are heavily regulated and repressed, dynamic small and medium-sized enterprises may be starved for finance. These types of firms represent the major potential for effective innovation in developing and middle-income economies in the Asia Pacific region. Thus, deregulating the domestic markets and opening the country to domestic and foreign venture capital firms can provide needed finance. Indeed, US venture capital firms are active in southern China (see Maskus 2000). Similarly, freeing banks to finance entrepreneurial development and product marketing programs can liberate enterprises from capital constraints.

Even as they develop, however, private capital markets may fail to fund all socially promising technologies. There is a substantial free-rider problem at the level of basic research in biology, medicine, and information technology engineering, which makes private firms unwilling to undertake and finance it. As well, such research can generate significant spillovers of new knowledge into applications in many commercial sectors. There is scope for publicly funded research programs at universities and government laboratories, and for tax advantages for R&D spending, where private markets fail to fund them adequately.

Governments need also to establish efficient methods for commercialising the applied fruits of research performed under public grants, paying heed to the need for public access to the research. Innovative programs in China, Taiwan and Hong Kong aim to promote commercialisation of information technologies through joint university–private sector enterprises, but this process is still being improved. In this regard, public funding is joined with private participation, but sorting out ownership claims is an important task. IPRs are the basic mechanism for this purpose, with government deciding the extent to which it wishes to retain rights for further research.

It should be noted that there is virtually no research on the issue of the role of intellectual property protection in determining how R&D is financed in developing countries. In the United States, events studies show that, at least in some sectors, share prices are sensitive to announcements of patent grants and expiration; this seems also to be true in the pharmaceutical sector in Korea.[6] Moreover, during the 'dot.com' craze in the United States, venture capital and equity finance flowed dramatically into internet firms, in part on the basis of patentable ideas. Of course, patents failed to prevent capital from abandoning those firms when the bubble burst, for patents are of little value when the underlying product or service is not demanded. In developing economies, however, we know relatively little about how information technology projects are financed – bank lending, equity, venture capital, joint ventures, or government subsidies – in relation to the structure of intellectual property protection. This is an area in which additional research would be informative.

## INTELLECTUAL PROPERTY RIGHTS IN THE NEW ECONOMY

It is useful to describe briefly the main components of intellectual property protection that are of particular significance for the technologies under discussion. Some of these technologies evolve rapidly and bear characteristics that pose challenges for traditional forms of IPRs.

### Computer software

Under the Agreement on Trade-Related Aspects of Intellectual Property Rights (TRIPs), the required global minimum standard for protecting computer programs is copyright. Developers of popular software face considerable appropriability problems because the high margins between protected software prices and costs of duplication create large markets for unauthorised copies. Illegitimate copies of programs such as Windows 98 and Office 2000 are sold over the counter and loaded onto hardware systems in most countries, though at the highest rates in developing nations. This activity is illegal under copyright protection; the TRIPs standard should be sufficient to reduce the problem, though in many locations adequate enforcement is years away.

There is evidence that an environment of weak copyright protection in developing countries limits the development of large and globally competitive software firms (Maskus 2000). Uncompensated copying discourages the development of software platforms in China, for example, where the legitimate programming industry concentrates on business applications with limited market reach. Many talented programmers prefer to work for multinational software firms, which have established resources for dealing with piracy. This situation constrains the development of software for local needs and also prevents the attainment of large-scale economies in program production and sales that can fund further innovation.

Copyright protection for software is a minimum standard and is widely followed in the Asia Pacific region. However, in the United States and Japan, software with commercial utility may be patented, a policy that is under consideration currently in China. Typically, patents preclude reverse-engineering of programs. This policy is restrictive in a development sense, because decompilation is an important source of follow-on program innovation and permits interoperability of programs in an open environment. It also potentially accords considerable market power to software firms; such power could be exercised in numerous user industries and through computer networks.

Different Asian countries have different interests in protecting software because their circumstances are different. All countries are obligated to enforce copyrights and reduce piracy, which can have positive impacts on local software industries in even poor countries. Countries in which software firms tend to produce applications programs have an interest in maintaining as open a framework as possible for learning the functional aspects of computer code. In that context, copyright protection with a liberal treatment of reverse-engineering and an explicit fair-use exception for research and education can be vital for development prospects. For such countries, however, it seems advisable to erect high barriers to patent eligibility for computer programs. Countries with advanced information technologies and software firms that produce platform programs or applications that could find global markets may wish to consider patents, albeit under rigorous standards for novelty and non-obviousness.

### Recorded entertainment

Film and music production is one of the more dynamic industries in the United States, and increasingly in China, Taiwan and Hong Kong. The industry depends critically on advanced technology to achieve special effects and sound quality. It also invests considerable amounts in talent development and performers. Thus, there are substantial investment costs at the creative end. Moreover, product marketing at the distribution end is costly. The industry's creative products are protected by both copyrights and trademarks.

Assessing the ability of copyrights to enhance music and cultural creation in developing countries is not straightforward. Literal copying provides a considerable consumer benefit, albeit at the cost of lower-quality copies. Firms engaged in copying also enjoy some profits, though the sector is highly competitive. However, there is no innovation or technology transfer associated with piracy. In a dynamic sense this can generate a misallocation of resources. For example, fieldwork by the World Bank in Lebanon, Jamaica and Senegal suggests that growth of the local film and music industries has been severely stunted by weak copyrights (Maskus 2000). In Senegal, for example, an outdated copyright law and virtually no resources in enforcement have discouraged the development of rights collection societies that would transfer income to musicians when their compositions are aired on the radio. Endemic copying of domestic musicians limits their ability to sell recordings beyond small localities. In this regard, the cultural protection offered by copyrights can be important in developing entrepreneurship.

### The internet and electronic commerce

Analysts of the 'new economy' often use the term to mean electronic transactions over the internet. The internet provides considerable scope for domestic and cross-border marketing of services and goods. For example, Forrester Research predicts that global e-commerce revenues will exceed $6.8 trillion by 2003, or some 9 per cent of worldwide sales.[7] E-commerce sales in Europe, Asia and Latin America are expected to spread rapidly and to account for over 50 per cent of global internet sales within five years. The global share of internet users in the Asia Pacific region and Latin America could soon approach 35 per cent.[8]

Internet use offers considerable opportunities for consumers and enterprises in developing countries.[9] For example, the internet is increasingly used to market and distribute artisan products, such as traditional music and handicrafts, even in the poorest countries. Increasingly, the medium is becoming a conduit for technical innovation and technology transfer as researchers and scientists communicate their results on a global scale. Business-to-business use is the largest and fastest growing component of the internet; it is export-intensive, with 38 per cent of total e-commerce revenues in the Asia Pacific region coming from exports (Mann and Wilson 2001). Building customer relationships and taking advantage of global online trading offer large potential cost savings for enterprises in developing nations. Mann et al (2000) employ estimates from the United Nations Conference on Trade and Development (UNCTAD) to compute that a cut in production costs of 0.3% associated with adopting more business use of e-commerce would raise real GDP by 1.2 per cent ($30 billion) in Asia and 1.0 per cent ($15 billion) in Latin America in the long run.[10]

In the United States and the European Union (EU), the development of e-commerce has been associated with, among numerous other factors, reliance on several types of IPRs: trademarks, copyrights, database protection, and patents for computer programs that effect a technique for doing business electronically. Trademarks are critical for assuring consumers of the ultimate origin of a good or service, which helps reduce market failures associated with information asymmetries and lowers consumer search costs. The internet extends the transactions reach of both producers and consumers across space and time. However, this extension requires investments in consumer trust and confidence. Thus, transparent and rapid trademark registration procedures, effective enforcement against misappropriation of marks on the web, and discouragement of 'cyber-squatting' on domain names are important structural planks for countries wishing to build access to the internet.

Electronic transmissions over the internet pose difficult questions for copyright regulation of the uses of digital products.[11] Because standard copyright principles apply under the TRIPs Agreement, duplication rights and distribution rights are held by the copyright owner. However, enforcing these rights is difficult, for software and recorded music may be easily downloaded with no deterioration in quality. Under most jurisdictions, such copying for personal use is acceptable within copyright law. However, the greater difficulty stems from the ease with which electronic files may be distributed and shared among computers.

Technology for downloading and distributing electronic products continues to improve, leading to calls for technical means to deter unauthorised copying. The Copyright Treaty and the Performances and Phonograms Treaty (agreed at the World Intellectual Property Organization in December 1996) allow countries to make it illegal to use technical means to circumvent electronic measures to control copying. They also make it easier to collectively manage copyrighted materials on the internet and they further clarify the rights of performers and music producers to authorise electronic transmission of their works. The treaties also encourage nations to define appropriate standards of fair use in order to avoid penalising researchers and educators. Several Asia Pacific nations, including China, have signed the treaties and are considering legislation to put them into effect.

### Electronic databases

Internet transactions often require construction of extensive databases, such as customer lists and preference profiles, news compilations, travel information, and price lists. Under TRIPs, databases must be protected at a minimum with copyrights. For developing countries, this form of protection is probably sufficient to induce international web service providers to make their products available. Countries need to decide the extent of fair use they will allow in accessing and employing database information and whether they will specify rules for unfair competition in this area.

Through its 1996 Directive on the Legal Protection of Databases, the EU has extended protection well beyond copyright; the United States has similar legislation pending. In some contexts, the directives set out conditions that could throw significant and costly barriers into the path of scientific researchers and educational institutions.[12] For example, they would extend copyright protection to data compilations that require nothing more than arranging publicly available data into a particular order, thereby protecting materials that should not be copyrightable. In the EU, protection lasts for fifteen years and may be indefinitely extended by simply improving the database. Developing countries would be advised not to award such protection to databases. It would be preferable to extend standard concepts of fair use to database copyright protection.

### Telecommunications equipment and information technology hardware

It is evident that access to global information technologies requires investments in numerous physical technologies: cable or wireless transmission systems, satellite signal receptors, personal computers, and so on. Such equipment is protected largely by patents, trade secrets, and trademarks. Information technology equipment is one of the manufacturing sectors with an intense use of patents; pharmaceuticals and biotechnology also have a high use of patents. Even so, equipment manufacturers tend to protect their proprietary technologies through trade secrets and management of technical personnel more heavily than patents (Maskus 2000).

For developing countries, selection of a sensible set of standards in this area is complex. While patents and trade secrets must be respected, TRIPs provides some flexibility. Thus, policy-makers may wish to preserve some ability for their firms to learn foreign technologies and adapt them to local circumstances. In this regard, poor countries with limited innovative capabilities may wish to couple their patents with rapid disclosure requirements, utility models, and requests for technical assistance in building a basic information technology structure. Middle-income economies with strong imitative capacities may prefer to link somewhat stronger protection with rapid disclosure, effective patent opposition procedures, and access to innovation systems. In considering an IPR strategy, however, it should be remembered that access to technologies supporting the public's attachment to global information and electronic markets bears strong potential benefits. Actions that restrict foreign firms' willingness to enter these countries and transfer technologies may be self-defeating.

## EVIDENCE ON CAPITAL AND INTELLECTUAL PROPERTY RIGHTS

The argument that IPRs help attract investment is self-evident. By providing exclusive market rights, intellectual property protection reduces the uncertainty attached to competition in new products and technologies. Moreover, the profits associated with patents on inventions that meet market demand provide both a return to investment and an ability to fund further innovation. Investors should respond to this incentive. An early recognition of this notion was provided by Shoven (1988), who pointed out that venture capital is highly sensitive to changes in marginal capital taxation in the United States. He interpreted this to mean that capital allocation is sensitive to profit incentives and that, for countries looking to increase investment in innovation, strong patents are advisable.

There is good evidence, at least using aggregate data, that the structure of patent rights across countries has an important effect on domestic investment and FDI. One econometric study (Park and Ginarte 1997) found that, while there was little direct correlation between an index of patent strength and growth rates, there was a significantly positive impact of patents on both physical capital investment and R&D spending across a large group of developed and developing nations. In another (Lee and Mansfield 1996) it was reported that multinational firms that have their headquarters in the United States are sensitive to perceived weaknesses in IPRs when deciding on the location of subsidiaries, particularly where those facilities will engage in production or R&D. Yet another (Maskus 1998) showed that as industrialising economies adopt stronger patent rights they could expect to see significant increases in their US-owned stock of FDI assets. Finally, arm's-length licensing of technical information on production processes in manufacturing is sensitive to patent rights once the strength of patents has achieved a threshold level (Yang and Maskus 2001).

There is also substantial anecdotal evidence of investment and structural change being sensitive to intellectual property protection. Sherwood (1997), for example, cites a Mexican expert who claimed that small companies developing a process or product innovation found it easier to attract investment after 1991, when a new Mexican patent law provided greater certainty about exploiting that innovation. Some of those interviewed indicated that there was a significant rise in domestic biotechnology applications also and that larger Mexican enterprises stepped up their internal R&D programs. Evidence mentioned earlier from China and Lebanon also suggested that entrepreneurship and product development are responsive to trademark and trade secrets protection in developing countries.

While these results are instructive, they come from aggregate studies and may not be fully applicable to information technologies and the new economy. I am not aware of any systematic cross-country studies of how firms in those sectors acquire capital as a result of software copyrights and patents or copyright and anti-circumvention protection on the internet. An additional caution is that, if the intellectual property system excessively protects the acquisition of market power, the resulting capital investment would suffer from a dynamic distortion. It is conceivable, as mentioned earlier, that proprietary standards and network effects supported by the exercise of IPRs (restrictive licensing, patent pools, product tying between software and hardware, and the like) could establish considerable monopoly power, particularly within countries that have few substitute technologies available. Capital flowing into such arrangements may be devoted more to the maintenance of market power, say through pre-emptive litigation, than to additional innovation. In that environment, competitive entry by rivals with better ideas could be curtailed.

The evidence remains sketchy, but it supports the view that IPRs can play a positive role in resolving some of the uncertainty involved in investing in the development of new knowledge and entering markets. At the same time, the depth and competitiveness of capital markets are at least as important as intellectual property protection. If countries retain heavy impediments to the development of local capital markets and restrict inward investment, reform in the intellectual property system is unlikely to have much long-term impact on domestic innovation or technology acquisition.

## CONCLUDING REMARKS

There is much that we do not know about innovation processes in developing economies as they link themselves more closely to the new economy. Some countries in the Asia Pacific region, most notably South Korea and Japan, wholeheartedly have embraced information technologies, at least as regards consumer access to the internet. Others, such as China and Malaysia, have young and vibrant software sectors that, unfortunately, retain limited potential because of capital constraints and weak intellectual property protection. Still others, such as Vietnam and Indonesia, have achieved little domestic attachment to the relevant technologies, because of poverty. A full regional entry into the new economy is not just around the corner and requires further policy reform.

The analysis reviewed here suggests a series of general conclusions about effectively promoting the use and development of new-economy technologies. First, it should be market-based as much as possible. Rigid and non-transparent controls over use of telecommunications equipment, software, the internet or e-commerce could severely limit incentives for adoption. Second, IPRs play an important role in supporting the use of new products and access to the internet. However, countries at differing levels of economic development should be flexible about the precise standards of protection to be provided. Third, a predictable and stable regulatory environment, covering both intellectual property protection and other issues such as privacy and consumer protection, is important to permit firms to inspire trust on the part of new and far-flung users.

Finally, there is an important synergy between IPRs and capital allocation. Overall, freeing up capital markets should provide the larger boost to the acquisition and development of important sectors of the new economy, but adequate intellectual property protection sets part of the framework within which this development may proceed.

## NOTES

1 'The new economy: theory and measurement', Chapter 3, this volume.
2 Richard Cooper discusses some macroeconomic implications in 'What's new in the new economy?', Chapter 2, this volume. See also Mann et al (2000) on growth in electronic commerce and OECD (1996) for discussions of structural change favouring knowledge industries.
3 See 'Zhongguancun: China's Silicon Valley,' *The China Business Review*, May/June 2001.
4 See Mann et al (2000) for an extensive discussion of factors determining growth of the internet and electronic commerce.
5 Lew Evans ('A force for market competition or market power?', Chapter 9, this volume) discusses the issue in depth from a management perspective.
6 See Lacroix and Kawaura (1996) and Maskus and Wong (2001).
7 Forrester Research, 'Global eCommerce Approaches Hyper-growth,' www.forrester.com/ER/Marketing/1,1503,212,FF.html, 29 September 2000.
8 See Mann and Wilson (2001) for these and related figures.
9 See Mann et al (2000), Mann and Wilson (2001) and OECD (1999) for discussions of the market-expansion impacts of electronic commerce.
10 United Nations Conference on Trade and Development, *Building Confidence: Electronic Commerce and Development*, 2000, available at www.unctad.org/ecommerce/docs/building.pdf. The authors of the UNCTAD study assumed that service input costs would fall by 0.3 per cent and used this assumption to compute increases in GDP from a computable general equilibrium model.
11 See WTO (1998) and Mann et al (2000).
12 See Reichman and Franklin (1999) for discussion.

## REFERENCES

Lacroix, S. and Kawaura, A. (1996) 'Product patent reform and its impact on Korea's pharmaceutical industry', *International Economic Journal*, 10: 109–124.
Lee, J-Y. and Mansfield, E. (1996) 'Intellectual property protection and U.S. foreign direct investment', *Review of Economics and Statistics*, 78: 181–186.
Mann, C.L., Eckert, S.E. and Knight, S.C. (2000) *Global Electronic Commerce: A Policy Primer*, Washington DC: Institute for International Economics.

Mann, C.L. and Wilson, J.W. (2001) 'Facilitating trade in developing Asia: what role for electronic commerce?' manuscript, World Bank.
Maskus, K.E. (1998) 'The international regulation of intellectual property', *Weltwirtschaftliches Archiv*, 134:186–208.
—— (2000), *Intellectual Property Rights in the Global Economy*, Washington DC: Institute for International Economics.
Maskus, K.E. and Wong, E.V. (2001) 'Searching for economic balance in business methods patents', *Washington University Journal of Law and Policy*.
OECD (Organisation for Economic Co-operation and Development) (1996) *The Knowledge-Based Economy*, Paris: OECD.
—— (1999) *The Economic and Social Impact of Electronic Commerce*, Paris: OECD.
Park, W.G. and Ginarte, J-C. (1997) 'Intellectual property rights and economic growth', *Contemporary Economic Policy*, 15: 51–61.
Reichman, J.H. and Franklin, J.A. (1999) 'Privately legislated intellectual property rights: reconciling freedom of contract with public good uses of information', *University of Pennsylvania Law Review*, 147: 875–970.
Sherwood, R.M. (1997) 'The TRIPs Agreement: implications for developing countries', *IDEA: The Journal of Law and Technology* 37: 491–544.
Shoven, J.B. (1988) 'Intellectual property rights and economic growth,' in Walker, C.E. and Bloomfield, M.A. (eds), *Intellectual Property Rights and Capital Formation in the Next Decade*, Lanham MD: University Press of America.
WTO (World Trade Organization) (1998) *Electronic Commerce and the Role of the WTO*, Geneva: WTO.
Yang, G. and Maskus, K.E. (2001) 'Licensing and intellectual property rights: an econometric investigation', *Weltwirtschaftliches Archiv*, 137: 58–79.

# 9 A force for market competition or market power?

*Lewis Evans*

## BACKGROUND

The acquisition, analysis and use of information are fundamental to the operation of all markets. The 'new economy' is characterised by the high level of information processing, storage and distribution made possible by modern computer, electronics and fibre-optic technology, and by the rapid development of new technologies relevant to production and distribution in all sectors of the economy. As long ago as 145–87 BC, the importance of the 'invisible hand' and prices for conveying information for the organisation of economic activity were recognised.[1] In this chapter I shall review the principal themes of the economic analysis of information over the past 50 years and examine the impact of new-economy technology on the organisation of firms, the organisation of markets and competitive and cooperative relationships between firms.

The costs associated with the writing and monitoring of contracts and the organisation of markets depend on the cost of acquiring, analysing, developing and using information.[2] It follows that changes in the technology associated with the acquisition, storage and use of information may change transaction costs and thereby the relative costs of different economic institutions associated with economic activity. Coase (1937) emphasised that transactions costs are a critical element in determining the size and scope of a firm. Changes in transactions costs therefore affect assessments of whether an activity should be carried out within a firm or contracted out. Changes in transactions costs may also affect the operation and even the feasibility of organising markets. It is conceivable that the information technology of the new economy may so lower the costs of the organisation and participation in a market that new markets may be created. An obvious example is the emergence of competitive markets for activities that have until recently been considered to be natural monopolies best internalised within state-owned or regulated firms such as electricity.

Information is also central to cooperation between firms. Economists have long recognised that cooperation between firms may be efficient (for example, when joint ventures reduce industry-wide costs for particular types of infrastructure) or inefficient (for example, when explicit or tacit cooperation increases market power). Firms are now better able to analyse and share information. This may help them to cooperate to achieve efficiencies, but it also has the potential (in the eyes of competition authorities at least) to make it easier for them to create and exercise market power.

The new economy poses a challenge for the assessment of market power because the speed of technical change substantially reduces the relevance of traditional static assessments of efficiency. Rapid technical change focuses competition on new investment in research and development (R&D) and the implementation of new technologies. In this environment, the dynamic efficiency of markets (in the sense of the incentives to invest in new technologies as a source of future competitive advantage) will be more important than static efficiency in improving consumer welfare into the foreseeable future.

Innovation in private goods and services in the new economy may not have the distinguishing externality features of information, but the innovations reflect many sources of change and affect the organisation of firms and markets and competition in many often subtle ways.

In this paper I suggest that the greater availability and utilisation of information associated with the new economy has had the net effect of increasing rather than reducing competition. The combined effect of the creation of new markets, the reorganisation of firms (as a result of lower transactions costs) and the introduction of new technologies has increased the scope for competition and reduced the feasibility of cooperation. Cooperative agreements work best with low levels of uncertainty about the path of future technical change and the payoff to new investment, but the new economy has increased uncertainty on both counts. More information may reduce transactions costs, but it will increase the potential for anti-competitive cooperation only if it reduces the uncertainty associated with economic activity and the path of technical change.

## INFORMATION NETWORK TOURNAMENTS AND THE PACE OF CHANGE

Information has three special characteristics that set it aside from many goods and services. The first is that it is an 'experience' good, in which it is difficult to assess utility arising from it without consuming it. This is a characteristic of services that have attached to them intangible qualities that can only be accurately assessed after they have been used: for example, the services of a plumber or a trip to Seattle have to be experienced to be assessed. This is the basis of asymmetric information in which different parties to a transaction have different information: the plumber will know the effort and materials she uses but the consumer will not, at least not before the job is done. Information is the classic experience good: a company with an idea may have to give it away before the recipient can assess whether to pay for it.

The second key characteristic is that information has the non-rivalrous quality of a public good in that any amount of consumption of information does not diminish additional consumption possibilities: for example, the contents of a book can be consumed innumerable times. Non-rivalry renders it very difficult to measure consumption of information and difficult to get consumers to reveal how much they are prepared to pay for its creation.

The third characteristic is non-excludability. Once produced, information is generally available for all to consume. Encryption and intellectual property right laws can protect information, but the techniques may not be effective and they come at some cost. Indeed, in the new economy it is difficult to protect information, because of its abundance and because techniques such as 'cookies' can be used to acquire it. In short, the intrinsic nature of information is that it must be consumed to be accepted and understood, although it is non-rivalrous and difficult to exclude other consumers.

These characteristics of information are not new; but they are a key to the acquisition and revelation of information that is so important for transactions costs and for the new economy. They raise issues about defining the property rights that are important if there is to be efficient investment and trade in information goods.

The new economy is characterised by networks. In addition to the established physical infrastructure networks, the new economy has actual and virtual networks arising out of network externalities that attend some of the new technologies. One effect of a network is that its value increases as the number of people on it increases. This effect has also been applied to new-economy products for which there are compatibility requirements. Communication of information is a cornerstone of e-commerce and e-government; the degree to which devices can communicate with each other is a feature of competition. Apple Macintosh's operating system came a distant second to the IBM personal computer (PC) because Apple chose not to open the Macintosh's architecture to any firm, and thereby limited the new-product race associated with that computer. Opening up the architecture of the PC allowed it to better compete in the new-product race and stopped PC manufacturers from being able to 'hold up' downstream users as they would be able to do if there were a single manufacturer. The consequent dominance of the PC architecture meant that it set the standard ensuring compatibility among PCs for software, hardware and communications. Network externalities then stemmed from advantages in owning a PC (relative to a Mac) because of the relatively large numbers of PCs and the attendant benefits of their general usage.

There continues to be debate in economics about the extent to which these externalities are distinctive (see Katz and Shapiro 1994 and Liebowitz and Margolis 1994). Economics has been founded on network models; indeed, economists since Adam Smith have stressed the role that prices play in conveying information in a decentralised but nevertheless coordinated economy. Nowhere is this more evident than in the modelling that is used to implement electricity spot markets, many of which are standard general equilibrium models. Further, the private sector has solved a wide variety of network externality effects in various innovative ways (Shapiro and Varian 1997, p.17). Network effects have occurred in the past and have been a critical element of economics; they may just be more extensive in the new economy.

A second element of the PC–Macintosh tale is of a product race in which the winner takes almost all. These sorts of races fall in the category of tournaments. They are an efficient mode of competition where actual and potential innovations and products are striving for economic success. Tournaments tend to work well where the prize for first place is very large relative to other prizes. The presence of network effects accentuates the prize differential because acceptance of the winning technology brings with it demands stemming from its broader utilisation. Many new-economy products can be produced very rapidly and have very low marginal costs of production: taking an extreme example, software can be reproduced almost instantaneously at virtually zero marginal cost. In addition, the dissemination costs of many of these products are very low: again taking an extreme example, new software products can be downloaded over the internet very quickly. Rapid, low-cost manufacture and delivery mean that innovative products reach consumers and firms very quickly, limiting the catch-up time for other technologies. Success in the tournament can therefore reach markets worldwide and render financial returns very quickly. Successful innovation leads to greater financial returns when products can be made fast and at low cost and when they can be quickly and cheaply disseminated across a vast global market. Moreover, network effects protect a leading position against alternative products, thereby raising the prize for the winner.

Agarwal and Gort (1999) suggested that competing products now appear in the market more quickly than in the past. They estimated that in the United States the average time between the introduction of a new product and the appearance of competitors has shrunk from thirty-three years in 1900 to only three to four years in 1986. The pace of change has almost certainly accelerated since then. For example, the software controlling some telecommunications networks is now replaced each month as new capability and standards are developed.

Several factors have allowed the increased pace of change. Skilled labour is more mobile; communication is better, so technical information and products spread more rapidly; there are more potential entering firms; and the market is larger. In the past, patents have helped people to recoup investment in successful product development; today, they offer limited protection, largely because of the pace of technological change. Today, returns (prizes) are available from dissemination across space (global markets) rather than time (through protection through patents). Agarwal and Gort (1999) concluded that the high pace of competition in the new economy is reducing the potential for regulations to be effective.

## ORGANISATION OR MARKET

The implications of new-economy changes for industry performance vary between different markets, but all industries have been affected by reductions in their costs, changes in the availability and quality of information, the creation of new markets, and the opening of competition within and between markets. [3,4] I will illustrate the following discussion of the organisational implications of the new economy with examples taken from the physical network industries, because they continue to attract a disproportionate amount of public policy analysis.

The new-economy changes have affected the optimal organisation and governance structures of network industries in two ways. First, they have challenged traditional public policy by making competition feasible where natural monopolies existed before. This has occurred through cheaper technological solutions to interconnection between networks, and the potential for competition in the provision of core facilities within networks, in the provision of new products and in the ability to create markets where none existed before. Second, the changes imply changes in transactions costs that affect the organisation of firms. There are several contributing factors.

### Changed costs

Technological change has lowered costs in networks and thereby increased the profits of incumbent firms.[5] These markets have attracted new entrants whose activities have been instrumental in the introduction of new products and lower prices for basic services. Technological change has also altered industry cost characteristics, influencing, for example, the nature and extent of any economies of scale and scope. These changes are, in turn, leading people to change their view about whether certain industries and firms act as natural monopolies.[6]

Industries and networks affected by rapid technological change may experience lower prices and an expansion in output. Substantial growth in industry output will affect the functional economies of scope defined by Stigler (1951) and produce specialisation, because functions that do not enjoy increasing returns are split off. The equipment businesses of telecommunications companies have been shed as output has grown, and this may be an example of such specialisation. Another example of output-induced structural change is infrastructure firms that contract out customer services. The Stigler argument is one of transactions costs: as output grows, performance gains can be achieved by efficiently consummating transactions to do with the core (growth) aspect of the business. Indeed, voluntary separation of vertically integrated businesses is one consequence of the benefit of focusing on a line of business and competitive pressure that limits the benefit of coordination in a vertically integrated structure.

The lowered total production costs and the use of one network by another network industry have provided new sources of competition and certain changes in economies of scope. A prominent example is the need of modern electricity companies for electronic communications networks, and the fact that if a new gas network is established the extra cost of laying cable for telecommunications is negligible. The telecommunications networks of electricity companies can be used for bypass, and thus compete with telecommunications companies. The separation of banking and insurance transaction networks is now also artificial.

## Unpredictability and competition

The timing of the arrival of new technology is unpredictable, as are its costs and characteristics. The prospect of new technology affects investment decisions by incumbent firms and by potential entrants (Choi 1994). It affects strategic decisions and decisions about the best time to invest, and it is affected by the intensity of prospective competition. Although formal analysis is complex, the more intense the competition, or the more likely is competition, the more rapidly it pays a firm to invest in new technology.[7] There are competitive dangers in waiting for additional 'good news' about new-technology investments.

Uncertainty will also affect pricing and investment and hence market performance. The uncertainty inherent in the arrival of new technology, its future cost and its future characteristics all combine to complicate investment and pricing decisions. In telecommunications, regulation is often based on a calculation of average incremental costs that depends on expectations of the future – the economic life of capital equipment, and output. However, estimates of average incremental cost and output can vary substantially, even when they are based on the same information. Furthermore, the investment decision is not just based on the calculation of expected returns and costs: as mentioned, uncertainty (for example, measured as variance) about the future also directly affects the investment decision. Investment in changed technology is often, of necessity, substantial. Thus, new technology investment decisions often entail considerable size and risk.

Cooperative behaviour among existing firms is affected by the high likelihood of the arrival of new technology and the often consequent threat of new entrants to the market. In particular, where entry is economically feasible, the arrival of technology can be viewed as a shock to established arrangements. In this environment, the strong possibility that a new technology will emerge is likely to restrict the advantages of cooperative (collusive) arrangements among firms. In competitive network industries, there is a tradeoff between cooperation – through inter-network access agreements – and competition. The prospect of technological surprises will tend to advance non-cooperative behaviour, and thence competition.

The rapid, uncertain appearance of new technology reduces a firm's strategic planning horizon for specific investments and encourages a portfolio approach to strategic investments. A shorter planning horizon is one result of not being able to anticipate new developments. A portfolio approach is one way of managing the risk, although it represents a departure from specialisation.

One common way to reconcile specialisation and portfolio options is to participate in joint ventures. Joint ventures are often explained as a response to a coordination problem or the need for standardisation (Carlton and Frankel 1995; McMillan 1997). In network industries, joint ventures probably also result from asymmetric information and complementarities within the network. For example, a private independent firm may not be able to participate in a joint venture if there is too great a risk of hold-up and other forms of opportunism. If firms entered into joint ventures to share the risk attached to large investments, the activities of the partners would usually be complementary. Efficient risk sharing usually occurs in capital markets, but joint ventures established for portfolio reasons may be economically efficient if there are functional complementarities such as internet services and content. It is no accident that joint ventures in network industries have become common or that they have formed and dissolved as technology and economic circumstances have rapidly changed. When participating companies compete in products and functions that are not part of the joint venture, the venture is especially likely to retain an internal tension that will preserve competition outside the venture.[8]

This tension is enhanced when information that firms claim as their own is essential to the joint venture.[9] Such information may be an element of competitive advantage that firms do not want to disclose. In consequence, joint ventures may not form or may be restricted in their operation. Information disclosure is an important element of joint venture electricity spot markets. In the New Zealand wholesale electricity market, firms can join the spot market or enter private bilateral contracts. If they join the spot market, they submit contractually to a set of rules that defines what is public information and what is private.[10] For example, generator offers and purchaser bids are held to be private under the contract that enables the spot market; this has been a source of tension, because some participants have argued that offers should be revealed. Instead, transformed offers are revealed in the supply curve 'offer stack' that is public information.[11]

### Asymmetric information: rapid standardisation

The nature, pace of product development, manufacture and distribution of new products – particularly in information and electronically related products – have rendered industry costs more transparent than before and reduced the problem of specialist knowledge that has always bedevilled the management of large organisations.

Newer technologies combine many functions and much power in components whose functions can be well understood by non-specialists, even if an understanding of the technology represented in the components requires very advanced knowledge. In addition, the components have become standardised, compatible with different technologies, very reliable and easy to replace. They are often sourced from various suppliers, so their characteristics are well known by all companies. Of course, specialist personnel are critical to the design and operation of network companies, but the role of technological specialists and their importance in managing the company has changed.

In the past, a knowledge of engineering was very useful for most employees in a telecommunications company, if only for one-off problem solving and communication. In part, this was because of internally produced equipment and internal, company-specific solutions to problems. Such a detailed knowledge of components is no longer required for good management.

This development has a number of implications for organisational governance. Where specialist knowledge is important for the operation of a company, there is a communication issue: certain employees know more about aspects of the business than do managers. In economists' jargon, there is an asymmetry of information that gives the specialists, perhaps at the expense of managers, more decision-making power. This affects company performance, because more time and other resources have to be allocated to the task of communication itself: for example, more employee monitoring is required. Within any organisation there must always be employees with different knowledge, but communication and monitoring are cheaper if knowledge differentials and organisation-specific knowledge do not matter so much.

The reduced importance of organisation-specific knowledge, skills and specialisms reduces transactions costs and enables executives to be drawn from a wider, more competitive pool of people; this should improve management.[12] The more vigorous competitive environment generated by technological advance will demand that chief executive officers give companies strategic directions and leadership, rather than specific direction on technical issues.

The availability of standardised, reliable, powerful, multi-task componentry affects not only competition between firms but also the decision to contract out. The reduced transactions costs of standardised componentry and less asymmetric information are likely to affect the size of the firm because they make it easier to contract out or even purchase functions in a market – for example, via business-to-business (B2B) exchanges. Such changes in organisational structure will of themselves increase competition in the affected services. There may also be a direct effect on competition, because competitors can more accurately estimate other firms' costs.[13] When firms have a better knowledge of their competitors, there is less scope and fewer incentives for incumbent firms or potential entrants to strategically misrepresent their costs.[14] An incumbent's costs depend upon past investment and may be more difficult to calculate than those of an entrant. Various scenarios are possible, but when all companies have a better knowledge of costs there will probably be greater vigour and competition in the industry. Different scenarios may arise because, where repeated interaction takes place, better knowledge can provide that basis for cooperation rather than competition per se. This issue is considered below.

### New markets and products

In the new economy, some of the most evident economic outcomes of technological change are new products and markets. Technological change relaxes existing constraints and opens up possibilities and alternatives, thereby generating competition. It has directly provided markets where none existed before: spot markets in electricity, for example, would not be economically feasible without recent developments in electronics and telecommunications.

## MARKET: COMPETITION OR OLIGOPOLISTIC POWER

I have shown that in the new economy even natural monopolies will usually have reduced power unless the monopoly is preserved under statute. This occurs because of the forces of change within the new economy and because of the ways in which information development, processing, storage and transmission affect transactions costs. The rapidly changing new economy means that market power is typically transitory, but essential to provide incentives for a dynamically efficient economy. Mergers in this environment do not generally pose issues of public policy concern, although in the United States some mergers (which entailed high switching costs and intellectual property held by an incumbent) have been prevented on the grounds that they would have lessened competition (Shapiro and Varian 1997, p.37)

Setting aside the question of monopolies, I want to discuss more deeply how reduced transactions costs affect the ability of firms to coordinate their activities (B2B exchange) and how any improved coordination affects competitive conduct. Some coordination is required for economic efficiency: even in a tournament, the economically efficient evolution of new-technology standards will generally require communication among firms.[15] I will conclude this section by considering limited evidence about B2B exchange and business-to-consumer (B2C) exchange (the final effect of commerce in the new economy).

### B2B market maker

To assess B2B exchange, it is useful to understand what determines the way in which trade is conducted – whether it is by privately negotiated agreements (perhaps through dealers or brokers) or through transactions at publicly observable prices posted on some 'market maker' exchange. Potentially, all factors that are important in contract enter the calculation. In particular, the transactions costs of search, negotiation, monitoring and enforcement all affect the ultimate determination. Private agreements will specify the exchange of goods and the concomitant exchange of money, and will provide for surety in various ways and to various extents. Typically, private agreements are not directly replaceable with an exchange that simply posts prices. Such an exchange must provide enforceable financial surety to the same extent as the private agreement if it is to duplicate B2B private or broker agreements. In short, we cannot bypass the economic principles of contracting if B2B exchanges are to provide better services than those of brokers or direct B2B search and transaction.

Rust and Hall (2001) studied the effect of introducing a monopolist market maker on search and the equilibrium of a commodity market. Businesses may purchase or sell at the market maker exchange, which covers its costs via the difference between what it pays and receives – the 'bid-ask spread' – or themselves search for the best deal. The market maker does not hold inventories; rather, it acts as an information gatekeeper. Posted prices mean that there is no negotiation by the market maker. Rust and Hall showed that, if the market maker has a lower marginal cost of processing transactions than the corresponding cost of the least efficient broker, the market maker is profitable, is economically efficient, generates higher trading volumes and reduces price dispersion. The market maker generally does not drive all brokers out of the market: some low-cost brokers remain and they limit the power of the market maker. Rust and Hall made the point that, although the market maker is a monopolist, it is disciplined by the entry of low-cost brokers, which should not be excluded from the market by regulation. Rust and Hill concluded that the entry of market makers on the World Wide Web has had unambiguously positive welfare effects, because both consumers and producers gain, and that the technology has significantly reduced the costs of market makers relative to brokers.[16]

Rust and Hall (2001) pointed out that there are market maker exchanges for metals such as aluminium ingot, and precious metals such as gold or silver (COMEX and the London Metal Exchange, respectively), and for heterogeneous products such as fish (gofish.com) or wheat and pork bellies (Chicago Mercantile Exchange). They noted that gofish posts the history of transactions and incorporates an auction for certain types of fish, that it provides the credit history on any potential buyer, that buyers obtain a line of credit with gofish, and that fish are graded for quality. These examples indicate the wide range of products that are traded on B2B exchanges: durable homogeneous goods, heterogeneous goods (where the heterogeneity is indicated) and perishable goods.

The issue of product quality leads to the second detailed examination of B2B exchanges that I shall consider. The decline of transactions costs accompanying the new economy suggests that firms will transact with more suppliers, but the evidence is that firms are working more closely with fewer suppliers (see Bakos and Brynjolfsson 1993). People have been concerned about businesses sponsoring their own exchange so they can more efficiently transact with their suppliers. The concern arises because such sponsorship could be a mechanism for tacit collusion in input markets or for buyers acting monopsonistically.[17] It is claimed that this would be facilitated by repeated transactions and the increased transparency – less asymmetric information – provided by the buyer-sponsored exchange.[18] This is a very simple statement of the issues, because the characteristics of the products and the institutional setting are also important factors. The conditions for effective tacit collusion are typically very stringent.[19] The question is whether the tacit collusion is facilitated by the lower transactions costs of the new economy that enable potentially efficient market makers to emerge.[20]

Advances in information transmission, storage and creation lower the search costs of both consumers and sellers. The claim that market makers will improve welfare is a consequence of their adoption of modern information technology: were it not for such technology, a market maker exchange would not be economic because it would have higher costs relative to the highest cost brokers. The market maker approach does not take into account the effect of the new technology across all buyers and sellers. The technology might lower brokers' costs and also provide lower-cost information as to the activities (prices) of other sellers. Furthermore, it might enable any seller to respond almost instantaneously to a rival's actions and thereby facilitate coordination. If a seller knew that its price cut would be followed instantly by price cuts by its rivals, it might engage in non-price competition or simply not compete on price (Varian 1999). The outcome would depend on various relevant circumstances, but such an environment would be no different from certain oligopoly markets (for example, markets for homogeneous products such as oil)[21] in which competition is observed. The ability to tacitly collude will be affected by the ease of entry and by the range and extent of uncertainty that attends markets in the new economy. Given the present uncertainty, it is unlike that sellers benefit from the new information technologies to such an extent that competition is inhibited.

Bakos and Brynjolfsson (1993) have explained the reduction in suppliers at the time of the introduction of a buyer-owned exchange. I use their explanation to illustrate that observed outcomes must be interpreted in the context of subtle contracting issues and the uncertainty that attends technology in the new economy. Bakos and Brynjolfsson pointed out that contracts are necessarily incomplete and that suppliers that are responsive to the variations in demands experienced by the purchaser in the new economy generally must invest in capital specific to the purchaser firm. If this is to occur, the purchaser cannot, ex post, expropriate the quasi-rent arising from these investments. A supplier will obtain more benefit from its specific investment if there are fewer suppliers, because it will have greater ex post bargaining power and therefore greater incentives to make non-contractible investments, such as those required for quality, response time and innovation. The buyer does not want to be 'locked in' to one supplier, but nor does it want many suppliers. This approach does not argue that supplier numbers should fall as transactions costs fall, but the dynamic nature of change in the new economy would suggest a fall in the number of suppliers in a competitive market – even, perhaps, with declining transactions costs.

In repeated transactions (games), even auctions, knowledge of the players (a breakdown of the anonymity presumption) and of their strategies can lead to more collusive outcomes. But the extra information must reduce uncertainty if this is to be attributed to an increase in information. Uncertainty is inimical to the coordination required for tacit or explicit collusion, so additional information is unlikely to reduce competition unless it reduces uncertainty.[22] The improved information of B2B exchanges is occurring at a time of increased competitive uncertainty, so B2B exchanges in this environment do not raise the spectre of increased tacit collusion.

## B2B and B2C: some evidence

The proportion of transactions that are conducted through B2B exchanges, whether private or public, continues to grow rapidly, but the early euphoria about public exchanges, at least on the part of investors, has evaporated. International Data Corporation carried out market research on 1,000 B2B exchanges launched between 1999 and 2001 and reported[23] that only 100 were handling any genuine transactions. Instead, active exchanges were typically supported by private enterprises owned by large firms and joint ventures formed by them. The problem with most public B2B exchanges has been that, while they enabled buyers and sellers to be matched, they did not duplicate the key contractual requirements of transactions. Indeed, the B2B exchanges run by large, well-known firms supply the reputation, liquidity and even the product and financial surety that is required for B2B to substitute for other methods of transaction.

It is not obvious that joint venture B2B exchanges will emerge as the norm. Large software companies have developed B2B exchange software that can be installed in private company-specific B2B exchanges, so the large natural monopoly element that is frequently the basis of joint ventures may not be present: the capital may be reproducible at low cost. B2B exchanges currently require broadband networks that may be specific to B2B tasks, but these networks will probably become the norm for internet usage so there will be exchanges that have no B2B-specific network characteristics. Potential joint venture firms compete at other levels and may be jealous of their proprietary intellectual capital, especially if it relates to information products, so the costs of joint venture B2B exchanges may outweigh the benefits of company-specific exchanges. If so, the new economy is no more likely than the old to exhibit tacit collusion.

As with B2B exchanges, the reduced transactions costs of the new economy imply that B2C exchanges – such as Amazon.com – will become economically efficient. The reasoning of Rust and Hall (2001) can be applied directly, although market makers will enter in competition and sellers can better monitor other sellers. The introduction of B2C exchanges will be economically efficient and can be expected to increase the volume of transactions, to lower final prices relative to wholesale prices (that is, reduce the 'bid-ask' spread) and to reduce price dispersion. In addition, the price elasticity facing any particular B2C exchange will probably be high because of the low search costs of B2C users. Electronic menu costs of price adjustment are also likely to be lower, so price may be adjusted more frequently on B2C exchanges than in traditional distribution channels.

Price studies that have compared conventional outlets with B2C have resulted in conflicting conclusions. They have been explained by the use of pre-1997 data, when there was little B2C competition and B2C was in its infancy, and by the difficulty of interpreting comparisons between heterogeneous goods. In a comparison of B2C car auction exchanges with car yard sales that involved heterogeneous goods, Smith et al (1999) found prices that seemed to reflect auction theory rather than search theory, between which there would have been a tradeoff. Brynjolfsson and Smith (2000) reported that books and CDs delivered to residences were cheaper on the B2C exchanges and that these exchanges made price adjustments in much smaller increments than conventional outlets (supporting lower B2C price-adjustment menu costs). They also found that price dispersion was generally lower on B2C exchanges, although it remained surprisingly high. They conjectured that price dispersion arose from heterogeneity in B2C brand or reputation and concomitant trust. Varian (1999) suggested that 'shopbots' – computerised search engines – affected market outcomes and may well have resulted in deliberate price dispersion (randomisation) that enabled online firms to sort among shopbot searchers and others. Loyalty programs are another device for discriminating among customers using different methods of search. Furthermore, the market maker model assumes there is a product with defined characteristics that the buyer knows and seeks to buy. In fact, people may search deeper than this. Online bookstores, for example, offer potential buyers the opportunity to search for what they want to buy, as well as the opportunity to search for a bargain. This too could affect pricing strategies.

Goolsbee (2000a) reported that variations in retail price resulting from variation in local sales tax seemed to have a large effect on consumers' online buying patterns. Goolsbee (2000b) estimated that the online purchase of computers is very sensitive to retail prices and vice versa, suggesting that there is such competition between the two activities that they are the same market.

This evidence helps in assessing the effect of B2B and B2C exchanges, but it does not measure their total effect on welfare. The presence of the exchanges will alter the behaviour of the supply chain and competing (traditional) sellers, and thereby affect the performance of traditional selling. This will occur directly through these outlets' use of new technology and indirectly through their response to the online competition. B2C and B2B exchanges are in their infancy, but their numbers are growing rapidly and they are being increasingly used. The available evidence generally supports the transactions cost argument that they are forces for improvement in economic wellbeing.

## CONCLUSIONS

In this paper I suggest that the greater availability and use of information associated with the new economy is increasing rather than reducing competition. New markets are being created, firms are being reorganised as a result of lower transactions costs and new technologies are being introduced. The combined effect is to increase the scope for competition and reduce the feasibility of cooperation. Cooperative agreements work best with low levels of uncertainty about the path of future technical change and the payoff to new investment, but the new economy has increased uncertainty on both counts. Increased information and cheaper dissemination have been accompanied by uncertain, rapid change, with higher discount rates for decision makers and less coordination. More information may reduce transactions costs, but it will increase the potential for anti-competitive cooperation only if it reduces the uncertainty associated with economic activity and the path of technical change.

Further, in many industries, reduced transactions costs are enabling increased contracting out and more decentralised decision making. This too increases competition. The increased competition should foreshadow less industry regulation, but many regulatory regimes are struggling to release industries to competition.

The increased information of the new economy, with the attendant new information and information-gathering products, should allow better assessments of competitive effects as the new economy continues to evolve.

Finally, the dynamic technological and concomitant economic changes in the new economy are resulting in the links between welfare and common static measures of market structure being much more tenuous than they would be in a more stable world. The characteristics of the new economy suggest that dynamic, not static, economic efficiency should be the yardstick of competition law.

## NOTES

The paper has benefited from the comments of Neil Quigley, Bronwyn Howell and participants in the PAFTAD 27 conference.
1   Young (1996) argued that Adam Smith's 'invisible hand' proposition (the Wealth of Nations 1776) was anticipated by the Han Dynasty historian Sima Qian, who, unlike Smith, linked the 'invisible hand' directly to prices.
2   The lowered search costs are exemplified by the organisations using computers to automatically check and sort among products available on the web to locate the best bargains.

3   The substance of this discussion is unaffected by the sources of technological change. I would argue that there is no unique source or cause, and that technological advance and diffusion are influenced by institutional arrangements. See Archibugi and Michie (1998) for some discussion of institutional and neoclassical approaches to understanding technical change.
4   By competition between markets, I mean that the gaps in the chain of substitution possibilities between goods (Tirole 1988, pp.12–13) have narrowed or vanished, leading to fewer, larger markets; an example is the convergence of modes of communication.
5   See Norsworthy and Jang (1991) for an analysis of rapid cost reduction in the computing industry and implications for the measurement of productivity.
6   For example, empirical work by Shin and Ying (1992) concluded that in telecommunications the structure of local service costs is not that of natural monopoly.
7   See Dixit and Pindyck (1994, pp.16–19), where the nature of increased competition is made specific.
8   Work on joint ventures has, however, focused on the potential for the joint venture partners to invoke rules that reduce competition and consumer welfare (Carlton and Salop 1996; Carlton and Frankel 1995; Evans and Schmalensee 1996). Economides and Salop (1992, p.107) argued that the efficiency implications of joint ownership of a firm producing an input for a network involves a trade-off between the welfare gains from vertical integration of complementary products and the losses from horizontal integration of competing products. Carlton and Salop (1996, pp.330–335) outlined three types of competitive harm that may result from the activities of a joint venture.
9   Information property rights remain an important issue for the new economy.
10  For a review of the operation and rule enforcement processes of the multilateral contract that is the New Zealand Electricity Market (NZEM) see Arnold and Evans (2001).
11  Electricity markets offer a tool for checking on the effect of subtle differences in market design. Different markets handle in different ways the fact that revealing bids or offers in a repeated game in which anonymity is hard to preserve may yield a collusive outcome.
12  See Friedlander et al (1992) for a discussion and evaluation of the importance of the background training and experience of chief executive officers for the operation of railways.
13  In telecommunications, costs can be estimated very accurately (Richard Simnett, Bellcore, personal communication).
14  The incentives and the effects of exchanging information will depend upon whether the unknown factors have a common value or private values (see Phlips 1995, Chapter 5).
15  Shapiro and Varian (1997) noted cases where communication among firms about new-product standards that have the potential to benefit consumers have led to prosecution by competition authorities in the United States.
16  Lucking-Reiley and Spulber (2001) reported that 'Lehman Brothers finds that a financial transaction is $1.27 for a teller, $0.27 for an ATM and $0.01 for an on-line transaction'.
17  The COVISINT exchange for autoparts is a joint venture of the three largest US auto makers.
18  For a discussion of lack of anonymity of bidders and the role of information in auctions see Maskin (1992); for a discussion of competition among oligopolists see Phlips (1995, Chapter 5).

19  See *Thresholds for the Scrutiny of Mergers and the Problem of Joint Dominance*, a review by Lewis Evans, Neil Quigley, Frank Mathewson and Patrick Hughes, available at http://www.iscr.org.nz.
20  Following Rust and Hall (2001), in general where there is sufficient buyer and seller competition, brokers will emerge that will discipline the monopoly market maker.
21  It is likely that the uncertainty that attends the wholesale oil market reduces the ability for companies to cooperate.
22  See Rassenti et al (2001) for experimental economics examples of the effect of demand uncertainty on coordination in electricity.
23  *The Economist*, August 2001.

## REFERENCES

Agarwal, R. and Gort, M. (1999) 'First mover advantage and the speed of competitive entry', Research Paper, New York: Department of Economics, Buffalo State University of New York.

Archibugi, D. and Michie, J. (1998) 'Technical change, growth and trade: new departures in institutional economics', *Journal of Economic Surveys*, 12, 3: 313–332.

Arnold, T. and Evans, L. (2001 ) 'Governance in the New Zealand electricity market: a law and economics perspective on enforcing obligations in a market based on a multi-lateral contract', *The Antitrust Bulletin*, 46, 3: 611–643.

Bakos, Y.J. and Brynjolfsson, E. (1993) 'From vendors to partners: information technology and incomplete contracts in buyer–supplier relationships', Center for Coordination Science Technical report, MIT School of Management.

Brynjolfsson, E. and Smith, M. (2000) 'Frictionless commerce? A comparison of internet and conventional retailers', *Management Science*, 46, 4, April: 563–585.

Carlton, D.W. and Frankel, A.S. (1995) 'The antitrust economics of credit card networks', *Antitrust Law Journal*, 63: 643–668.

Carlton, D. and Salop, S. (1996) 'You keep knocking but you can't come in: evaluating restrictions on access to input joint ventures', *Harvard Journal of Law and Technology*, 9, 2: 319–352.

Choi, J.P. (1994) 'Irreversible choice of uncertain technologies with network externalities', *RAND Journal of Economics*, 25, 3: 382–401.

Coase, R.H. (1937) 'The nature of the firm', *Economica*, 4: 386–405.

Dixit, A.K. and Pindyck, R.S. (1994) *Investment under Uncertainty*, Princeton: Princeton University Press.

Economides, N. and Salop, S.C. (1992) 'Competition and integration among complements, and network market structure,' *Journal of Industrial Economics*, 40, 1: 105–123.

Evans, D. and Schmalensee, R. (1996) 'A guide to the antitrust economics of networks', *Antitrust Magazine*, 10, Spring: 36–40.

Friedlander, A.F., Berndt, E.R. and McCullough, G. (1992) 'Governance structure, managerial characteristics, and firm performance in the deregulated rail industry', in M.N. Bailey and C. Winston (eds) *Brookings Papers on Economic Activity: Microeconomics* Brookings Institution: 95–186.

Goolsbee, A. (2000a) 'In a world without borders: the impact of taxes on internet commerce', *Quarterly Journal of Economics*, 115, 2: 561–576.

Goolsbee, A. (2000b) 'Competition in the computer industry: online versus retail', Mimeo, University of Chicago Graduate School of Business.

Katz, M. and Shapiro, C. (1994) 'Systems competition and network effects', *Journal of Economic Perspectives*, 8, 2, Spring: 93–115.

Liebowitz, S. and Margolis, S. (1994) 'Network externality: an uncommon tragedy', *Journal of Economic Perspectives*, 8, 2, Spring: 133–150.

Lucking-Reiley, D. and Spulber, D.F. (2001) 'Business to business electronic commerce', *Journal of Economic Perspectives*, 51, 1: 55–68.

McMillan, J. (1997) 'Rugby meets economics', *New Zealand Economic Papers*, 31, 1: 93–114.

Maskin, E. (1992) 'Auctions and privatisation', in H. Siebert (ed.), *Privatization: Symposium in Honor of Herbert Giersch*, Tubingen: Mohr (Siebeck): 115–136.

Norsworthy, J.R. and Jang Show-Ling. (1991) 'Productivity growth and technological change in the United States telecommunications equipment manufacturing industries', in M-A. Crew and P. Kleindorfer (eds), *Competition and Innovation in Postal Services*, Norwell, Mass: Topics in Regulatory Economics and Policy Series, and Dordrecht: Kluwer Academic: 121–139.

Phlips, L. (1995) *Competition Policy: A Game Theoretic Perspective*, Cambridge: Cambridge University Press.

Rassenti, S.J., Smith, V.L. and Wilson, B.J. (2001) 'Controlling market power and price spikes in electricity networks: demand side bidding', Mimeo, Tucson: Economic Science Laboratory, University of Arizona.

Rust, J. and Hall, G. (2001) 'Middle men versus market makers: a theory of competitive exchange', Mimeo, Department of Economics, Yale University.

Shapiro, C. and Varian, H. (1997) 'US Government information policy', Mimeo prepared for the Department of Defense.

Shin, R. and Ying, J. (1992) 'Unnatural monopolies in local telephone', *RAND Journal of Economics*, 23, 2: 171–183.

Smith, M.D., Bailey, J. and Brynjolfsson, E. (1999) 'Understanding digital markets: review and assessment', in E. Brynjolfsson and B. Kahin (eds), *Understanding the Digital Economy*, Cambridge, MA: MIT Press.

Stigler, G.J. (1951) 'The division of labour is limited by the extent of the market', *Journal of Political Economy*, LIX, 3: 185–193.

Tirole, J. (1988) *The Theory of Industrial Organization*, Cambridge, MA: MIT Press.

Varian, H. (1999) 'Market structure in the network age', paper prepared for a conference on 'Understanding the Digital Economy', Washington DC: Department of Commerce.

Young, L. (1996) 'The tao of markets: Sima Qian and the invisible hand', *Pacific Economic Review*, 1, 2: 137–45.

# 10 Internet providers: an industry study

*Haryo Aswicahyono, Titik Anas and Dionisius Ardiyanto*

INTRODUCTION

The 'network revolution' promises to make a substantial contribution to economic development in the developing world. First, the new network technology can improve economic efficiency and competitiveness. Second, it can be used for more efficient and effective education, health care and public administration. Third, it creates opportunities to exploit low factor costs in the international market. Fourth, it provides opportunities to increase social capital. Finally, new information technology creates opportunities to bypass failing domestic institutions.

The network revolution also has the potential to widen the development gap. In most developing countries, it may expose known weaknesses such as inadequate legal and commercial frameworks, shortfalls in education and knowledge, and weak network services and infrastructure. By increasing tradability in services, it will expose weak domestic services such as finance, business services, professional services and retailing to the full force of international competition. The increased volatility and instability of world markets that may accompany the network revolution also carry risks to the developing world.

This chapter addresses the question of how Indonesia can best adopt new information technologies. We argue that in Indonesia the high price of internet services is the main deterrent to the diffusion of information technology in general, and of the internet in particular. We also argue that the best way to reduce the price of the services is to make the market for internet services more competitive. By way of background, we briefly review the dynamics of the 1990s networking revolution that paved the way for the convergence of communications, information technology and the multimedia industry. We also briefly discuss the challenges and opportunities posed by the networking revolution for developing countries.

## BACKGROUND

### Opportunities

New information technology is believed to bring various opportunities to developing countries. The World Bank (2000) has listed several ways in which the internet is revolutionising business, creating opportunities for the poor, and improving governance and the delivery of government services. The Centre for International Development at Harvard University[1] has emphasised the benefits of the internet to developing countries because it creates new opportunities, eliminates barriers to development, and promotes the efficient production and delivery of goods and services.

Canning (1999) proposed three areas in which information technology brings economic benefits. First, he suggested that the productivity of service sectors that rely on information will increase, because of increases in the power of storing, computing, retrieving and processing information. Second, he suggested that the market will grow as the internet makes communications easier and that expansion of the market will, in turn, increase competition and allow production to be more geographically dispersed. Finally, he suggested that information technology may increase the rate of technological progress and diffusion, because it revolutionises education and innovation.

The internet is often said to have brought about the 'death of distance', in the sense that it overcomes physical and virtual isolation. As a result, businesses in developing countries can link into global supply chains, extend market reach, and exploit local comparative advantage (World Bank 2000). The tremendous growth of e-commerce is the clearest example of how this affects the economy. Forrester Research (an independent research firm that helps companies assess the effect of technological change) estimates that e-commerce in the Asia Pacific region will grow at 98 per cent annually, reaching US$1.6 trillion by 2004 (see Table 10.1). This growth rate is much faster than that of other regions. As a result, the Asia Pacific region's share of world-wide e-commerce is expected to increase from 8 per cent in 2000 to 24 per cent in 2004. In developed countries, most e-commerce is in the form of business-to-consumer (B2C) transactions. In Asia, by contrast, business-to-business (B2B) transactions have the most potential, in both the manufacturing and service industries. The research firm GartnerGroup[2] predicts that Asian B2B e-commerce will top US$272 billion in 2003, with overall Asian e-commerce reaching US$340 million.

The internet also offers developing countries new opportunities for trade in information-based products such as software, engineering, architecture, education/training, medicine and law products. Developing countries have a comparative advantage in software trade because of their generally lower costs of labour. An example is India's increasing share of software trade. Table 10.2 shows Indian software labour costs compared to those of other countries. Software exports increased from a mere US$4 million to US$54 in just seven years, from 1980 to 1987, an annual average growth rate of 47 per cent. The

Table 10.1 Worldwide e-commerce growth

|  | Value (US$ billion) |  | Share (%) |  | CAGR(%) |
|---|---|---|---|---|---|
|  | 2000 | 2004 | 2000 | 2004 | 2000–04 |
| *North America* | 509 | 3,456 | 78 | 51 | 47 |
| United States | 489 | 3,189 | 74 | 47 | 46 |
| Canada | 17 | 160 | 3 | 2 | 56 |
| Mexico | 3 | 107 | 0 | 2 | 102 |
| *Asia Pacific region* | 54 | 1,650 | 8 | 24 | 98 |
| Japan | 32 | 880 | 5 | 13 | 94 |
| Australia | 6 | 208 | 1 | 3 | 106 |
| Korea, Republic of | 6 | 206 | 1 | 3 | 106 |
| Taiwan | 4 | 176 | 1 | 3 | 112 |
| All other | 7 | 180 | 1 | 3 | 94 |
| *Western Europe* | 87 | 1,533 | 13 | 23 | 77 |
| Germany | 21 | 387 | 3 | 6 | 80 |
| United Kingdom | 17 | 289 | 3 | 4 | 76 |
| France | 10 | 206 | 2 | 3 | 84 |
| Italy | 7 | 142 | 1 | 2 | 82 |
| Netherlands | 7 | 98 | 1 | 1 | 72 |
| All other | 26 | 411 | 4 | 6 | 74 |
| *Latin America* | 4 | 82 | 1 | 1 | 87 |
| *Rest of world* | 3 | 69 | 0 | 1 | 85 |
| Total | 657 | 6,790 | 100 | 100 | 60 |

*Source*: Calculated from Forrester Research, Inc., electronically accessed in April 2000 (http://www.forrester.com/ER/Press/ForrFind/0,1768,0,00.html).
*Note*: CAGR = compound annual growth rate.

growth rate remained very high between 1989–90 and 1999–2000, at an average of 44 per cent per annum.

The internet has revolutionised manufacturing by streamlining product and service delivery, increasing the transparency of operations, and reducing transaction costs. All these promote efficiency. Better information flows mean that firms may reduce transaction costs by outsourcing functions that had previously been carried out expensively or poorly within the firm. Conversely, firms can increase revenues by providing functions outsourced by a larger partner. The internet has also had a big impact on the efficiency of market intermediaries: firms have better and more direct interaction with customers and suppliers, and there are fewer inefficient traditional market intermediaries.

These efficiency gains should result in improved productivity throughout the economy. However, it is hard to find evidence for this, because the internet economy is a new phenomenon. Many people believe that the recent acceleration in US productivity growth was due to the internet. In the four

*Table 10.2* Indian software labour: cost comparison (US$ '000)

|  | Switzerland | USA | Canada | UK | Ireland | Greece | India |
|---|---|---|---|---|---|---|---|
| Project leader | 74 | 54 | 39 | 39 | 43 | 24 | 23 |
| Business analyst | 74 | 38 | 36 | 37 | 36 | 28 | 21 |
| Systems analyst | 74 | 48 | 32 | 34 | 36 | 15 | 14 |
| Systems designer | 67 | 55 | 36 | 34 | 31 | 15 | 11 |
| Development programmer | 56 | 41 | 29 | 29 | 21 | 13 | 8 |
| Support programmer | 56 | 37 | 26 | 25 | 21 | 15 | 8 |
| Network analyst/designer | 67 | 49 | 32 | 31 | 26 | 15 | 14 |
| Quality assurance specialist | 71 | 50 | 28 | 33 | 29 | 15 | 14 |
| Database data analyst | 67 | 50 | 32 | 22 | 29 | 24 | 17 |
| Metrics/process specialist | 74 | 48 | 29 | 31 | – | 15 | 17 |
| Documentation/training staff | 59 | 36 | 26 | 21 | – | 15 | 8 |
| Test engineer | 59 | 47 | 25 | 24 | – | 13 | 8 |

*Source*: Adapted from Rubin (1996).

*Note*: Dashes indicate data not available.

years ending in the fourth quarter of 1996, productivity growth averaged 0.9 per cent per annum; but in the four years ending in the third quarter of 1999, it averaged 2.7 per cent per annum.[3] Productivity growth in well-measured gross domestic product (GDP) accelerated more rapidly than that in total GDP. It increased from 2.24 per cent annual growth in 1990–95 to 4.65 per cent in 1996–98 (Nordhaus 2001).

Recent news on dot.com performance in the United States portrayed a very bleak picture because of the failure of a few large dot.com companies. However, numerous large dot.coms performed relatively well. A recent report on internet economy indicators in the United States[4] stated:

> From the first to the second quarter of 2000, the total revenues associated with these dot coms increased a healthy 18.7 percent, while gross profit margin increased 1.3 percent from 33.1 percent to 34.4 percent. By contrast, a 'top-ten' list of traditional companies (chosen from retail, financial services, hardware, software, telecommunications and aerospace sectors) experienced a 5.1 percent reduction in gross margin from 34.5 percent to 29.4 percent from the first to the second quarter of 2000, indicating a possible slowdown in the overall economy.

## Threats

For developing countries, the main threat posed by the internet is the risk of exclusion for economies that do not adjust to its use. This is like a weak port infrastructure reducing the attractiveness of a country to merchandise trade. It is estimated that half the difference between Africa's manufactured exports as a share of GDP and East Asia's share could be accounted for by the weak communications networks in Africa.

*Internet providers: an industry study* 241

Developing countries must build the necessary information infrastructure. They must also extend internet usage as widely as possible, by making it more accessible and more affordable to the general public. In the following sections, we discuss the main barriers to internet use in Indonesia and the policy environment that is most likely to foster internet development there.

### Barriers to internet diffusion

Access price is the most important factor in internet diffusion, although the pace and depth of diffusion also depend on factors such as internet content and public awareness of the benefits of internet access. Figure 10.1 shows the inverse relationship between internet access costs and proliferation in different countries. The lower the costs of accessing the internet (relative to income), the faster the internet diffusion.

In 1999, Indonesia's internet access rate (relative to average income) was lower than that of Thailand, India and the Philippines but much higher than that of Malaysia and Korea. The price elasticity of internet penetration is −1.4 per cent, so a 1 per cent reduction in the access price per capita gross national product (GNP) would lead to a 1.4 per cent increase in the penetration rate.

We need to know not only the total access price but also the cost of the individual components. Table 10.3 shows dial-up internet access prices in the Asia Pacific region. Most Indonesian internet access costs are quite low compared

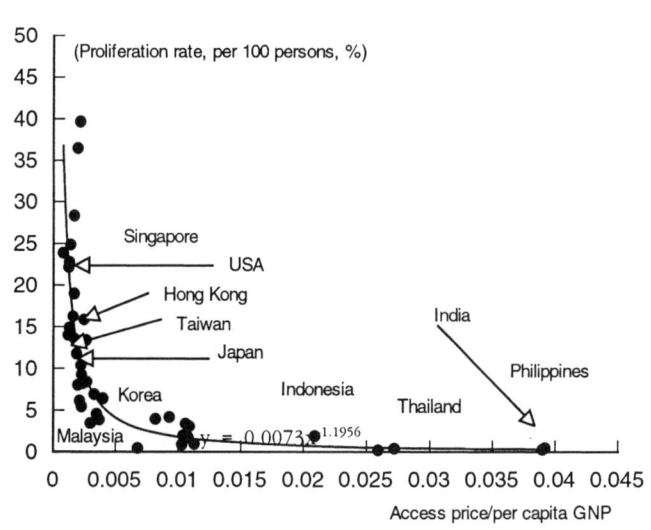

*Figure 10.1* Proliferation of the internet and access prices

*Note:* Access prices include monthly subscription charge, call charges for 20 hours, and montly line rental charge of major ISP in each country as of 1998. ($y = 0.0073x^{-1.1956}$ where y equals proliferation rate, x= access price/per capita GNP).

*Source:* ITU (1999) *Challenges to the Network*; The World Bank *World Development Indicators.*

to average access costs for lower-income groups, upper-income groups, developed countries and the Asia Pacific region. However, the telephone call charge for accessing the internet is comparable to that of developed countries – the highest in the region. It is higher than the average telephone call charge in developing countries and much higher than that of the Asia Pacific region generally. In Indonesia, about 57 per cent of internet access costs were telephone call charges.

In order to increase the internet penetration rate, Indonesia must maintain the current low total internet access charges and, if possible, reduce the high telephone call charge. In the next section, we argue that competition should be the single most important principle for telecommunications reform in Indonesia and that it can play a critical role in sectors previously thought to be natural monopolies.

## Telecommunications reform in Indonesia

The information revolution has been made possible by cheaper information transmission, increased computing power, and the shift from analogue to digital technology. Bond (1997) has estimated that transmission costs per bit per second per kilometre have declined by a factor of 10,000 in twenty years as a result of the development of fibre-optics, cheap electronics and smart wireless applications. Fibre-optic cables allow thousands of telephone conversations to be carried simultaneously, so the cost per voice call has declined considerably. Expensive electromechanical switches have been replaced by microprocessor chips, allowing existing telecommunications infrastructure to be converted into 'intelligent networks', with improved capacity and cheaper switch maintenance (Bond 1997).

The second driver of the information revolution has been cheaper and faster computer power. Computer power per dollar invested has risen by a factor of 10,000 in twenty years. Several factors contributed to this. First, integrated

*Table 10.3* Internet access costs, October 2000 (US$)

|  | Sign-up fee | Monthly fee | Free hours | Excess time | Total charge | Telephone call charge | Total charge |
|---|---|---|---|---|---|---|---|
| Indonesia | 8 | 8.46 | 15 | 6.99 | 15.45 | 21.21 | 36.66 |
| Lower-income | 23 | 17.43 | 13 | 29.09 | 46.52 | 12.37 | 58.89 |
| Upper-income | 2 | 17.49 | 7 | 3.82 | 19.35 | 23.50 | 42.85 |
| Developed |  | 21.19 | 40 |  | 21.19 | 21.01 | 42.20 |
| Asia Pacific | 15 | 18.01 | 16 | 19.44 | 36.35 | 16.33 | 52.66 |

*Source:* ITU (2000).

circuits can pack millions of transistors into a single chip. In fact, a microchip can be seen as an entire computer. Second, transistor density has increased considerably. In 1971, the Intel 4004 packed only 2,300 transistors in a single chip. By 1997, the Pentium II had already integrated 7.5 million transistors in a single chip. The extra manufacturing costs have risen only slowly, so the investment cost per unit of computer power has declined considerably (Bond 1997).

The third driver of the information revolution has been the conversion from analogue to digital technology. This development was crucial to the convergence of computer technology, communications technology and media industries into the multimedia industry; as Bond (1997) put it:

> as costs have fallen and digitalisation has replaced analog technologies in telecommunications, information technology and media industries are merging into a 'bit industry' that manipulates voice, image, video, and computer data in binary form.

Before convergence, each information service – telephone, internet or television – was delivered to end users through a different type of infrastructure. After convergence, generalised information infrastructure – whether wire-based or wireless, packet-switched, coaxial or satellite – delivered voice, pictures and data to end users.

In the past, telecommunications were viewed as a natural monopoly with economies of scale and scope, because marginal costs declined continuously and monopoly was the most efficient way to produce the service. Having such services in private hands risked the abuse of monopoly power, so governments operated telephone systems. Potential competitors were excluded on the grounds that they would wastefully duplicate existing facilities or inhibit the government's ability to provide universal services.

However, there was a marked discrepancy between theory and reality. A company that is shielded from competition has no incentive to be efficient, so the gains from economies of scale may be outweighed by the losses in efficiency. For countries with chronic under-investment, it is often argued that the losses in efficiency may be offset by high monopoly profits. However, competition seems more likely to stimulate investment, because it creates incentives to invest to meet demands: companies that do not invest will risk losing market share.

Convergence and new information technology have made existing natural monopoly models for the telecommunications industry obsolete. First, telecommunications technology has changed. Wireless, cable television and other technologies that have a flatter cost curve challenge the conventional copper wire-based technology used by existing local loop providers. With a flatter cost curve, size has no real cost advantage. Network providers can now compete without raising the network's overall cost by much. Moreover, the new information infrastructure permits competition even across different modes of signal transmission. Internet services can be delivered through telephone lines, cable television, microwave or satellite – even power lines in the not-too-distant future. Convergence also allows competition within the industry. For example, broadcasting can be challenged or complemented by narrow-casting, which provides individuals with information tailored to their needs.

## COMPETITION IN INTERNET SERVICE PROVIDER INDUSTRIES

The literature on strategic management and standard industrial organisations provides an analytical framework to assess the competitive environment faced by internet service providers (ISPs). According to Porter's standard 'five forces' model of competition (Miller and Dess 1992), the level of competition in the ISP industry depends on the threat of new entrants to the market; the bargaining power of a firm's customers; the threat of substitute products; the intensity of rivalry within the industry; and the bargaining power of a firm's suppliers.

Of course, every firm in an industry will act strategically to reduce the level of competition. They will do so by creating barriers to entry, differentiating their products, refusing to divulge strategic information to customers, undertaking vertical integration, and adopting many other strategic behaviours to raise their market power. Regulators have the opposite objective: to foster (or at least not to hinder) competition in the market by eliminating barriers to entry; fostering technological change that provides substitute products; fostering competition among suppliers, buyers and existing firms; and informing buyers about their rights. We will use this framework to assess the competitive environment in the internet access market in Indonesia.

### Threat of new entrants

The threat of new entrants depends on how easily a firm can enter the market. If new entrants can enter the market relatively easily, there will be a higher level of competition than otherwise. The most important deterrent to new entrants is economies of scale. Economies of scale exist when average production costs decline as the size of a firm increases. Economies of scale are quite low for ISPs, because their cost structure is dominated by variable cost. To increase the customer base while maintaining quality, ISPs require a fixed proportion of inputs. For example, they must maintain the ratio of bandwidth to access lines – at 5 kilobits per second (kbps) per access line – and the ratio of access lines to subscribers (20:1 for analogue lines, 15:1 for digital lines).

The size of the incumbent may give a good indication of the presence or absence of economies of scale. If an industry has large economies of scale, firm size will be uniform and the market will be dominated by few firms. In contrast, firms can choose any size they like in an industry that is characterised by constant returns to scale. Table 10.4 shows the size distribution of firms in the Indonesian ISP market: firm size varies considerably, from 12,000 subscribers of RadNet to 90,000 subscribers of Telkomnet (CastleAsia and AC Nielsen 2000), suggesting that economies of scale do not represent a barrier to new entrants.

The second deterrent to new entrants is product differentiation. If the incumbent's product is unique in the eyes of its customers, new entrants must overcome brand loyalty. Brand awareness does seem to be important in Indonesia's ISP industry, in which new entrants have been disadvantaged by lack of brand awareness. However, Linknet, a new player in the market, built its brand awareness by being the first company to offer free internet connection, resulting in rapid subscriber expansion. Having obtained its brand awareness, the company introduced a fee-paying dial-up service which maintained the quality of its services.

Other deterrents to new entrants could be their cost disadvantage (independent of size), capital requirements and the cost of a customer switching from one supplier to another. However, these do not seem to be important in deterring new entrants to the Indonesian market.

We can conclude that the threat of new entrants is therefore quite high in the ISP market. One important potential entry that may increase the level of competition in the ISP industry is foreign investment in the domestic ISP market. The Indonesian government previously tried to prevent foreign direct investment (FDI) in the multimedia industry, but revoked the relevant regulations after strong reaction from media and telecommunications observers. The temptation to protect domestic firms is ingrained in Indonesia, but it is important to adhere to the principle that government should protect competition, not competitors.

## Bargaining power of customers

ISPs cannot exercise their market power if their customers have a strong bargaining position. Current internet services are quite standard and undifferentiated; customers can easily shop around for the most favourable terms. However, Indonesia's ISPs rarely disclose to customers the quality of their services (the ratio of bandwidth to subscribers or the ratio of the number

*Table 10.4* Market share of Indonesian internet service providers

| *Internet service providers* | *Market share (%)* |
|---|---|
| Telkomnet | 25 |
| Linknet | 22 |
| Indosatnet | 14 |
| CBN | 10 |
| Indonet | 7 |
| Centrinnet | 7 |
| Wasantara | 6 |
| Radnet | 3 |
| Dnet | 3 |

*Source*: CastleAsia and AC Nielsen (2000).

of lines to the number of subscribers). In the absence of accurate information, internet services become experience goods that can be evaluated only after purchase. This may reduce the bargaining position of buyers to some extent.

The bargaining power of customers also depends on their price sensitivity. Incomes are low in Indonesia, so internet users are very sensitive to price increases. For example, home internet users limit their access time so they can enjoy low rates by not exceeding the set time limit for these low rates (typically 20 hours per month). They may also use the internet in the office or obtain cheaper rates in internet cafes through 'warnets'.[5]

Warnets have become an important ISP customer as they play an increasingly important role in providing and distributing internet access. Currently, there are around 3,000 warnet outlets in Indonesia; the number is expected to reach as many as 15,000 by the end of 2005. The telephone charges for internet connections are very expensive in Indonesia, so in the last few years people have increasingly used warnets to access the internet. Figure 10.2 shows this situation: between 1996 and 2000, user numbers grew much more than subscribers numbers, and there was an increasing ratio of users to subscribers. At the time of writing, 50 per cent of Indonesian internet users connected from warnets; this number is expected to triple in the coming five years. Warnets that are bulk buyers of internet services may be very price sensitive. Moreover, unlike millions of individual customers, it is easier for warnets to organise collective action so as to increase their bargaining position and pressure suppliers for better terms.

Standardised products, elastic pricing, low switching costs and the increasing use of warnets provide customers with relatively great bargaining power. Bargaining power can be further increased if ISP performance is accurately monitored and evaluated. This can be achieved through regulations that require ISPs to disclose information or by creating a market for ISP assessment. In developed countries, such information is typically supplied by consumer protection or rating agencies.

### Threat of substitute products

Products such as fax, mail and telephones pose little threat to the internet, because email is much cheaper and quicker than even fax. As more and more services are available through the internet (browsing, sound, pictures and data), users become more familiar with the technology and the price of internet services declines relative to other information technologies, which become less and less suitable as alternatives to the internet.

### Rivalry within the industry

With high customer bargaining power and little threat of substitute products, there is likely to be intense competition within the ISP industry, resulting in rivalry among competitors. The intensity of the rivalry is determined by the size distribution of firms in the market, the growth rate of the market, the

*Figure 10.2* Indonesia: internet subscribers and users, 1996–2000 (million)

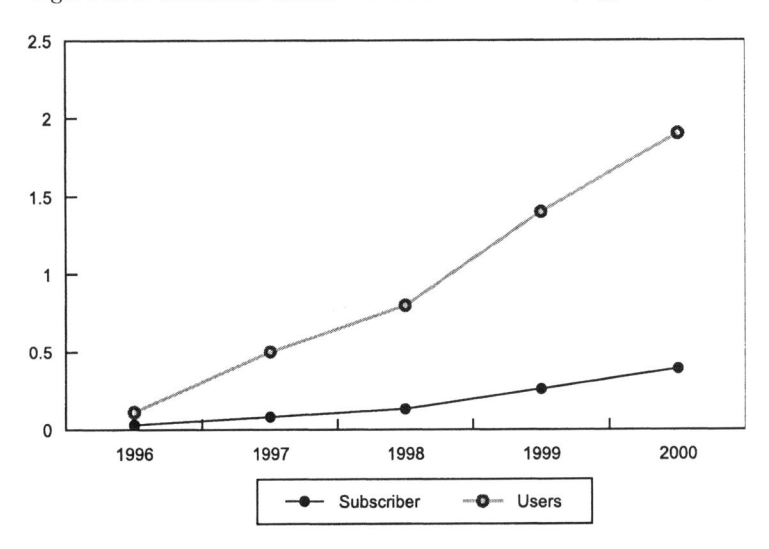

*Source*: CastleAsia and AC Nielsen (2000).

proportion of fixed costs to total costs, the ease of product differentiation and the switching costs faced by customers (Miller and Dess 1992).

In Indonesia, the ISP industry is an oligopoly in which four firms have about 72 per cent of the market. However, it is a loose oligopoly, because the firms' ability to increase market power is limited by ease of entry into the market, competition from fringe ISPs and the substantial bargaining power of users. In this environment, it is not easy for firms to exercise market power.

Three factors may reduce the intensity of rivalry among competitors in Indonesia. First, the market for internet services is expanding rapidly there. In the next five years, dial-up subscribers are forecast to increase by 117 per cent (from 227,000 in 2000 to 493,000 in 2005). Moreover, the number of cable/broadband subscribers is expected to grow from an insignificant 2,000 in 2000 to 253,000 in 2005, equivalent to 12,000 per cent growth. With rapid growth in demand, individual ISP sales can increase without taking market share away from rivals. Second, ISPs service different geographic markets. In some areas, their market share may be higher or lower than that shown in Table 10.5. Third, increases in overall ISP production capacity are possible only in large increments.

The intensity of competition can be gauged by price or quality competition among firms in an industry. In Indonesia, price competition is clearly evident from Linknet's successful attempts to gain market share. There is a positive correlation between price and quality (bandwidth per access line): the higher

Table 10.5 Indonesian internet service provider profiles

| | Bandwidth (Mbps)[a] | Ratio of number of subscribers to access lines | Phone lines | Number of subscribers | Price (Rp/hour) | Estimated access lines | Bandwidth per access line (Mbps) |
|---|---|---|---|---|---|---|---|
| Indosatnet | 70 | 16 | 3,300 | 50,000 | 5,250 | 3,125 | 22 |
| CBN | 32 | 20 | 2,600 | 37,000 | 4,000 | 1,850 | 17 |
| Idola | 18 | 12 | 400 | 5,000 | 3,000 | 417 | 43 |
| Wasantara | 5 | 20 | n.a | 20,000 | 3,150 | 1,000 | 5 |
| Radnet | 10 | 15 | 800 | 12,000 | 2,750 | 800 | 13 |
| Indonet | 8 | 20 | 1,400 | 25,000 | 4,400 | 1,250 | 6 |
| Visionnet | 4 | 9 | n.a | 1,000 | 4,600 | 111 | 36 |
| Centrinnet | 32 | 10 | 4,000 | 24,000 | 3,293 | 2,400 | 13 |
| Linknet | 30 | 8 | n.a | 80,000 | 2,750 | 10,000 | 3 |
| Telkomnet | 35 | 20 | 5,000 | 90,000 | 4,000 | 4,500 | 8 |
| Dnet | 9 | 20 | 1,000 | 12,000 | 4,000 | 600 | 15 |
| Biznet | 1 | 4 | 46 | 60 | 5,250 | 15 | 67 |

Source: Calculated from CastleAsia and AC Nielsen (2000).

Notes
a Includes monthly fixed charge. Mbps is megabits per second.

the quality, the higher the price (CastleAsia and AC Nielsen 2000). However, the correlation between price and the number of subscribers per access line is less clear (CastleAsia and AC Nielsen 2000). Some ISPs have low-quality access (with a higher than average number of subscribers per access line) but charge a high price; others have the reverse strategy, with high-quality access but a low price. This anomaly may be due to limited access to PT Telkom's lines: even if companies were willing to maintain quality by maintaining their subscriber base, they may be prevented from doing so.

We conclude that in Indonesia rivalry among ISPs is quite high and that ISPs should offer either low prices or higher quality to gain market share.

### Bargaining power of suppliers

For ISPs, the most important inputs are bandwidth and telephone lines. About 40 per cent of ISP costs are to connect to a foreign 'backbone'. In the past, the bandwidth market was dominated by two big players: Indosat and Telkom. As a result, bandwidth costs were quite high. In 2000, Indosat provided access to a foreign backbone at a cost of $42.000 per megabyte per second per month, almost four times higher than direct access via Lora and Panamsat satellites, which costs $12.000 per megabyte per second per month).[6] The regulations do not make clear whether Indosat still monopolises access to the foreign internet backbone.[7] ISPs do not seem to be aware of the changes in the status of Indosat monopoly rights to access a foreign internet backbone.

Currently, most ISPs buy direct access to the foreign backbone; however, some small ISPs form consortiums to buy bandwidth collectively and distribute it flexibly among their members according to demand. A more recent development is the launching of PalapaNet, the price of which is claimed to be 30 per cent lower than that of other backbone providers in the Asia Pacific region.

Telephone lines are another essential input for ISPs. They provide the link to the international backbone, and connections for end users. To connect to the foreign backbone, ISPs require E1 (the European format for digital transmission, similar to the North American T-1), which is in short supply; an incoming call line, which costs Rp 88.000 per month, is required to connect to end-users. As mentioned earlier, end-users pay high costs for telephone lines (around 60 per cent of total internet access costs). Given the high price elasticity of internet costs, it is clear that phone call charges are the main barrier to internet connection. Table 10.6 shows local call costs in various countries. The Indonesian telephone tariff is higher than that of the Philippines and India, but lower than that of Japan, South Korea, Taiwan and Thailand

PT Telkom currently monopolises the local loop in Indonesia, so its bargaining power is quite high. However, it will not have a monopoly after 2003.

## DRIVING FORCES FOR COMPETITION

In the previous section, we described the competitive environment faced by ISPs in Indonesia. We now turn to the regulation of intermodal competition and to competition arising from new technologies for the provision of services (intramodal competition)

*Table 10.6* Local call costs (US$/3 minutes)

| Country | Cost (US$) |
| --- | --- |
| Indonesia | 0.02 |
| India | 0.01 |
| Japan | 0.08 |
| South Korea | 0.03 |
| Malaysia | 0.02 |
| Philippines | 0.00 |
| Singapore | 0.02 |
| Taiwan | 0.05 |
| Thailand | 0.07 |

*Source*: *Koran Tempo*, 5 June 2001.

Some studies have suggested that the number of internet users is growing much more quickly than the number of conventional telephone lines. ITU (1999) showed that between 1991 and 1998 there was an 87 per cent increase in the number of internet hosts throughout the world compared with about 6 per cent for telephone lines. For Indonesia, CastleAsia and AC Nielson (2000) forecast an annual internet growth of 30 per cent in the next few years, compared with 12 per cent for telephone lines.

The huge increase in internet usage will require an adequate telecommunications infrastructure. Telephone lines will be required not only for voice communication but also for internet access. Telkom has a limited capacity to provide sufficient infrastructure. The unreliability of the current network has led to the adoption of new technologies such as microwave, fibre-optic cable, cable television and satellites for internet users who require fast connection, high-quality data transmission and cheaper prices. However, many subscribers, particularly households, will continue to rely heavily on dial-up connections. Some regions will continue to require telephone line access because they lack access to alternative technologies.

How can Indonesia accelerate the growth of fixed lines and how can it make internet access cheaper? The answer to the first question is to allow more participants into the market. The answer to the second question is to require the price of domestic local calls and connection to reflect the true cost of providing such services.

Allowing more participants into the market will break the incumbent's monopoly power. Is it possible? Smith (1995) has shown that competition is technically and economically feasible and that it is increasingly accepted by investors.

Introducing competition to the telecommunications industry should bring about lower costs, lower prices, greater innovation, increased investment and better service, particularly for countries like Indonesia whose telecommunications development is in its infancy. When the Philippines government allowed new entrants to provide local telephone services, there was a sharp increase in investment in fixed lines (Petrazzint 1996).

Introducing competition does not hurt incumbents very badly and there is little evidence that it puts universal service at risk. Previous research suggested that competition leads to an increase in network penetration and service availability. It revealed that former monopoly operators are not as vulnerable to entry of competing service providers as initially expected (Petrazzint 1996). For example, when China allowed a second carrier to enter the market, there was a dramatic improvement in the rate of network and service deployment. In fact, new entrants faced difficulties in gaining market share. For example, when Malaysia introduced competition in long distance and international services in 1993, none of the new entrants took a significant market share from the incumbent. There was a similar result in China.

## Regulation affecting intermodal competition

Competition requires effective regulation to guarantee competition in price, access and technology. Local carriers need to be free to set prices according to their cost. In a competitive market, both users and ISPs must be able to easily switch from one local carrier to another (intermodal competition) by simply dialling a certain code number. Such interconnection schemes cushion market entrants from the incumbent's competitive advantage. The Chilean model, which introduced a reasonable interconnection scheme, allowed new entrants to gain 16 per cent of the international market after only seven months of operation. Competition in technology allows innovative undertakings to benefit from their innovations.

In Indonesia, Law 36/1999 on telecommunications and Law 5/1999 on anti-monopoly and unfair business practices can be used to liberalise the telecommunications industry. Law 36/1999 will significantly change the structure of the Indonesian telecommunications industry. Fixed local wire lines will be exclusively operated by PT Telkom until 2003, but new entrants will then be welcome. In 2004, Indosat will provide domestic local services. Law 36/1999 provides that telecommunications networks and services will be provided by state-owned companies, private companies and cooperatives; that network providers can provide telecommunications services; and that network providers can use and/or rent telecommunications networks. It makes government responsible for telecommunications policy, regulation, supervision and control but allows for the establishment of an independent body responsible for regulation, supervision and control of the industry. The law provides that private firms, state-owned enterprises, regional government-owned companies and cooperatives will be able to provide fixed lines, although Telkom will have exclusive rights to provide them until 2003. Under the law, network providers will be free to negotiate all aspects of interconnection, with the government acting as arbitrator. The tariff will be determined by the market through a cost-based approach based on a government pre-determined formula (article 28).

Law 38/1999 was enacted almost a year ago but it has been poorly supported by ministerial decrees and government regulations, most of which are still being formulated.

## New technologies and intramodal competition

Convergence and technological progress have made it possible to provide internet services through mediums other than phone lines. Figure 10.3 shows how new technologies such as microwave, cable and power lines can be used to break the PT Telkom monopoly.

### *Microwave*

Microwave technology allows internet access through wide area networks (WANs), promising considerable cost savings to users. However, the installation

Figure 10.3  Technology as monopoly buster

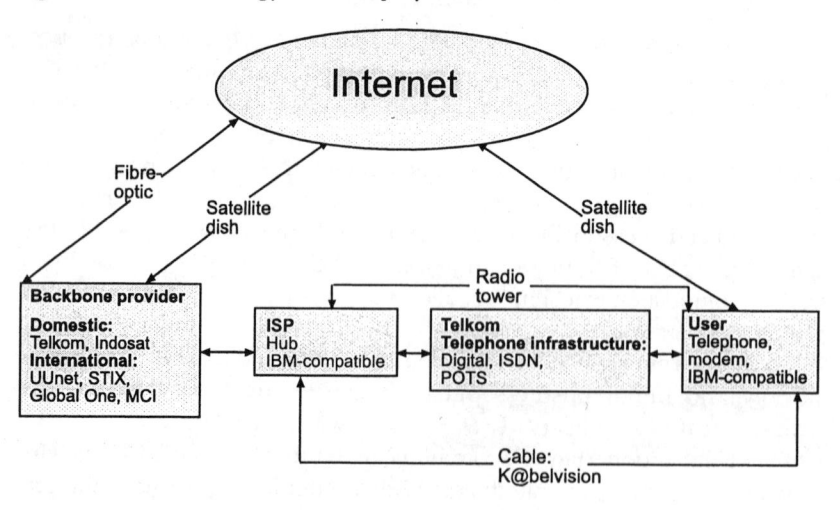

[Figure: Diagram showing Internet connected via fibre-optic, satellite dishes, and radio tower to:
- Backbone provider (Domestic: Telkom, Indosat; International: UUnet, STIX, Global One, MCI)
- ISP Hub (IBM-compatible)
- Telkom Telephone infrastructure: Digital, ISDN, POTS
- User (Telephone, modem, IBM-compatible)
- Cable: K@belvision]

Note: ISP = internet service provider.

costs are relatively high, so WANs are more suitable for high-usage customers – for example, warnets, especially in Yogyakarta, Surabaya and Bandung – than for low-usage customers.[8] WANs are currently restricted to a maximum of 2.4–2.8 gigahertz (Ghz), which may stifle internet penetration, since it makes it more difficult and costly for ISPs, warnets and schools to use alternative technology. In December 2000, the Director General of Post and Telecommunication (MOCT) regulated the use of the 2.4–2.8 Ghz frequencies. It required hub users at these frequencies to register and to pay a licence fee. Negotiations between the internet association (APJII) and MOCT resulted in a compromise that gives a win–win solution. The fee will be reduced from Rp 17 million per year to Rp 1.7 million per year; ISPs will be the main hub for the WAN; and warnets will rent bandwidth from ISPs according to demand. With this solution, only ISPs that use this microwave technology need to have the licence; warnets do not need to have the licence to use the frequency.

*Cable*

The second alternative to telephone lines is a cable network. Currently, a cable television provider (Kablevision) and six ISPs (LinkNet, Indosatnet, CBN, Centrinet, M-Web and Uninet) cooperate in a revenue-sharing scheme to provide internet access through fibre-optic and coaxial cables. This technology entirely bypasses Telkom's network, because it uses direct satellite connection for uplink and uses fibre-optic and coaxial cable for downlink. Another cable television operator, Indonusa Telemedia, a subsidiary of Telkom, also plans to provide high-speed internet connection through fibre-optic cable. Moreover, PLN (the Indonesian electricity company Perusahaan Listrik Negara) owns fibre-optic cable that could be commercialised as an internet conduit in addition to potential revenue from right of way. Its subsidiary, Indonesia Comnet Plus, is already licensed as a network operator and expects services to be available in about March 2002.[9]

*Power lines*

Digital power line technology is not yet commercially available but is a promising technology for cheap, high-speed internet access. In North America, Digital PowerLine, developed by Northern Telecom and United Utilities, can transmit data at 1 megabit per second (Mbps) over existing electricity infrastructure. However, its use has been held back because of issues about the number of users per transformer. In Indonesia, digital power line technology would be more suitable for densely populated areas such as Java than for sparsely populated areas outside Java.

## Anti-competitive behaviour

Although there is room for competition in the ISP industries, players in the industry may exhibit potentially anti-competitive behaviour. This usually (but not always) occurs if a player has a large market share and therefore a high market power, if a player controls a strategic input for other players, or if players are integrated vertically.

*Market power*

A monopoly can exploit its market power by setting prices well above the marginal cost of producing goods and services; in doing so, it extracts economic rent at the expense of consumer surplus. The ability to raise prices above marginal costs is greater when demand for the services is inelastic. Table 10.7 shows the estimated price elasticity of telephone services in Indonesia. From the first quarter of 1998 to the second quarter of 1999, demand for domestic calls became more inelastic (from –2.06 to –0.25). Clearly the current price is not a maximising profit price since the monopolist will never operate in the elastic part of the demand curve. Therefore there are immense opportunities to extract monopoly rent by raising telephone charges.

Ministerial Decree 19/2001 recently revised the formula used to calculate tariff adjustments, but the new formula was widely criticised by consumers. The main complaint was that tariff adjustment would only result in a huge monopoly profit to the incumbent. PT Telkom has made a profit but still wants to increase its price. The government is examining several options to amend the decree, but is determined that fixed line tariffs should be raised.

*Control of strategic inputs*

The law also guarantees that all users have equal opportunity to access the network at non-discriminatory prices. In the past, one ISP has been allocated

Table 10.7 Indonesia: price elasticity of telephone services

|  | Q1 1998 | Q2 1998 | Q3 1998 | Q4 1998 | Q1 1999 | Q2 1999 |
|---|---|---|---|---|---|---|
| Price elasticity | −2.06 | −0.56 | −0.41 | −0.32 | −0.13 | −0.25 |

Source: Prabowo (2000).

telephone lines when another has not or PT Telkom has given one ISP but not another a discounted rate on leased lines. The implementation of non-discriminatory principles will require an independent body with the power to impose administrative charges in cases of violations or the establishment of some other transparent means of allocating limited resources through auction.

### Vertical integration

If an ISP is a subsidiary of a fixed line provider, it may have an unfair advantage over its rivals. This can occur in two ways. First, the fixed line provider may subsidise its subsidiary so that the vertically integrated ISP enjoys cost advantages over its competitors. A recent conflict between APJII (Asosiasi Penyelenggara Jasa Internet Indonesia) and Telkomnet in Indonesia illustrates this possibility. PT Telkom's ISP, Telkomnet, announced that it would not raise its price when the telephone charge was increased. APJII alleged that this was because PT Telkom intended to implicitly subsidise Telkomnet. It argued that, if the telephone charge were raised, Telkomnet would be unlikely to remain profitable with the current ISP charge. APJII accused Telkom of practising price discrimination and threatened to ban PT Telkom from using IIX, a domestic backbone under APJII's management.

The second way in which vertical integration can disadvantage competitors is if a fixed line provider raises the telephone charge to its subsidiaries' competitors. On April 2001, PT Telkom in Kalimantan announced that for incoming calls it would raise the ISP telephone charge from Rp 35.000 to Rp 300.000. This would be anti-competitive behaviour unless Telkom paid the same tariff.

### Preventing access to alternative technology

Government may hinder the development of substitute products by preventing the use of alternative technology or setting a high licensing fee for using the technology. As discussed earlier, the attempt to set exorbitant fees for using WANs may discourage firms from using substitutes for fixed lines. The ban on voice over internet protocol (VOIP) in the past can also be seen as non-neutral technology policy that protects the incumbent fixed line provider. Landing rights for satellite transmission, rights of way in deploying cable, and technical standards can also be used to stifle competition.

## Preventing entry

Finally, FDI can be prohibited for nationalistic reasons. For example, Indonesia previously prohibited FDI in multimedia. It is possible also to partially discourage FDI by imposing unnecessary costs through divestment and joint venture with domestic partner requirements.

## CONCLUSION

In Indonesia, the uncompetitive fixed-line telecommunications industry has made internet access expensive. Statistical information and analysis of the forces that affect competition in the ISP industry suggest that PT Telkom's monopoly power has led to high telephone call charges, making internet services expensive for both ISPs and consumers. Indonesian telephone call charges are higher than those of most other Asian countries.

We suggest that internet access could be made cheaper if there was competition in the telecommunications industry, particularly for the domestic local loop connection. This could be achieved by allowing more players into the market, eliminating the incumbent's monopoly power and eventually reducing prices and providing better quality service. The new Indonesian telecommunications law includes provisions for competition, but there have been no major telecommunications developments since the law was enacted almost a year ago.

## NOTES

1 'Readiness for the networked world: a guide for developing countries', http://www.cid.harvard.edu/ciditg/resources/guide.pdf.
2 Paul A. Greenberg, 'Asian B2B E-Commerce To Top $272B By 2003', in Ecommercetimes, 14 January 2000, available at http://www.ecommercetimes.com/perl/story/2220.html.
3 Alan S. Binder (2000) 'The internet and the new economy', Brookings Institution Policy Brief 60, available at http://www.brook.edu/comm/policybriefs/pb060/pb60.htm.
4 *Internet Economy Indicators*, 23 October 2001 (see www.internetindicators.com/jan_2001.pdf).
5 Warnets are basically internet kiosks. A warnet could give a cheaper price for internet connection because it has economies of scale in producing internet connection. In warnets, an internet connection from one telephone line can be used to serve ten people using the internet simultaneously. Some warnets are beginning to adopt the latest non-dial-up technology such as microwave or internet cable, but most Indonesian warnets are still using dial-up technology.
6 Kontan-online at http://www.kontan-online.com/05/10/bisnis/bns1.htm.
7 Law No 36/1999 on telecommunication. The law suggested that entry is open. However, it also indicated that special rights that had been given to a particular company remained in place until the granted due date, unless the government and the particular company agreed to terminate them.
8 Bisnis Indonesia, 29 January 2001.
9 Hukumonline, 31 May 2001.

## REFERENCES

Bond, J. (1997) 'The drivers of the information revolution: cost, computing power, and convergence', Public Policy for the Private Sector No. 118.

Canning, D. (1999) 'Telecommunications, information technology and economic development', Consulting Assistance on Economic Reform Discussion Paper 53, available at http://www.hiid.harvard.edu/projects/caer/pubs.html.

CastleAsia and AC Nielsen (2000) *The Internet in Indonesia,* CastleAsia and AC Nielsen, Indonesia.

ITU (International Telecommunication Union) (1999) *Challenges to the Network: Internet for Development,* Geneva: ITU.

Miller, A. and Dess, G.G. (1992) *Strategic Management,* McGraw-Hill.

Nordhaus, W.D. (2001) 'Productivity growth and the new economy', Cambridge, MA: National Bureau of Economic Research Working Paper 8096.

Petrazzint, Ben (1996) 'Competition in telecoms: implication for universal service and employment', World Bank, Public Policy for Private Sector, No 96, 1996. Available at http://www.worldbank.org/html/fpd/notes.

Prabowo, A. (2000) *Telecommunication Sector: Entering a New Paradigm,* Danareksa, Indonesia.

Rubin, H.A. (1996) *Worldwide Benchmark Project,* Pound Ridge, NY: Rubin Systems.

Smith, Peter (1995), 'End of the line for the local loop monopoly? Technology, competition and investment in telecom networks', World Bank, Public Policy for the Private Sector, No. 63, Available at http://www.worldbank.org/html/fpd/notes.

World Bank (Global Information and Communication Technologies Department) (2000) *The Networking Revolution: Opportunities and Challenges for Developing Countries,* The World Bank Group. Available at http://www.infodev.org/library/working.htm.

# 11 Automobiles: an industry study

*Roger Farrell and Christopher Findlay*

## INTRODUCTION

Rapid change is occurring in the regional patterns of trade in the automotive industry. There is evidence, for example, of Association of Southeast Asian Nations (ASEAN) export success into markets in Japan, not only for motorcycles and standard components, but also for complex components (Farrell and Findlay 2001a). Another significant change is the emergence of China as a competitor in those markets. These results illustrate the scope for market access in Japan for developing country suppliers, such as those based in ASEAN.

Farrell and Findlay (2001a) have pointed out that there is interesting further work to be done on the management of these international transactions. One issue is whether the forms of the transaction (for example, arm's length versus another intermediate form) vary significantly according to product type or the characteristics of the buyer, and Farrell and Findlay (2001a) also stressed that the management of the procurement process is changing rapidly, with greater use of new information and communication technologies (ICTs).

The emergence of global electronic markets and online auctions in the automotive sector has been a commonly cited example of the new applications of ICTs. Farrell and Findlay (2001a) identified the following priority topics for further work: the scope for producers based in developing economies to benefit from the diffusion of ICTs; the application of ICTs in the trading relationships with developed economy markets, like those in Japan; and the manner in which the application of ICTs might vary between product type and between type of purchaser. It is these questions which are taken up here.

We begin with a broad discussion of changes in the organisation of the automotive industry and the requirements for ICT application. We then review the different ways in which ICTs may affect the automotive sector and the different models that may emerge. We note some key differences between US and Japanese approaches to the application of ICTs in this sector. Finally, we examine the implications for developing country suppliers of automotive components.

## 'BUILD-TO-ORDER'

Vehicles will acquire new features as a result of the application of ICTs. Cars will become 'informative and intelligent'. Toyota, for example, has indicated that it intends to 'transform the automobile and its chief role of providing transportation into a mobile information processing and communications platform...by equipping it with information functions' (Toyota 2000). General Motors (GM) has a similar goal, using its OnStar service, and Nissan chief executive Carlos Ghosn recently announced that Nissan would wire its luxury vehicles to the internet.[1]

These are all interesting possibilities, but do consumers want them? The question of how to use ICTs to apply information about customer preferences to the management of automotive production and distribution systems is even more important than using ICTs to design new features. In this respect, the following goal of Toyota illustrates the impact of information technology (IT) in the industry:

> All of Toyota's operational processes from sales to production, distribution, procurement and development should fully utilise IT and be seamlessly integrated into one organic whole. This makes it possible to grasp what a customer needs and wants and to provide products and services that perfectly meet any request. Obtaining this type of one-to-one business model is our goal, and one that can only be achieved using IT to create a 'market connected environment'. Accomplishing this will serve as the completing link in Toyota's 'value chain management'. (Toyota 2000)

Helper and MacDuffie (2001) outline the way in which this approach, also called 'build-to-order' (BTO), will change the nature of businesses in the industry and the linkages between them. They suggest that the use of BTO will lead to the greater use of modular designs, which contain standardised or common parts that can be substituted across models; a larger role for suppliers in designing, building, delivering and even installing modular parts; and a new role for dealers in providing a conduit for information, providing samples for test drives and providing a contact point for complementary activities, such as servicing (which could be relatively high margin activities compared to production and sales of new vehicles).

The modularisation of parts occurs when parts suppliers assemble individual parts into large units for supply to vehicle assemblers. The aim is to reduce the number of parts in assembly, lower the cost of parts assembly, and improve assembly times while maintaining quality. The major Western automobile manufacturers are also moving towards a greater reliance on more modular parts that can be supplied across model and platform types and are accepted by different original equipment manufacturers (OEMs). GM, for example, is aiming to reduce the number of its suppliers from several thousand to about 600. Its plan to move to more modular parts and components is intended to reduce parts costs by 30 per cent, with in-house production falling from 70 to 40 per cent of total production (Takayasu and Toyama 1997).

When an assembler designates a primary supplier to provide modularised parts, the supplier is assured of large-volume orders. A primary supplier company must be able to participate in the assembler's development and design process from the platform design stage. The company typically needs to have the capacity to control its own suppliers and to supply parts on a global basis.

The consequence is that the supply model in which an OEM is supplied by firms specialising in particular components is being replaced by one in which major component firms are becoming module suppliers to the OEMs. Both special and common components are put into modules, which are becoming increasingly sophisticated. The module suppliers, rather than the OEMs, will be procuring the individual components, and will be making the choices about how to do so.

As this process occurs, the OEMs will shift to focus on the design of vehicles while module makers concentrate on research and development, within the framework of current technology, and on the processes of production. As we noted already, the competitiveness of the OEM will depend on its capacity to capture and process information on consumer preferences. It will also depend on providing consumers with sufficient options, while maintaining the scope to gain from large-scale production.

Helper and MacDuffie (2001) paint a picture of 'the most radical outcome' for the industry:[2]

> With a new dominant design built around fuel cells creating an opportunity for a full modular design, OEMs eager to shrink their asset base and diversify their risk could outsource much design to suppliers and virtually all production to contract assemblers. Automakers could then focus solely on determining the over-arching or meta-design rules that would guide a modular product architecture, on developing and extending their brands, on differentiating their product line with respect to customisation and on developing and personalising a full array of 'mobility services'. (p.3)

The application of information technology may also be associated with changes in the distribution system. Morita and Nishimura (2000) consider that there appear to be three main strategic patterns: firstly, the consolidation of dealer networks by manufacturers; secondly, dealer consolidations by retailers; and thirdly, the emergence of 'infomediaries'.[3] Helper and MacDuffie (2001) argue, however, that dealers will continue to be linked to particular OEMs because of the value of managing and collecting the information on consumer preferences.

The application of ICT is critical to these developments. That technology facilitates the integration of a number of functions, including electronic data interchange (EDI), supply chain management methods and electronic payment systems through the internet. This is possible even for relatively small factories.

The new methods can be applied in design and production processes, supply networks and purchasing of parts and components, communication within and between industry participants, and the marketing and sale of vehicles and other equipment. The options are summarised in Table 11.1.

PricewaterhouseCoopers (2000) suggest that, as a consequence of the changes related to e-commerce, the cost of a new car could be reduced by $US2,000 over the next few years. The emergence of online buying exchanges and online retailing has the potential to create car production savings, they argue. Manufacturers can use lower cost suppliers, and administrative and purchasing costs fall. A.T. Kearney (2000), referring to examples of its clients, reported a 50 per cent reduction in transactions costs per purchase order; a 3–8 per cent reduction in purchasing cost price due to discounts induced by competition through online procurement; a 12–20 per cent reduction in purchasing price through consortium buying; a 50–70 per cent reduction in order cycle time; and a doubling (from 40 per cent to 80 per cent) of compliance with purchasing agreements.

E-commerce applications are therefore claimed to lower procurement prices. However, price competition, and the scope for purchasers to gain from the use of online auctions, is likely to be greater for generic products than for those which are specific to particular OEMs. In the automobile industry, the use of generic parts remains relatively low. About 5 per cent of the cost of inputs in a car are estimated to be for purely generic parts (Table 11.2). This share is much lower for an automobile than, for example, for a computer.

*Table 11.1* Introduction of information technology into the Japanese automobile industry (multiple responses)

| Area of introduced IT | Share (per cent) |
|---|---|
| Procurement | 41.8 |
| Sales | 48.8 |
| Basic research | 8.5 |
| Applied research | 15.0 |
| Distribution | 34.0 |
| Manufacturing control[a] | 60.1 |
| Managerial and control, including accounting | 75.2 |
| For information gathering | 60.8 |
| Sharing of information | 52.9 |
| Other | 4.6 |

*Source*: JBIC (Japan Bank for International Cooperation) FY 2000 Survey (2000), 'The Outlook of Japanese Foreign Direct Investment' *JBIC, 12th Annual Survey*, October.

Note
a Computer-aided design (CAD) and computer-aided manufacturing (CAM).

*Table 11.2* Types of components in a car and a computer (per cent)

| Share of cost | Japanese car | Dell computer |
|---|---|---|
| Original equipment manufacturers | 30 | |
| Specialist parts manufacturers | 60 | 10 |
| Contract supply | 5 | 10 |
| Outside suppliers | 5 | 80 |

*Source*: Fujitsu Research Institute, fieldwork interviews.

Dell's use of the internet to sell computers is a good example of how commodity-like products can be assembled and marketed. In contrast, the automobile is not yet a generic product. A bicycle is widely regarded as a generic product because all the parts can be sourced from various suppliers and then assembled, but a motorcycle is built from many specialised parts and components that involve proprietary knowledge of the manufacturer. Trucks (including commercial vehicles), on the other hand, use more combinations of standard components than passenger vehicles, because of the design of their body and frame and the manner in which they are combined.

An open procurement system is more relevant to generic parts than to specific parts, but the extent to which parts are generic currently varies between types of vehicles. If the gains from the application of IT are as significant as the industry reports suggest, there is an incentive to re-design automobiles and their production systems to increase the extent of standardisation – as long as consumer preferences for variety can still be satisfied. Doing so might be easier in some sub-markets than others. When consumers value effectiveness more highly than appearance or 'feel' then standardisation might be easier to adopt.

## GETTING THERE

The previous section outlined the directions of change in the evolution of relationships between firms in the automotive sector. The role of ICTs in that evolution was stressed. The next task is to focus on the detail of the ways in which this new structure and division of roles could be established. White (2001) lists seven requirements for the ICT systems that must be applied. He says there must be computer networks to enable people-to-machine and machine-to-machine communication; data and application servers that store information and programming instructions; and software applications that activate it. There must be web-based tools and browsers that provide global accessibility; digital identity and access control software to ensure that only the right people are authorised to execute transactions; and auditing and tracking tools to document business transactions and provide an audit trail. Finally, there must be a wide assortment of tools, middleware and interfaces that make it all work together.

The public internet offers an infrastructure in which specialist software suppliers might provide these services. At present, however, the public internet has some shortcomings in security, reliability and performance. Various groups of firms in the industry have cooperated to build their own e-marketplaces to overcome these problems.

White (2001) observed that e-marketplaces have generally started by focusing on basic services, such as product catalogues and auctions. He argued that these activities are less time intensive than others and that the impact of downtimes or interruptions is tolerable because of the savings possible compared to paper systems. However, he argued that the longer-run targets of these systems will be to introduce new activities such as collaborative design, which requires managing large files with security. These initiatives aim to preserve internet features but in a more controlled environment.

A.T. Kearney (2000) review a number of options for the development of these e-marketplaces. OEM groups could create some marketplaces and large suppliers could decide to create their own marketplace. However, a bigger issue is whether to create a proprietary marketplace, a collaborative one (a group of OEMs with fixed membership) or an open one. Further variations relate to the coverage of the marketplaces – commodity-specific, industry-specific, or cross-industry.

Figure 11.1 summarises the discussion so far. It shows the different sorts of firms in the industry (OEMs, module suppliers and various component suppliers) and the mechanisms through which they might be linked. The module suppliers appear between the component makers and the OEMs. Component makers can be generic parts makers or specialists; the former are more likely to be involved in online auctions. All elements of Figure 11.1 are connected by the IT systems which eventually run throughout the automobile life-cycle.

The use of e-marketplace services will also vary, depending on the strategies adopted in vehicle design and production. OEMs that require modules specific to their vehicles will seek a lot of interaction with suppliers, and services with internet-like capabilities will make this information exchange easier. Module makers who are responding to OEM-specific requests might also be seeking the production of specialised components. These sorts of firms will make more use of the system for sharing information and cooperative design. They will make relatively less use of the service for procurement, for example, via online auctions.

Using the system for information sharing can be more demanding in terms of service quality, and these high-level applications may appear later in the cycle of the development of these e-marketplaces. OEMs and module makers who put more stress on those information-sharing applications may therefore be slower to abandon their proprietary systems.

Such differences in perspectives on priorities make it more difficult for all members of an industry to agree on the design of a system and its applications.

*Figure 11.1* Automobile industry relationships

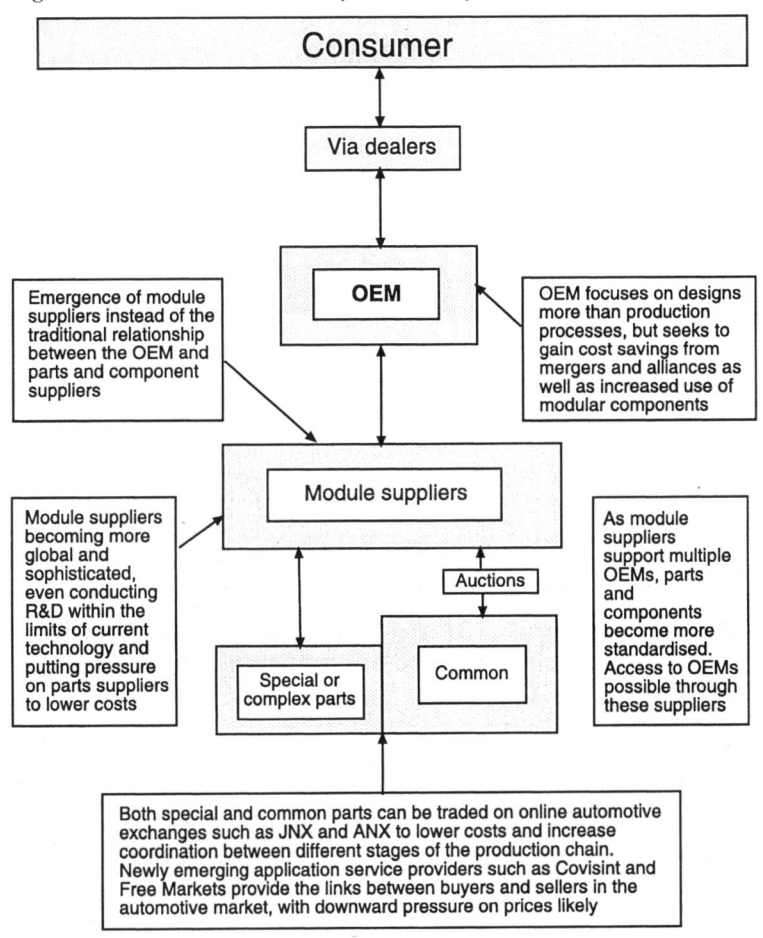

```
                          ┌─────────────────────┐
                          │      Consumer       │
                          └─────────────────────┘
                                    ▲
                                    │
                          ┌─────────────────┐
                          │   Via dealers   │
                          └─────────────────┘
                                    ▲
                                    │
                          ┌─────────────┐
                          │     OEM     │
                          └─────────────┘
```

- Emergence of module suppliers instead of the traditional relationship between the OEM and parts and component suppliers
- OEM focuses on designs more than production processes, but seeks to gain cost savings from mergers and alliances as well as increased use of modular components

**Module suppliers**

- Module suppliers becoming more global and sophisticated, even conducting R&D within the limits of current technology and putting pressure on parts suppliers to lower costs
- As module suppliers support multiple OEMs, parts and components become more standardised. Access to OEMs possible through these suppliers

Auctions: Special or complex parts | Common

Both special and common parts can be traded on online automotive exchanges such as JNX and ANX to lower costs and increase coordination between different stages of the production chain. Newly emerging application service providers such as Covisint and Free Markets provide the links between buyers and sellers in the automotive market, with downward pressure on prices likely

*Note*: OEM = original equipment manufacturer; JNX = Japanese automotive Network exchange; ANX = Automotive Network exchange.

Cooperation is valuable at this stage because of the shortcomings of the public internet. The difference in approach may be one reason why the e-marketplace has been slower to develop in Japan than in the United States. These differences are now outlined.

In the United States, OEMs have cooperated and collaborated on e-commerce initiatives, in the recognition that cost savings will be large for all competitors. These savings are expected to flow from linking their supply chains and distribution channels to e-commerce networks, from agreeing on common standards for the communications infrastructure and by creating joint ventures to improve the marketing of both vehicles and parts and components. These initiatives include the Automotive Network eXchange (ANX); Covisint (a joint venture marketplace for OEMs and their suppliers); dealer networks; and after-market initiatives (Jahn 2001).

ANX is a private virtual network that connects a large number of firms in the industry. Currently, there are almost 600 companies connected to it in the United States, including the 'Big Three' US OEMs and 75 per cent of automotive suppliers. Covisint is a procurement portal for the automotive supply chain that provides support for the marketing of goods and services in the industry.[4] Covisint will use both ANX and the public internet to enable communication between trading partners at different levels of confidentiality. White (2001, p.42) reported that the vision of Covisint is to 'have Internet technologies provide industry-wide, online services that span the life-cycle of a motor vehicle – from planning and design, through to procurement, supply chain management and distribution'. Helper and MacDuffie (2001) pointed out that even if this vision is not completely realised, and if BTO is not completely implemented, the online auction functions will still generate significant cost savings.

E-commerce complements the lean production systems pioneered by Japanese OEMs, as it can improve just-in-time (JIT) production and delivery of parts to assembly plants. However, Japan, has different patterns of corporate organisation than those of the United States. There has been a closer connection between OEMs and their suppliers in Japan, with each partner closely involved in the planning and design of a vehicle.[5]

For companies such as Toyota and Honda, an internet-based market for the common supply of parts to assemblers could disrupt traditional supplier–assembler relationships as well as risk proprietary knowledge about design, materials and process manufacturing. While the US companies have adopted a more global strategy for the procurement and supply of parts, firms in the Japanese industry rely to a greater extent on the already established infrastructure of supplying industries. Keller (1997, p.5) notes that:

> Japanese [producers] also have benefited from very close relationships with suppliers. In fact they were able to get new cars on the market every four years, mainly because their suppliers were linked to the automobile company in familial relationships that entrusted the supplier to a great deal of the engineering work for the manufacturer. In effect, the Japanese shifted a lot of their fixed costs onto their suppliers and became variable cost assemblers.

The traditional proprietary system also has its supporters. Katoh (2000) observed that:

> [b]y skillfully using the close relationship with their business partners (the keiretsu), the Japanese companies have not been bested by the Americans, but instead

have made a Business-to-Business/Business-to-Consumer Information System, which in some respects is even better.

Even so, new ICTs offer Japanese OEMs opportunities other than online auctions. Japanese OEMs and components manufacturers are using e-commerce links to streamline procurement and reduce costs in the supply chain. Toyota has converted about 70 per cent of its kanban JIT[6] ordering system to an online system that automates the process. This system has been tested with major suppliers, such as Denso, and Aisin Yokohama Rubber, the tyre maker, has established e-commerce procedures to connect its dealers with manufacturing lines in order to reduce inventories and shorten supply lead times.[7]

The Japanese automotive Network eXchange (JNX) was launched in October 2000 as a standard network for the car industry in Japan.[8] Its aim is to connect all auto manufacturers and parts makers via a secure 'extranet', or online marketplace, in which participants are able to efficiently exchange design information and negotiate prices.[9] Previously, each auto manufacturer has had its own transmission circuits and terminals for each of its suppliers. Under JNX, however, participants can use standardised communication methods so that they can communicate with everyone else. JNX is also expected to ultimately hook up with counterparts in America and Europe – ANX and the European Network eXchange (ENX) – to form a Global Exchange Network, which will make it even easier for auto makers to conduct business with their suppliers throughout the world.[10]

Nevertheless, there is a lack of agreement about JNX in the Japanese automotive industry. Interviews in the industry in Japan suggest that there is resistance to the wide use of JNX. A common system has been agreed and technical problems have been resolved, but 'business issues' between companies remain the main obstacle to the success of such an exchange, which would operate outside traditional networks between assemblers and suppliers. Some companies with their own internal EDI systems are not likely to replace them by external procurement of parts, components or services. As White (2001) remarks, 'getting separate technologies to work together can be a lot easier than reaching agreement among competing stakeholders' (p.42).

## OPTIONS FOR DEVELOPING ECONOMY SUPPLIERS

Suppliers based in developing economies may seek to communicate with the e-marketplaces and procurement systems set up by the OEMs or the module makers, but what capacities exist for that purpose?

Figure 11.2, taken from Sidorenko and Findlay (2001), shows three groups of indicators of the capacity to apply ICTs for procurement and sales. The top three rays of the star indicate key economic variables, namely population density, GDP per capita and the services share of GDP. On the right hand

266  *The New Economy in East Asia and the Pacific*

*Figure 11.2* Overall assessment: snapshot of the Asia Pacific economies (see text for explanation of shapes)

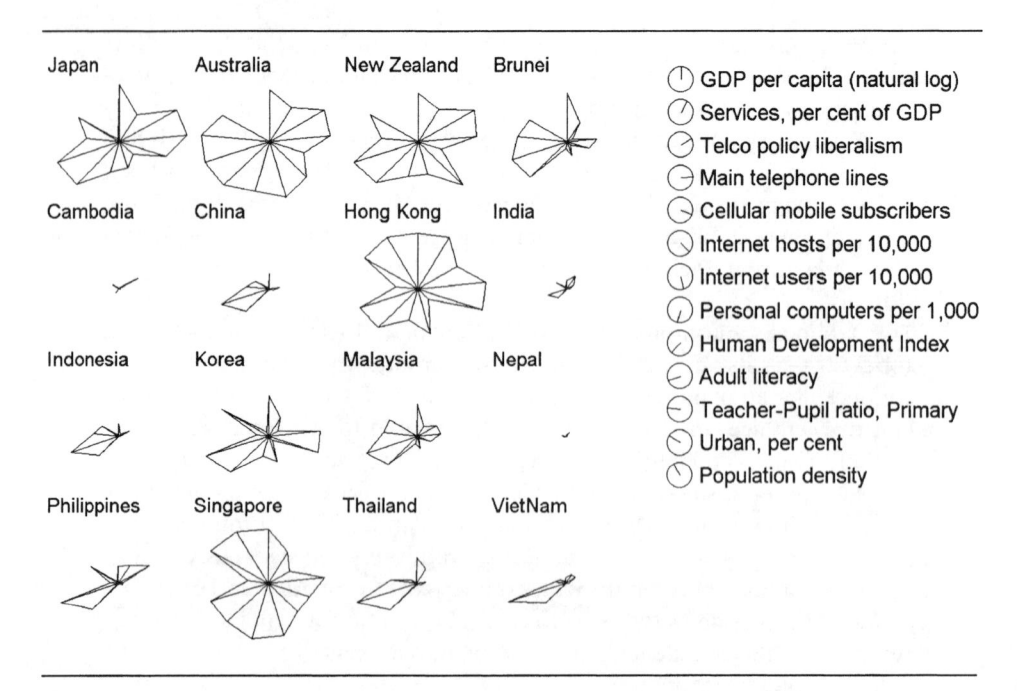

*Source*: Sidorenko and Findlay (2001).

side are some telecommunications and IT indicators, including telecommunications penetration rates, and computer and internet accessibility. On the left-hand side are a series of population and education indicators, including adult literacy, teacher:pupil ratio in primary education, the share of the urban population in the total, and population density.

Higher scores for all these variables are expected to be associated with greater capacity to make more effective use of ICT. A 'missing element' in the star graph for the country visually suggests that this indicator is relatively low in that economy compared to the average across the countries.

The size of the shapes in Figure 11.2 indicates which economies can be classified as more developed in terms of access to and use of ICTs. There are significant differences in capacity between Japan or Australia and the group that includes both China and the ASEAN economies (other than Singapore). If these capacities are important to the management of trade in the new economy, then the outlier economies appear to be at a disadvantage.

To examine the significance of these differences, Sidorenko and Findlay (2001) also applied cluster analysis to the data. They used the Euclidean measure of dissimilarity; the distance between two groups is the distance between the farthest members. The vertical distance along the y axis of Figure 11.3 represents a dissimilarity gap between the identified groups. Sidorenko and Findlay (2001) made four important observations about these data. First, Hong Kong and Singapore appear to be fairly similar to each other, but they are very different from the rest of the sample. Second, Japan and Korea are also similar, and they form a group with Australia and New Zealand. These two collections are however very different from other economies in the sample. Third, Brunei and Malaysia form a small similar group. Finally, differences among the rest of the economies, including the other ASEAN economies as well as China, are not large compared to their difference as a group from the more advanced countries.

The indicators chosen for this graphic presentation are not necessarily the best to characterise the degree of readiness for the new economy. APEC (2000, November) contains a much more elaborate set of indicators pertaining to the innovation system, human resource development, infrastructure and the regulatory environment. APEC used these to compare the APEC member economies and classify them into four groups. Furthermore, the results of Sidorenko and Findlay (2001) refer to national averages, which conceal important variations within economies: in China, for example, some coastal urban cities may report a status much closer to that of the city states of Singapore and Hong Kong. Nevertheless, the indicators reported above present a handy summary of some of the information available.

One option for suppliers from developing economies to participate in the global procurement strategies of the automobile companies is to construct new local e-marketplaces. These could operate at a number of levels. The entry level would provide information, help to establish a reputation (for example, past history, standards certification, statements from customers) and provide catalogues. At the next level, orders by purchasers could be inspected, and workflow control and billing and payment systems established. Online auctions are another option at this level. The final stage could involve collaborative product development and further integration of the supply chain.

These systems would most likely be established industry-wide, to share investment costs and the costs of managing content, and to reduce the costs of connecting all the interested parties, especially given the state of the existing infrastructure, as indicated in Figures 11.2 and 11.3. A significant proportion of the transactions might actually take place within ASEAN countries or China, for example.

The industry in these economies may decide not to build this marketplace, because the cost is higher than in other economies, as suggested by the indicators in Figures 11.2 and 11.3. Or its members may be unable to cooperate at sufficient scale to do so.

However, this does not mean that these suppliers will be excluded from the world market for components. Another interface could be set up between

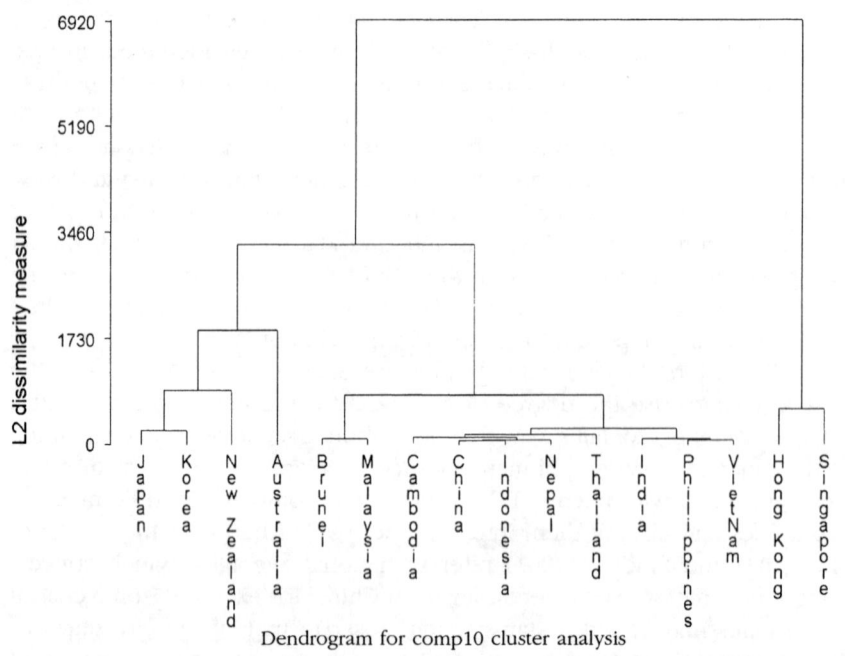

Figure 11.3 Cluster analysis of the Asia Pacific economies

[Dendrogram showing L2 dissimilarity measure on y-axis (0, 1730, 3460, 5190, 6920) for economies: Japan, Korea, New Zealand, Australia, Brunei, Malaysia, Cambodia, China, Indonesia, Nepal, Thailand, India, Philippines, Vietnam, Hong Kong, Singapore]

Dendrogram for comp10 cluster analysis

*Source*: This figure was supplied by Alexandra Sidorenko and prepared as part of her research undertaken for the drafting of Sidorenko and Findlay (2001).

them and the e-marketplace systems: this could take the form of a service provider who connects the e-marketplace system with a paper system, for example. Given the availability of other suppliers world-wide (who have direct access to OEM and module maker procurement systems), the suppliers would need to bear the costs incurred in the construction of this interface, and also bear the costs of lower quality outcomes (eg errors in the fulfilment of orders). Bearing these costs is the alternative to investing in their own e-marketplace.

As the quality of the public internet develops, and the e-commerce indicators improve, less coordination of the development of these services for particular industry members will be required. One group of users will have less control of the interaction, standards are likely to be more common and more widely applied, and the outcome is likely to be an open and cross-industry system. Users may still procure specialist services, for example, those provided by the developers of software relevant to each stage of the life-cycle of the car – or other software services on the list prepared by White (2001) and quoted above. The web sites of these software or system providers may replace those of the e-marketplaces as the sites most frequently visited by transactors in this environment.

Once there is an open and cross-industry system, developing economy component makers with internet access will be able to 'log on' direct to the OEM or module maker sites. Until then, a commitment to establish a secure environment in which to conduct these transactions (for example, an ASEAN e-marketplace) is required in order for these firms to participate directly in the new systems of procurement.

## CONCLUSION

Information technology has a wide range of applications in the automobile industry, including applications that go far beyond the possibility of a 'talking car'. The use of ICTs is likely to have significant impacts on the costs of production and procurement, and also on the corporate structures evident in the industry. For component makers, the impact of ICTs could be different for different types of product. The emergence of ICTs may also change the typical mix of component types in a vehicle, leading to a trend to standardise individual components. At present, however, there remains some contention among industry members, particularly in Japan, on the most likely trajectories of change.

One visible application of this technology is the development of e-marketplaces. However, they are not the only application; for some brand name companies other forms of information sharing may be more important. Tailor-made systems will continue to be important while the public internet suffers from security or capacity problems.

From a component maker's point of view, e-marketplaces that are the result of cooperative initiatives by their customers can lower the costs of transactions, and of access to exports. They also introduce further competition in local markets where there may previously have been few bidders for any contract.

There may be concerns that the cooperative behaviour of the establishment firms extends beyond making the market institutions to making the prices. These competition policy issues would be less significant if a number of such markets competed with each other, as would effectively be the case if the public internet provided the relevant services.

Developing-economy component makers face new competitive challenges in this environment, but they also have new opportunities to sell into world markets. National indicators suggest that ASEAN countries and China, for example, are at some disadvantage in terms of access to digital technologies. This does not mean they will be excluded from world markets. Their long-term relationships, especially with the Japanese OEMs, continue to be important. Other interfaces with digital systems can be established. Costs are incurred one way or the other, either in establishing a new secure ASEAN or China e-marketplace or in using other routes until the public internet delivers higher standard services.

In the last decade, exports from ASEAN countries and China have performed strongly in the Japanese market. Despite these positive results to date, in terms of international adjustment and auto sector export performance, policy-makers in those economies will continue to be concerned about the quality of the access to digital technologies and its impact on the sector's outlook.

Sidorenko and Findlay (2001) have recently examined policy responses to the 'digital divide'. They pointed out that a number of policy areas are relevant to the efficient adoption and diffusion of any new technology, as well as investment decisions on research and development, the environment for entrepreneurship, the intellectual property regime and so on. Sidorenko and Findlay also argued that some other policies are more specifically relevant to digital divide issues at this time. These include initiatives which help capture the network externalities present in ICTs, which recognise the two-way connections with human capital development and the use of ICTs, and which provide access to world markets for ICT products and services. While none of these proposals is limited to automobile production and distribution, a policy-maker who is driving with that sector in the new economy will probably want to watch out for them on the road ahead.

## NOTES

1. 'Drive Distractions', by Charles Bickers, *Far Eastern Economic Review*, 21 June 2001, pp.34–37. Features which might be included in a 'telematic car' include navigation, toll collection, web-browser email (perhaps audible, as might be the navigation system), television, rear seat movies and games, and, reflecting consumer interests in security, crash auto-help and 'find me' options. On 5 March 2001, the Japanese government announced that researchers from academic and business circles would join forces on a government-sponsored project to develop an 'internet car' which will be capable of sending information on traffic and weather conditions and will also provide drivers with wide-ranging access to the internet. See also *Japan Times*, 6 March 2001, p.12.
2. See Steiger and Oberg (2001) for further commentary on powertrain technology, which is an example of a radical technological change.
3. On 1 January 2000 Toyota Motor Corp elevated the status of its e-commerce department to a division, and it plans to make it an independent entity in the future, as a public corporation. The automobile industry, one that includes automotive parts makers and auto distribution networks, is one of the broadest-based industries in Japan. Toyota is aiming to build up a network of automotive parts suppliers jointly with General Motors. More specifically, Toyota plans to spin off its Gazoo Division, an auto distribution network launched in April 1998. At present the division has a membership of about 400,000 and provides its members with information concerning new models and used cars through computer terminals installed at affiliated car dealers and convenience stores or through the internet.
4. Covisint is an e-business exchange developed by DaimlerChrysler, Ford, General Motors, Nissan and Renault. Its current product and service offering focuses on procurement, supply chain and product development solutions, including catalogues, auctions, quote management and collaborative design.

5  The transfer of technical information and data between Japanese companies typically involves parts information such as product specifications and information on parts make-up, together with information on the design of parts and products, as well as computer-aided design information.
6  Kanban is a Japanese process of time-based management like JIT but with some other features.
7  *AutoAsia*, 10 February 2000.
8  JNX uses a standard internet communication technology. This will allow parts suppliers to connect and communicate with all automobile manufacturers with a single link and a single protocol. It will also reduce communication and operation costs. JNX is designed as a reliable, secured and high-performance infrastructure for supply chain management, which will improve the information flow and reduce marketing time ('Japan automated Network eXchange is launched', Statement by Covisint, 20 October 2000).
9  The JNX Centre is run by the Tokyo-based Japan Automobile Research Institute. By March 2001, the institute expected to have 200 member companies, including the thirteen Japanese car, truck and motorcycle companies. See http:www.japanauto.com.
10 See Japan Automobile Manufacturers Association (JAMA), 'JNX Network Opens', *Japan Auto Trends*, Volume 4, No. 4, December 2000. The JNX web site is at http://www.jnx.ne.jp.

## REFERENCES

APEC (Asia-Pacific Economic Cooperation) (2000, November) *Towards Knowledge-Based Economies in APEC*. Singapore: APEC.
A.T. Kearney (2000) 'E-fulfilment 2000: corporate initiatives in automotive B2B', in *Global Automotive Manufacturing and Technology*, World Markets Series Business Briefing, London: World Markets Research Centre.
Farrell, R. and Findlay, C. (2001a) 'Japan and the ASEAN4 automotive industry', mimeo.
—— (2001b) Japan and the ASEAN4 automotive industry: developments and inter-relationships in the regional automotive industry', Kitakyushu: International Centre for the Study of East Asian Development (ICSEAD).
Helper, S. and MacDuffie, J. P. (2001) 'How the automotive industry might evolve – the internet and consumer and supplier relationships', in *Global Automotive Manufacturing and Technology*, World Markets Series Business Briefing, London: World Markets Research Centre.
Jahn, K. (2001, February) 'New economy initiatives: Global developments and their implications for the Australian automotive industry', in *Outlook 2001, Manufacturing, Tourism and Services*, Canberra: Australian Bureau of Agricultural and Resource Economics.
Katoh, H. (2000) 'An IT network revolution changing the framework of business', *The JAMA Forum*, 18, 2.
Keller, M. (1997) 'International automobile production: how will firms compete in the 21$^{st}$ century?', *Occasional Paper No. 34*, Center on Japanese Economy and Business, November.
Morita, M. and Nishimura, K. (2000) 'Information technology and automobile distribution: a comparative study of Japan and the United States', mimeo, University of Tokyo, November.
PricewaterhouseCoopers (2000) *The Second Automotive Century Executive*, Summary, available at www.pwcglobal.csm/auto.

Sidorenko, A and Findlay, C. (2001) 'The digital divide in East Asia', *Asian-Pacific Economic Literature*, 15, 2, November: 18–30.

Steiger, W. and Oberg, H.J. (2001) 'Outlook for alternative powertrain systems', in, *Global Automotive Manufacturing and Technology*, World Markets Series Business Briefing, London: World Markets Research Centre.

Takayasu, K. and Toyama, A. (1997) 'Business development of the big three US automakers in Asia accelerating their global strategies' *RIM Pacific Business and Industries*, Vol III, No. 37.

Toyota (2000), *IT, Interactive Toyota, Toyota's IT Operations*, corporate brochure, p.5.

White, N. (2001) 'A quiet revolution?', in *Global Automotive Manufacturing and Technology*, World Markets Series Business Briefing, London: World Markets Research Centre.

# 12 The digital divide in East Asia

*Emmanuel C. Lallana*

## INTRODUCTION

Information and communication technology (ICT), as the advertisement reminds us, will 'change the way we live, work and play'. Already it has made possible the emergence of what Manuel Castells (1996) identified as a 'global economy' – an economy 'with the capacity to work as a unit in real time on a planetary scale'.

Economists are still debating the precise effect of ICT on productivity, but the G-8's Digital Opportunity Task Force, in its final report, concluded (DOT Force 2001):

> ... when wisely applied [ICTs] ... offer enormous opportunities to narrow social and economic inequalities and support sustainable local wealth creation, and thus help to achieve the broader development goals that the international community has set.

The DOT Force also reported (DOT Force 2001) that ICT 'can provide new and more efficient methods of production, bring previously unattainable markets within the reach of local producers, improve the delivery of government services, and increase access to basic social goods and services'.

Policy-makers and opinion makers in developing countries are encouraged to see ICT as an 'enabler of development' (Accenture, Markle Foundation and UNDP 2001). ICT is said to help in enhancing rural productivity by allowing solution sharing among communities and providing timely market information to farmers. It is also being touted for its ability to help developing economies make real progress in health and education. For reformers, the appeal of ICT also comes from its potential to create an informed (and therefore empowered) citizenry. They are also attracted to the idea of a more efficient and transparent government brought about by widespread ICT use.

However, the bad news for developing countries is that the rapid uptake and global spread of ICT has been unequal. Bridges.org reports that, while all countries are increasing access to and use of ICT, '...the "information have" countries are increasing their access to and use at such an exponential rate that, *in effect*, the divide between countries is actually growing'.[1]

## DEFINING THE DIGITAL DIVIDE

The Organisation for Economic Co-operation and Development defines the digital divide as the 'gap' between individuals, households, businesses and geographic areas with regard to (a) their opportunities to access ICTs and (b) their use of the internet for a wide variety of activities (OECD 2001). The Digital Divide Network defines it as the 'gap between those who can effectively use new information and communication tools, such as the Internet, and those who cannot'.[2] For Bridges.org, 'Simply put, "the digital divide" means that between countries and between different groups of people within countries, there is a wide division between those who have real access to information and communications technology and are using it effectively, and those who don't'.[3] Underlying these definitions is the belief that lack of access to ICT goods and services will translate to social and economic disadvantages.[4]

While many studies of the digital divide make the important distinction between 'access' and 'use' of ICT, many (particularly the early studies) tend to focus on the question of physical access to ICT services. Access is usually measured by looking at telephone density (teledensity). For instance, the OECD (2001, p.7) argued that:

> The most basic and the most important indicator of the digital divide is the number of access lines per 100 inhabitants. It is the leading indicator for the level of universal service in telecommunications and a fundamental measure of the international digital divide.

Until very recently, teledensity was measured in terms of availability of fixed or wired lines. But with the recent explosion in cellular phone use, the teledensity picture of most developing countries will not be accurate if cellular phone ownership and use are not taken into consideration. Indeed, in a number of East Asian economies (such as Hong Kong and the Philippines) there are now more mobile phones than 'wired' phones.

Aside from measuring telephone density, most digital divide studies also look at personal computer (PC) deployment or penetration. The logic is that the PC is the main way to access the internet. But there are limitations in using PC penetration to study the digital divide. The first is that in many developing countries several people may use the same PC (in schools or internet cafes). The other is that the PC is not necessarily the only way people can access the internet. In Japan, a significant number of people access the net through their iMode (mobile) phones. Furthermore, a number of studies project that in the next three years the number of internet-capable mobile phones will exceed the number of internet-connected PCs.[5]

The number of internet users is another indicator of the digital divide. Michael Minges (2000) distinguished between 'internet users', 'internet subscribers' and 'internet coverage'. He noted that comparative estimates of the number of 'internet users' must be used with care as 'there is no standard definition of frequency of use (e.g., daily, weekly, monthly) or services used (e.g., e-mail, World Wide Web)' in these studies. He also noted that private market research outfits typically ignore developing countries.

Minges (2000, p.5) believes that 'internet subscribers' is a more precise indicator of access than 'internet users', even as he recognises 'that the number of subscribers measures those who are paying for a subscription and not the number of users'. While 'subscriber' is a more accurate measure, it does not give an accurate picture of the digital divide. In Southeast Asia, many people access the internet through internet cafes, because of the prohibitive cost of a PC, a phone and an internet connection. Even if people can afford to buy a PC, have a phone at home, and pay internet access charges, they often prefer to access the web through internet cafes because of high telephone charges.

*Coverage*, the portion of the population of a country within easy access of a PC or internet access device, is for Minges an 'ideal indicator'. He said (Minges 2000, pp.6–7):

> This indicator would express the potential Internet user market and is the fundamental measure of universal access to the Internet. High coverage would denote a significant policy achievement in that technically, people can access the Internet even though they may choose not to do so for other reasons (e.g., high cost, lack of interest).

There are at least two problems here. The easier one is defining 'easy' access. As Minges pointed out, access can be defined in terms of time (how long it will take for a potential user to walk to the nearest internet access station) or distance (e.g. whether there is an internet access point within a two-kilometre radius). Another issue is that potential availability is not always translated into reality. And even if networks are in place, providing citizens with easy access to the internet is no guarantee that they will use the internet![6]

The OECD (2001, p.13) also noted that a 'good indicator of the international digital divide in relation to the Internet is the penetration rate of Internet hosts'. Minges (2000, p.2) cautioned about the use of internet hosts to compare internet development between countries. He noted:

> While hosts might be a useful infrastructure indicator of the number of computers in a nation that are connected to the Internet, it is a poor indicator of accessibility since it does not measure the number of users.

Tuvalu's '.TV' top-level domain name is being sold to television companies all over the world by *The .TV Corporation*, based in Los Angeles. Thus, the number of addresses with a '.TV' domain name is not a good indicator of internet use in Tuvalu. In the Philippines, there are plans to market the '.PH' top-level domain name not just to Filipinos or entities in the Philippines but also to phone companies everywhere. Another limitation of measuring internet hosts is that 'there is not necessarily any correlation between a host's domain name and where it is actually located'.[7] A host with a '.MY' domain name could be located in the United States or any other country as easily as in Malaysia. Furthermore, hosts under the top-level domain names such as '.EDU', '.ORG', '.NET', '.COM' and '.INT' could be located anywhere.

## MEASURING ACCESS

Among the internet access indicators, data for teledensity are easiest to obtain. According to the International Telecommunication Union (ITU 1998), there is a wide gulf that separates the country with the lowest teledensity (Congo, with 0.04 phones per 100 individuals) and the highest teledensity (Monaco, with ninety-nine). The ITU acknowledged that up to one-quarter of its members have a teledensity of less than 1 per cent. Once a country has reached a teledensity of 1 per cent, it would take fifty years for it to increase its teledensity to 50 per cent. The ITU conceded that the goal of a telephone in every home may prove impossible to achieve in developing countries.

In terms of internet access, the ITU reported that in 2000 there were 214 countries connected to the internet – up from 60 in 1993 and just eight in 1988.[8] In 2001, the ITU estimated total global internet users at 352 million.[9] On the other hand, Nua put the number of global internet users at 513.41 million in 2001.[10] Various estimates predict that by 2005 there will be between 977 million and one billion internet users worldwide. According to Nielsen/NetRatings, 41 per cent of all internet users come from the United States and Canada.[11] The Asia Pacific region accounts for 20 per cent of the total 'netizens' and Latin America for 4 per cent.

On the other hand, Pippa Norris (2000, p.3) has warned that, despite the rapid growth in internet use globally, 'only 4 per cent of the world's population is online'. Norris has also said (Norris 2000, p.5):

> ... in most developing nations the inequalities of resources that continue to produce disparities in health care, longevity and education are also, not surprisingly, evident in the virtual world.

The digital divide cannot be understood or solved if one looks at infrastructure or application issues alone (e.g., telephone access or internet access). It is important to look at content. Many Filipino adolescents and young adults patronise internet cafes, but the majority use these cafes for internet gaming. Is the lack of meaningful content part of the explanation for this phenomenon?

## BEYOND ACCESS

The Children's Partnership has argued that content is one aspect of the digital divide that has been neglected.[12] It identified the following four content-related barriers to greater internet uptake across society: local information barriers; literacy barriers; language barriers; and cultural diversity barriers.[13] The Children's Partnership documents what most internet users from developing countries discover when they go to the internet – that there is a great deal of information about many things, but hardly any about communities outside the major cities of the world. Providing farmers in Southeast Asia with internet access will not necessarily make a difference in their lives, because they are unlikely to find on the web anything of significance to them.

The lack of varied content is a function of the commercialised nature of the web. Commercial content providers focus only on markets with acceptable returns to their investment. Even when corporations use 'online communities' to generate business, there is no guarantee that low-income communities will get the information they need (Werry and Mowbray 2001). Compounding the problem is that not-for-profit and community-based initiatives to create content face sustainability problems.

Literacy is another barrier. According to the Children's Partnership:

> The vast majority of information on the Net is written for an audience that reads at an average or advanced literacy level. Yet 44 million American adults, roughly 22 percent, do not have the reading and writing skills necessary for functioning in everyday life.[14]

The literacy issue goes beyond basic and functional literacy. Technological literacy is also an issue. For the elderly, who are very literate, learning how to use a computer or get on the internet may be a very intimidating experience. An even bigger issue for developing countries is developing meaningful, up-to-date and inexpensive (if not free) content accessible to all, including illiterates. It is hoped that voice recognition technology will make internet access for all easier.

The language barrier compounds the literacy issue. Over half (68 per cent) of internet content world-wide is in English.[15] Only 3.87 per cent of internet content world-wide is in Chinese; French accounts for only 2.96 per cent, Spanish for only 2.42 per cent, and Arabic for a mere 0.04 per cent.[16] English is even more dominant in e-commerce. According to the OECD (2001, p.23), more than 94 per cent of links to pages on secure servers were in English. The only other languages to account for more than 1 per cent of detected links to secure servers were German, followed by French, Spanish, and Japanese.

For many non-English speakers who are able to gain access to the net, there are very few sites that they can visit. But this will change soon. By 2003, it is expected that at least 36 per cent of all internet users will prefer to use a language other than English.[17] Up to 75 per cent of users in Latin America, 52 per cent in western Europe and 54 per cent in Japan would demand local language sites.

The predominance of English also helps to maintain the lack of cultural diversity on the net. For instance, a majority of Philippine web sites are in English. Another contributory factor to the lack of cultural diversity is the fact that most of the internet content is provided by businesses. The result is information geared for consumers in the biggest market – North America.

## IS E-COMMERCE THE KILLER APP?[18]

The internet is supposed to change the way business is conducted. In a recent global survey, 78 per cent of North American executives, 64 per cent of Association of Southeast Asian Nations (ASEAN) chief executive officers and 46 per cent of European business leaders believed that the internet had fundamentally changed the way companies do business.[19]

Conservative estimates put the e-commerce market at over $1 trillion by 2003.[20] Worldwide business-to-business (B2B) purchases are expected to ignite the internet, rising from $96.9 billion in 1999 to $1,431.1 billion, or 87 per cent of all worldwide e-commerce, by 2003.[21] At the same time, global business-to-consumer (B2C) e-commerce is expected to grow from $33.5 billion in 1999 to $209.1 billion by 2003. As is to be expected, North America and Europe are the largest e-commerce regions. However, the Asia Pacific area and Latin America are experiencing robust e-commerce growth.

Also to be expected is that the underdeveloped information infrastructure of Asia is affecting the development of e-commerce in the region. E-commerce activity in Asia will be concentrated in B2B because of the issue of access. In 1999, 77 per cent of the US$6.63 billion in e-commerce revenues generated in Asia were in B2B transactions.[22] By 2003, the share of B2B transactions in total Asian e-commerce will increase to 87 per cent. Despite this rapid growth, Asia's share of global B2B e-commerce will remain small (about 5 per cent of the global total by 2005, according to eMarketer).

The growth of e-commerce in East Asia will be affected by at least two factors – access and language. Market analysts believe that, as the percentage of people online in a given market approaches a critical mass of about 10–20 per cent of the population, e-commerce becomes attractive to businesses.[23] As well, internet consumers are up to four times more likely to shop and purchase online from web sites that support their native language.[24]

## THE DIGITAL DIVIDE IN ASIA

Analysing data from the ITU, eMarketer, a private market research firm, has suggested that Asia can be divided into three major groups when looking at teledensity.[25] In the first tier are economies with about 50 per cent teledensity (Hong Kong, Taiwan, Japan, Australia and New Zealand). The second-tier markets are those with a teledensity of 8–49 per cent (Singapore, South Korea, Malaysia, China and Thailand). The third tier comprises economies with a teledensity of 2–5 per cent (the Philippines, India, Indonesia, Vietnam and Pakistan).

As is to be expected, the number of internet users is closely related to teledensity (Table 12.1). In Northeast Asia, Japan has the largest number of users, but Korea has a bigger percentage of the population online. In Southeast Asia, Singapore and Malaysia lead the pack.

How do East Asians access the internet? In Hong Kong, Singapore and Taiwan, over 80 per cent of people access the net from their homes,[26] which is not surprising given the economics of wired telephone deployment. The largest cost in installing telephone lines is the 'last mile', so it is cheaper to roll out telephone lines in densely populated areas than in thinly populated ones. City-states with few rural areas are easier to wire than continent-sized nations or archipelagos. For instance, a greater percentage of households are connected to the internet in Singapore than in the United States, the United Kingdom or France. South Korea presents an interesting case: 64 per cent of Koreans access the internet from the home, and 43 per cent from internet cafes. In Hong Kong, the corresponding figure is 91 per cent from the home and 6 per cent from internet cafes.

In the developing countries of Southeast Asia, the number of people who access the web using internet cafes is likely to be higher. For instance, in Indonesia in 1999 about 250,000 internet users accessed the net using internet cafes and other public access points; there were 250,000 internet subscribers in the same year.[27] In the Philippines, it is estimated that there are two million internet users and 1,500 internet cafes.[28] In this estimate, Cebu (a major city in central Philippines) has 200,000 internet users, 400 internet cafes and only 43,000 subscribed internet accounts. The popularity of internet

*Table 12.1* Internet users in East Asia

| Region | Users (,000) | Users per 10,000 inhabitants |
|---|---|---|
| *Northeast Asia* | | |
| China | 22,500 | 173.70 |
| Japan | 47,080 | 3,709.45 |
| South Korea | 19,040 | 4,025.37 |
| *Southeast Asia* | | |
| Brunei | 30 | 913.78 |
| Cambodia | 6 | 4.58 |
| Indonesia | 2,000 | 94.30 |
| Laos | 6 | 11.04 |
| Malaysia | 3,700 | 1,590.03 |
| Myanmar | 1 | 0.21 |
| Philippines | 2,000 | 261.44 |
| Singapore | 1,200 | 2,986.78 |
| Thailand | 2,300 | 379.49 |
| Vietnam | 200 | 25.05 |

*Source*: ITU (2001) (2000 data), see also www.itu.int/itu-D/ict/statistics/at-glance/internet00.pdf.

cafes in developing societies is not surprising given the low telephone and PC penetration and the relatively high cost to connect to the internet. Internet access costs (as a percentage of average monthly income) have been estimated at 1.2 per cent in the United States, 16 per cent in Hungary, 100 per cent in Uganda, and a whopping 191 per cent in Bangladesh.[29]

What do Asians do when they are online? According to Netvalue:

> Web surfing is still the predominant online activity, in terms of size of data and files transferred, accounting for 83.0% of all Internet users in China, 75.6% in Hong Kong, 72.2% in Taiwan and 69.5% in Korea. Pure web usage goes down to 53.3% in Singapore, where non-web activity audio-video takes on a huge percentage of market (30.4%).[30]

In China, Taiwan and Korea, local web sites are popular. On the other hand, international web sites (for example, Yahoo or Lycos) are popular in Hong Kong and Singapore.

A good sign, in terms of bridging the digital divide, is the high literacy rates in East Asia (Table 12.2).

## HOW ASIA IS DIFFERENT

The proliferation of mobile phones across much of Asia represents the biggest opportunity for widespread adoption of the internet in the region. Mobile phones are easier and cheaper to roll out than fixed lines. Already, there are more cellular phones than fixed line phones in a number of countries.

Taylor Nelson Sofres Interactive (TNS) estimates that 57 per cent of the adult population (defined as those between ages 15 and 65) in the Asia Pacific region have a mobile phone.[31] In terms of the actual number of users, the Asia Pacific region had 34.4 million regular mobile internet users by the end of 2000, up 29 per cent from September of that year. Japan led the region, with 26.8 million users, followed by South Korea, with seven million.[32]

Asia is already the 'largest and fastest-growing' wireless internet market in the world. Even more importantly, Asia (or at least Japan) is leading in the deployment of third-generation cellular services. Third-generation phones provide colour screens and faster access to the net.

Another bright spot on the Asian scene is the narrowing gender divide.[33] In terms of a gender-balanced internet user population, Asia as a whole has yet to catch up with the United States and Canada, but its situation is better than some European countries and most Latin American ones. eMarketer reports that in the United Kingdom, France, Germany and Spain, for example, men continue to account for up to two-thirds of internet users. The Latin American internet user population is 60 per cent male and 40 per cent female. In South Korea, the Asia Pacific region's second largest user base after Japan, women already make up 45 per cent of the internet population; Australia, the region's fourth largest user base, has almost achieved gender parity.

*Table 12.2* Adult literacy rates and total enrolments in East Asia, 1999

| Region | Literacy rate (%) | Combined enrolment ratio (%)[a] |
|---|---|---|
| *Northeast Asia* | | |
| China | 70.2 | 73 |
| Japan | 99 | 82 |
| South Korea | 97.6 | 90 |
| *Southeast Asia* | | |
| Brunei | 91 | 76 |
| Cambodia | 68.2 | 62 |
| Indonesia | 86.3 | 65 |
| Laos | 47.3 | 58 |
| Malaysia | 87 | 66 |
| Myanmar | 84.4 | 55 |
| Philippines | 95.1 | 82 |
| Singapore | 92.1 | 75 |
| Thailand | 95.3 | 60 |
| Vietnam | 93.1 | 67 |

*Source*: UN Human Development Index (2001) (1999 data) in www.undp.org/hdr2001/indicators/indic_11-_1_.html.

*Note*: a Combination of total enrolment in the primary, secondary and tertiary levels vis-à-vis total population.

Asian men still spend more time online than women: 14.5 hours per user per month, on average, in comparison with 12 hours for women. Yet analysts point out that Asian women exhibit more 'efficient' behaviour in that they go online with a specific goal in mind and log off after this is accomplished.

## NORTHEAST ASIA

Japan is Asia's leading internet market. In 2000, there were about 47 million Japanese internet users.[34] By 2005, the number of internet users in Japan is expected to double to 87.2 million. What makes the Japanese case interesting is that about half of its internet users use mobile phones to access the net. Given the recently announced policy for Japan's internet use to catch up with that of the United States, broadband connections in Japan are projected to reach 25 million households in 2005.

Internet users in Japan are predominantly young; email and browsing appear to be the most popular internet applications. Some 50.4 per cent of internet users use email once or twice a day; 41 per cent do so three times a day. Some 44.2 per cent of Japanese 'netizens' browse once or twice a day.[35] The top reason given for subscribing to iMode (mobile internet phone) is email. The top three categories of broadband content are paid movies and music videos (40 per cent), promotion and live events (33 per cent) and video sampling and home security (25 per cent each).

If Japan leads in wireless access to the net, South Korea is a world leader in broadband internet access. Close to 60 per cent of Korean households with internet access have broadband connections. The United States comes in a distant second, with 11 per cent, followed by Hong Kong (8 per cent) and Singapore (7 per cent).[36] Another study estimates that South Korea has 4.3 million broadband connections, or ten for every 100 people, far outpacing runners-up Canada (four connections per 100 people) and the United States (three per 100 people).[37] Even more impressive is the plan to provide ultra-high-speed (20 Mbps) internet access to 13.5 million households (84 per cent of the total household population) by 2005. It is planned to increase wireless access to 2 Mbps by the same time.[38]

Internet use among South Koreans is on the rise, according to the country's Ministry of Information and Communication. As of March 2001, 48.6 per cent of the population over the age of seven, or 20.9 million people, accessed the web at least once a month – an increase of 3.9 per cent from December 2000. Some 19.56 million people over the age of seven were frequent users, logging on more than once a week.[39] Where do these South Koreans go when they are online? Media-related sites, pornography sites, and art and cultural sites topped the list.[40] More South Korean netizens visited pornography sites than did other Asians: 71 per cent of South Korean netizens compared to 62.6 per cent for Asia as a whole. At the same time, more South Koreans visited arts and culture sites (according to NetValue, 59.5 per cent of South Korean surfers, topping the 51.6 per cent figure for the entire region).

China is at the other end of the digital divide – close to most Southeast Asian nations. However, its sheer size ensures that it will soon be a major internet player. The difficulty of counting internet users is reflected in the recent debate over how many Chinese are online. According to the 'Semi-Annual Survey Report on the Development of China's Internet' of the China Internet Network Information Centre (CNNIC), 26.5 million people had access to the web by midyear 2001.[41] This figure is higher than the estimates of the private International Data Corporation (IDC), which puts the number of mainland China internet users at about 17 million, and Interactive Audience Measurement Asia Ltd. (IAMAsia), a Hong Kong company engaged in internet surveys, which puts the number at about 15 million.[42] The difference, of course, lies in who gets counted.

A NetValue survey of household internet users in Beijing, Shanghai, Guangzhou, and Shenzhen showed that household users logged onto the internet for an average of 9.6 days a month.[43] Another survey, by CCIDnet.com, revealed that the average age of a web surfer in China is between 22 and 27. The most popular uses are email, reading news, chatting with friends, and downloading software and documents. Most Chinese netizens are technical personnel, students or managers with an average monthly income of 900 renminbi (US$108.70) – 73 per cent higher than the average income of people living in cities and towns in 1999.[44]

## SOUTHEAST ASIA

The ten member states of ASEAN can be divided into three groups: the leaders (Singapore and Malaysia); the middle (Thailand, the Philippines, Brunei and Indonesia); and the laggards (Cambodia, Laos, Myanmar and Vietnam). A study conducted in 1999 by an ad hoc group of private corporations showed that most ASEAN members have low teledensities (less than 5 in 100).[45] Only one country had a teledensity above 40, with three in the 10–40 range. Most ASEAN countries indicated that at least two to three people shared one internet service provider (ISP) account to access the internet. In one country, an internet account was shared by more than seven people. Only Singapore and Malaysia indicated that 25–50 per cent of their businesses accessed the internet directly; in most ASEAN countries this number was less than 10 per cent.

In most ASEAN countries, the internet is used for email, random surfing, and information on products. In three member countries, it is also used for low-value transactions. Most ASEAN governments use internet technologies for providing information to the public. Singapore and Malaysia lead in e-government, in that they use the internet to transform government.

Singapore is the most connected country in Southeast Asia. According to Singapore's official statistics agency, 50 per cent of Singaporean homes had internet access at the end of 2000, and 60 per cent had PCs.[46] Singapore is also where e-commerce, even of the B2C variety, is flourishing. B2C e-commerce transactions in Singapore grew from $200 million in 1999 to $1.17 billion in 2001.[47] Singaporeans use the internet for instant messaging, web surfing, and chatting. When surfing, a number of Singaporeans visit pornography and gambling sites.[48] According to a study by NetValue, the percentage of surfers visiting adult web sites increased from 32 per cent to 36 per cent in the first three months of 2001 (despite an official ban on porn sites). At the same time, those visiting gambling sites declined from 36 per cent to 33 per cent. Adults over the age of 35 were the only demographic group to visit gambling sites more often than sex sites.

In Thailand, the internet population is expected to hit to 4.6 million in 2001 (up from 2.3 million in 2000).[49] A survey of Thai students (primary, secondary and university) showed that 16 per cent went online daily and nearly 28 per cent went online once or twice a week. They used the internet for games and entertainment (36 per cent), chatting (32 per cent), research (16 per cent), and email (11 per cent). Malaysia was expected to have three million internet users in 2001, and double that by 2005. In Malaysia, 68 per cent of the online population is male, and the vast majority of users (84.7 per cent) are aged 19–34.[50] Indonesia has a much larger population than these countries, but in 2000 had only an estimated two million internet users. A survey by the Association of Indonesian Internet Service Providers (APJII) predicts that the number will continue to double on an annual basis.[51]

## THE STATE AND THE DIVIDE

The concern with the digital divide is not simply political; studies have shown that technological diffusion is a 'slow and costly matter' and that developing countries 'cannot assume that relevant new technologies will flow easily to them across international borders' (Lipsey 2000, p.19). The successful models being held up for emulation – the high-performing economies of East Asia – implement not only market-friendly economic policies but also aggressive technology adaptation strategies.

Governments have undoubtedly played a vital role in developing information infrastructures in general, and government funding has been important in developing the internet (Naughton 1999). But, instead of calls for massive government funding for ICT development, the consensus among policy-makers and advisers is for governments to promote 'open' and competitive ICT markets. As with other development issues, the 'proper' role for government is seen to be creating an enabling policy and a legal and regulatory environment for ICT development.

Most governments in East Asia have formulated, or are in the process of formulating, their national ICT development plans. Among the most aggressive is the e-Japan Strategy, adopted in January 2001, which aims to turn that country into the '[w]orld's most advanced IT nation within 5 years'.[52] Through this strategy, the Japanese government hopes to achieve four goals: (a) enable every Japanese to enjoy the benefits of IT; (b) reform economic structure and strengthen industrial competitiveness; (c) realise affluent national life and a creative and vital community; and (d) contribute to the formation of an advanced information and telecommunications network society on a global scale. Specific goals are to establish 24-hour connection to high-speed access networks to 30 million households and ultra-high-speed access for ten million households within five years. The private sector is seen as the driving force, with government's role limited to establishing an enabling policy environment and dealing with 'non-private' areas such as e-government and the dissolution of the digital divide.

In ASEAN, Singapore and Malaysia are leaders in creating an ICT-friendly policy, legal and regulatory environment. Both countries are implementing ICT development plans that would catapult these countries to the front of the ICT revolution. Already, Singapore is a global leader in e-government applications. Malaysia has the most (seven) 'cyber laws': the Computer Crimes Act 1997; the Digital Signature Act 1997; the Copyright Amendment Act 1997; the Telemedicine Act 1997; the Communications and Multimedia Act 1998; the Multimedia Convergence Act 1997; and the Electronic Government Act 1998.

Uniquely, Korea enacted a digital divide law in January 2001. This law aims to:

... ensure a universal, unlimited access to the telecommunications networks and use of the telecommunications services for low-income earners, rural residents, the disabled, the aged, women, etc., who have difficulties in accessing or using the telecommunications services for economic, regional, physical or social reasons.[53]

The law calls for the creation of a committee for closing the digital divide and a five-year master plan and annual implementation plans to bridge the digital divide.

Korea's digital divide law has provisions on setting up telecommunications service accessibility guidelines for the disabled; subsidising the acquisition of PC or other telecommunications equipment by the poor and the disabled; developing access-enabling technologies for the disabled; supporting information providers that provide content for fishermen, farmers, the disabled, the elderly and the poor; and establishing public access centres that offer internet access and learning opportunity to residents in need.

Economic reforms are also important in developing ICT infrastructure and competencies. One of the most important is telecommunications sector reform. Liberalisation increases competition in telecommunications, leading to growth in access (fixed or mobile), with implications for alternative access technologies, reduced prices, internet access and increased usage. The rapid increase in the Philippines' teledensity would not have happened without the 1996 policy to liberalise the telecommunications sector. Singapore's 'big bang' approach to telecommunications liberalisation is seen by that country's policy-makers as an important reason for the vitality of its ICT sector.

But is liberalisation enough? Clearly, a market-based approach will be the foundation of any effort to enhance existing national information infrastructures and upgrade to broadband digital networks. Yet such an approach will also mean that certain areas may not reap the benefits of the digital age. Governments will have to seriously consider a supplementary approach to their market-based telecommunications strategy in order to build broadband networks. Even in the United States, industry experts are suggesting that the roll-out of broadband networks to certain areas would have to be subsidised.[54]

Following a study of leading broadband countries, Ben Macklin of eMarketer has suggested (Macklin 2001):

> Broadband has prospered in countries where national and regional governments have set aggressive targets, invested in infrastructure and developed a regulatory environment conducive to competition.

Another study, which reviewed the broadband strategies of 14 countries, suggested that there are three public policy approaches to the issue: (a) a 'light touch' regulatory approach or a market-led approach; (b) a cooperative approach or an approach where programs target areas and groups where market forces will not adequately address disparity to broadband access; and (c) comprehensive national broadband plans or proactive government involvement in broadband roll-out (Savage n.d.).

The market-led approach is exemplified by New Zealand and Switzerland. The cooperative approach is being practised by the United States, Australia, Germany and the United Kingdom. Korea (Cyber Korea 21), Norway (eNorway), Malaysia (Multimedia Super Corridor), Singapore (Singapore ONE initiative), Taiwan (Silicon Island program) and Japan (Info-Communications Strategy for the 21$^{st}$ century) are practitioners of comprehensive national broadband plans.

More active government strategies are needed not only for the deployment of broadband networks but also for providing basic information and communication services in rural areas. A McKinsey study (Dhawan et al 2001) showed how cooperation among mobile operators, government, mobile service providers and local entrepreneurs could develop low-cost mobile networks that would provide access to underserved areas in developing economies. This study suggested that mobile operators could reduce the cost of rolling out this type of service if they built 'no frills' networks, increased reliance on prepaid cards and outsourced customer relations to handset retailers. Government measures to encourage this development would include the creation of a universal service fund and a regulatory environment that would increase competition and keep prices down. The study also suggested that value-added service providers and local entrepreneurs could share the costs of building and operating the networks if given an acceptable role.

Governments can also help bridge the digital divide by lowering or removing import duties and/or sales tax on IT goods and services. In terms of lowering tariffs, the WTO member countries from the Asia Pacific region are already signatories to the Information Technology Agreement (ITA). ASEAN members, by virtue of the e-agreement, are also committed to reducing tariffs on a wide range of ICT goods and services[55] and to facilitating ICT trade between and investment in ASEAN countries.

Governments' own use of technology to enhance efficiency, effectiveness and transparency (e-government) is critical to bridging the digital divide.[56] Already, numerous e-government projects are under way. These include improving the efficiency and effectiveness of the bureaucracy by computerising governmental processes and creating government portals to provide information to the relevant public.

It is not surprising that numerous initiatives to provide information online are under way. After all, governments are the biggest repository of information that is important to citizens, and information is a public good. A useful guide to the role of government in providing content is given in 'Principles for government provision of goods and services in a digital economy' (Stiglitz et al 2000). In this report, it was argued (pp.51–52) that providing public data and information is a proper governmental role; that governments should exercise caution in adding specialised value to public data and information; that governments should provide services online only if it would not be more efficient to provide them privately, with regulation or appropriate taxation; and that governments should be allowed to maintain proprietary information or exercise rights under patents and/or copyrights only under special conditions (including national security).

A 2001 survey of 2,288 government web sites in 196 countries conducted by World Markets Research Center and Brown University concluded:

> While some countries have embraced e-government, a number of other countries have not placed much information or services online, and are not taking advantage of the interactive features of the internet.[57]

Not surprisingly, the United States emerged as the top e-government country – that is, its government web sites have features important for information availability, citizen access, portal access and service delivery. Taiwan, Australia, Canada and the United Kingdom complete the top five countries in this category. However, Taiwan tops the list of countries whose governments offer online services (65 per cent of its government web sites offer online services). Singapore (47 per cent) is in sixth place, but ahead of Canada and the United States (tied for seventh place, with 34 per cent each). China (26 per cent) and Malaysia (16 per cent) were the other East Asian countries in the top twenty of this survey of web sites with online services. In terms of regions, this survey showed that 28 per cent of government web sites in North America offer online services. The Pacific Ocean islands region is second, with 19 per cent; Asia comes in third, with 12 per cent; and the Middle East is in fourth place, with 10 per cent. Among governments, Taiwan has the highest percentage of web sites that offer online services (65 per cent).

Apart from providing information and transacting business online, the welfare-enhancing e-government initiatives include e-learning and telemedicine/e-health initiatives.

The goal of most e-learning initiatives is to harness technology to help prepare students for the emergent knowledge-based new economy. For many developing countries with limited resources, distance education would allow students in remote areas to access the same information and other educational resources as those in more accessible locations.

Governments interested in bringing technology into classrooms can look at a number of successful best practice endeavours. Canada claims the distinction of being 'the first country in the world to connect its public schools… and public libraries to the Information Highway'.[58] Its SchoolNet is an integral part of 'Connecting Canadians' – a government strategy to keep Canada among the leaders in connecting its citizens to the internet. In Southeast Asia, Thailand's SchoolNet has linked 3,414 Thai secondary schools to the internet since it was launched in 1995.[59] World Links for Development (World Links) is a global learning network to link students and teachers in collaborative projects and to integrate technology into learning via the internet.[60] There are World Links SchoolNet projects in over twenty developing countries, where 130,000 students and teachers collaborate over the internet with partners in twenty-two industrialised countries. World Links is led by the World Bank, but it involves ministries of education, schools, parents and private sector partners at every stage of its SchoolNet projects.

The promise of telemedicine is that it moves information instead of patients. This would give people living in developing countries, particularly in rural and remote areas, health services comparable to those in developed and urban areas. Telemedicine projects are already under way throughout the world.[61] Satelife's HealthNet is an example of the kind of e-health activity that governments could fund.[62] HealthNet provides low-cost access to healthcare information to 19,500 health care workers in more than 150 countries through a link to basic email. Physicians and other health workers connect to the network through local telephone nodes to access services such as physician collaborations, medical databases, consultation and referral scheduling, epidemic alerts, and medical libraries. In China, medical schools and hospitals in Zinjiang, Qinghai, Gansu, Ningxia and Shaanxi Provinces will be linked to medical centres elsewhere in China and in the United States through the Northwest China Telehealth Service (NCTS).[63] The NCTS will provide continuing medical education, telemedicine and library services, and will allow users in China to access health services via telephone, fax, email or the World Wide Web.

But national governments alone will not solve the digital divide. Multilateral agencies must also participate as partners. Such agencies can supplement (in some cases, provide) the resources to allow governments, as well non-government groups, in developing countries to launch digital opportunity initiatives.

## GLOBAL AND REGIONAL INITIATIVES

In 2001, the United Nations created an ICT Task Force with the objective of finding new, creative and quick-acting means to spread the benefits of the digital revolution and avert the prospect of a two-tiered world information society.[64] The task force comprises representatives from the public and private sectors, civil society and the scientific community, and leaders of the developing and transition economies. It is positioned to build strategic partnerships and to meld diverse efforts.

The United Nations Development Programme (UNDP) has projects to boost internet connectivity and access in some of the poorest countries in the world.[65] These include an initiative to determine the extent of 'e-readiness' in developing countries, the United Nations Information Technology Service (UNITeS) (a volunteer corps to train groups in developing countries in the uses and opportunities of the internet and IT), and the Sustainable Development Networking Program (an initiative to kick-start networking in developing countries and help people share information and expertise relevant to sustainable development to better their lives). The UNDP Asia Pacific Development Information Program (APDIP) was launched in 1997 to promote and establish information technology for social and economic development throughout the region. It 'assists developing countries to bridge the digital divide and to benefit from the global information infrastructure in numerous ways'.[66] APDIP has a capacity-building and technical assistance program and a research and development program, which focuses on 'finding solutions to expand connectivity in places underserved by typical market-oriented services, and customising applications to answer the information needs of developing countries'.[67]

The World Bank also has a number of initiatives related to the digital divide. Its Global Information and Communication Technologies Department seeks to promote private participation in developing telecommunications services in the emerging markets.[68] It has three goals: to accelerate the participation of client countries in the global information economy; to promote private sector investment in developing countries, which will reduce poverty and improve people's lives; and to promote innovative projects on the use of ICTs for economic and social development, with special emphasis on the needs of the poor in developing countries.

The World Bank's Information for Development Program (infoDev) was created in 1995 to address the obstacles faced by developing countries in an increasingly information-driven world economy.[69] It is a global grant program that promotes innovative projects on the use of ICTs for economic and social development in developing countries. The Global Development Gateway is a development portal that provides access to information and knowledge on development activities.[70] Through this initiative, the World Bank hopes to make it easier to share experience and knowledge in development, and offers up-to-date information on projects, resources, best practices and expertise on such subjects as poverty, governance, gender, IT, development and environment.

The G-8, in its 2000 summit in Japan, adopted the Charter on Global Information Society and established the DOT Force, aimed at integrating efforts to bridge the digital divide into a broader international approach. The DOT Force had representatives from government (G-8 and selected developing countries), multilateral agencies, the private sector and the non-profit sector. Of the forty-three members who have participated in its work, only eight came from developing economies.[71]

After a year of work, the DOT Force issued a report which proposed a nine-point plan of action for the G-8 to adopt. There are nine action points for the G-8 countries. First, they are to help establish and support developing country and emerging economy national 'e-strategies'. Second, they are to improve connectivity, increase access and lower costs. Third, they are to enhance human capacity development, knowledge creation and sharing. Fourth, they are to foster enterprise and entrepreneurship for sustainable economic development. Fifth, they are to establish and support universal participation in addressing new international policy and technical issues raised by the internet and ICT. Sixth, they are to establish and support dedicated initiatives for the ICT inclusion of the least developed countries. Seventh, they are to promote ICT for health care and in support against HIV–AIDS and other infectious and communicable diseases. Eighth, they are to make national and international efforts to support local content and applications creation. Finally, they are to prioritise ICT in G-8 and other development assistance policies and programs and enhance the coordination of multilateral initiatives.

In October 2001, the Asia Pacific Economic Cooperation (APEC) forum adopted an e-APEC strategy and stated (APEC 2001) that it aimed to build APEC:

> towards a digital society, with higher growth, increased learning and employment opportunities, improved public services and better qualities of life by taking advantage of advanced, reliable and secure ICT and networks and by promoting universal access.

The e-APEC strategy has three pillars: to create an environment for strengthening market structures and institutions; to provide an environment for infrastructure investment and technology development; and to enhance human capacity building and promote entrepreneurship.

Another intergovernmental initiative, this time by developing economies, is the e-ASEAN Task Force (eATF). ASEAN created the eATF in 1999 to develop a broad and comprehensive action plan for an ASEAN e-space and to develop competencies within ASEAN to compete in the global information economy. Unlike other ASEAN advisory bodies, the eATF includes representatives from the public and private sectors of the ten ASEAN member countries. Through the eATF, the governments of Southeast Asia can receive advice on the most appropriate enabling policy, and on the legal and regulatory requirements for the development of a regional environment for ICT growth.

The eATF was instrumental in the formulation of the ASEAN Framework Agreement on Information and Communications Technology Products, Services and Investment, which was signed by the ASEAN heads of state in 2000. The agreement is a landmark in ASEAN efforts to create a coherent strategy for ICT development in the region. It includes provisions on connectivity, regional content, a seamless legal and regulatory environment, a common marketplace for ICT products and services, human development, and e-governance, forming the basis for a coordinated approach to implementing the e-ASEAN action plan. The e-agreement establishes the political commitment to undertake the necessary steps to achieve the goals of e-ASEAN and allows the horizontal treatment of cross-cutting issues; it therefore facilitates a coordinated approach to addressing these issues, given the many sectors and institutions concerned.

## PRIVATE SECTOR INITIATIVES

Given the market-friendly approaches to ICT development that governments are taking and multilateral institutions are supporting, the private sector – through its investments and economic activities – plays an important role in bridging the digital divide. Private sector involvement is also seen in support for not-for-profit initiatives aimed at spanning the digital divide.

The Benton Foundation is a non-profit organisation that works to realise the social benefits made possible by the public interest use of communications.[72] The Benton Foundation produces and coordinates the Digital Divide Network (DDN).[73] The DDN was created to serve as a 'catalyst for developing new, innovative digital divide strategies and for making current initiatives more strategic, more partner-based and more outcome-oriented, with less duplication of effort and more learning from each others' activities'. DDN sponsors include the Ford Foundation, the Bill and Melinda Gates Foundation, Lucent, Intel and AT&T.

Bridges.org is another international non-profit organisation devoted to spanning the digital divide.[74] It undertakes research and promotes best practices for sustainable, empowering ICT use. Bridges.org provides information and resources on the digital divide, advises decision makers on key issues, provides expert advice to grassroots projects and e-government efforts, and helps in teaching basic computer and internet skills.

For-profit organisations are also involved in spanning the digital divide. Not surprisingly, most IT companies are involved in human capacity development efforts. Oracle has the Oracle Academic Initiative, Sun Microsystems the Java Competency Centre, and Cisco Systems the Cisco Networking Academy. Given the great demand for IT skilled individuals, these training initiatives are also geared to ensure the longevity of these corporations. These activities might be considered as emanating from enlightened self-interest.

Bridging the digital divide is also the aim of a number of corporate philanthropy efforts. Microsoft's international community affairs program aims to bring the benefits of ICT to disadvantaged people in countries where it does business.[75] Grants (in cash or software) and technical assistance are made to non-profit and community organisations through Microsoft's subsidiary offices. There are four aims: to support learning through the effective use of technology; to support access to IT skills training, with a particular focus on helping to provide opportunities for employment; to improve the effectiveness and productivity of non-governmental, charitable organisations by providing software and technical assistance; and to support disaster and humanitarian crisis relief efforts. In China, Microsoft is supporting the efforts of Project Hope – a non-government organisation (NGO) – to create five computer labs, or cyberschools. The project will involve teaching computer skills to disadvantaged youth, who will have internet access and the highest quality teachers and curricula in China. In Indonesia, Microsoft is working with Pact Indonesia to create six computer centres and offer IT skills development for disadvantaged youth.

Another example of private sector involvement in bridging the digital divide is the ed.venture project of the Coca-Cola Export Corporation and the

Foundation for IT Education and Development, a Philippine non-profit organisation. The ed.venture initiative provides computers and internet connectivity, training and post-training support services to high schools in the Philippines. In 2001, the initiative involved fifteen pilot high schools, with plans to increase the number by thirty-five. Ed.venture will also help to establish best practices in technology integration in the public secondary school system in the Philippines and, more broadly, in national policy formulation on ICT and its role in youth and community development and poverty alleviation.

The selection criteria for the ed.venture program include buy-in from the community, be it the local government, the parents and teachers association, the local school board or local NGOs and civic organisations. Unless the school (with its local partners) commits to generating the necessary resources to continue operating the centre after the one-year pilot period, that school will not be selected for the program.

Ed.venture is not only a technology deployment initiative but also a capacity-building and curriculum integration program. Through a comprehensive training package for teachers, administrators, students and technical staff, FIT-ED hopes to ensure that the school and the community can optimise the use of the technology for their own educational, social and economic ends. Ed.venture is a private sector response to the very real limitations faced by government in responding to the digital divide. Ed.venture is not unique in Southeast Asia. Sekolah 2000 in Indonesia is a similar initiative. It is a non-profit association of ninety member ISPs (there are about 150 ISPs in Indonesia) and aims to connect 2,000 (out of a total of 13,000) high schools in Indonesia to the internet. By using schools as an entry point to the community, ed.venture and similar projects are laying the foundation for greater community participation in the digital universe.

## FUTURE PROSPECTS

The digital divide is a multifaceted issue that will only be addressed with appropriate government policy, private sector participation and strategic community initiatives, including initiatives to deal with access to information tools, the economic vitality of communities, content that is relevant to and produced by communities, and the establishment of societies devoted to lifelong learning (DDN n.d.). In 'Spanning the digital divide: understanding and tackling the issues', Bridges.org said:

> Governments, businesses, individuals, and organisations have studied the issues at stake in the digital divide and drafted a range of valuable reports – from statistical analysis to in-depth case studies. ... Unfortunately, there is significant duplication of effort in these studies and recommendations, and too few of the suggestions are followed up in practice. *There is a lot of talk, but not enough action.*[76] [author's italics]

This is certainly an alarming conclusion. It is easy to justify the current condition by suggesting that we are at the early stages in grappling with the digital divide, but only since 1999 has the issue received significant attention from multilateral agencies and intergovernmental organisations.

A look at the history of reforms in developing countries will give us an indication of what to expect of efforts to bridge the digital divide. Brian Levy and Pablo Spiller in their comparative study of telecommunications reform suggested (Levy and Spiller 1996) that success depends on 'how a country's political and social institutions (its executive, legislature, judicial system, and informal norms of public behaviour) interact with the regulatory processes and economic conditions'. This is also true of the broader development effort. William Easterly, a former World Bank economist, in his review of development efforts, suggested (Easterly 2001) that governments and development agencies must get the incentives right if there is to be broad and deep development. He believes that there are four requirements for development efforts to prosper in a given country. First, government incentives must induce technological adaptation, high-quality investment in machines and high-quality education. Second, the poor must get good opportunities and incentives. Third, politics must not be polarised between antagonistic groups, and there be a common consensus to invest in the future. Finally, government must energetically take up the task of investing in collective goods such as health, education and the rule of law.

Getting the politics right is easier said than done, but ICT can help to generate the political consensus needed to help digital divide initiatives to succeed. E-government is another important tool in making government institutions more effective institutions in delivering public goods.

ICT could also help to make international development agencies more effective in combating the digital divide. David Morrison, President of NetAid.org, a joint endeavour of Cisco Systems and UNDP, has argued (Morrison 2001) that 'retooling the existing development industry, including putting some of its key processes online, would lead to strong efficiency gains, leaving more money to combat poverty directly'. Coordination among non-government groups is also made easier and cheaper with the use of ICT.

The digital divide will probably be the development issue of the twenty-first century. As such, it is important that policy advocates and policy-makers take to heart the lessons of the development debates in the last century. Unfortunately, this might mean that the underdeveloped countries of today will constitute the information-poor in a world where information is vital for development. For East Asia, the future may not be so bleak. Its biggest economies – Japan and China – are well aware of what is at stake. The newly industrialised countries of East Asia are poised to be major players in what Alfred Chandler calls the electronic century (Chandler 2001). In Southeast Asia, governments of the region are making a major effort to assist each other in creating an environment conducive for ICT development. It may yet be that in the twenty-first century the faded dream of the Pacific century will be realised.

## NOTES

The author acknowledges the research assistance provided by Ms. Patricia Pascual and Ms. Shelah Lardizabal in preparing this paper.

1. Bridges.org ,'Spanning the digital divide: understanding and tackling the issues', p.3, available at www.bridges.org.
2. http://www.dividedividenetwork.org/content/sections/index.cfm?key=2.
3. Bridges.org has twelve indicators to 'real access'; see www.bridge.org.
4. While the digital divide within nations is as important as the divide between nations, the focus of this paper is to look at the latter.
5. 'The internet, untethered: a survey of the mobile internet', *The Economist*, 13 October, 2001, p.4.
6. Minges seems aware of the issue but does not tackle the problem of content.
7. See Internet Software Consortium at www.isc.org/ds/faq.html.
8. See http://www.emarketer.com/estatnews/estats/eglobal/20010420_global_usage_itu.html?ref=wn.
9. See www.itu.int.
10. See www.nua.com.
11. eMarketer's *eAsia Weekly*, Issue 19, 2001.
12. Children's Partnership, 'Online content for low-income and underserved Americans: the digital divide's new frontier'. Available at http://www.ChildrensPartnership.org/pub/low_income/index.html.
13. Ibid. While the study focused on online content for low-income and underserved Americans, the findings are also relevant to the global digital divide.
14. Children's Partnership, p.8.
15. CyberAtlas, http://cyberatlas.internet.com/.
16. In the 'real' world, there are about 1.2 billion Chinese speakers, 478 million English speakers, 437 million Hindi speakers and 392 million Spanish speakers.
17. This is up from 28 per cent in 1999. See 'Web site globalization: the next imperative for the internet 2.0 era' at http://www.etranslate.com/include/idcrep.pdf.
18. Short for 'killer application', a term used for a software application that is wildly successful because it has better features than its competitors.
19. See Steve Butler, *The Road Ahead for B2B eCommerce*, at http://www.emarketer.com/analysis/ecommerce_b2b/20010202_top_ten_b2b.html. Regional as well as national differences in the business uptake of B2B e-commerce provide a view of the digital divide from the 'production' side. This is an important perspective but one that will not be pursued in the subsequent sections of this paper. The focus of this paper is the digital divide from the 'consumption' side.
20. See www.etranslate.com/include/stat1010.pdf.
21. 'Web site globalization: the next imperative for the internet 2.0 era'. Available at http://www.etranslate.com/include/idcrep.pdf.
22. 'New eAsia report indicates China and India will outpace Japan in internet growth over the next 4 years'. Available at http://www.emarketer.com/about_us/press_room/press_releases/051800_easia.html.
23. See www.estranslate.com/include/stat1010.pdf.
24. 'Web site globalization: the next imperative for the internet 2.0 era'. Available at http://www.etranslate.com/include/idcrep.pdf.
25. Cellular penetration was not included. See http://www.emarketer.com/analysis/easia/20010522_asia.html?ref=asw.

*The digital divide in East Asia* 295

26 'Asian internet users come out of the closet: first ever look into five regional Asian markets released by Netvalue', Netvalue Press Release in http://hk.netvalue.com. Subsequent figures are from this survey.
27 Estimates by APJII (*Asosiasi Penyedia Jasa Internet Indonesia* in *Indonesian Internet Statistics*. See http://www.insan.co.id/internet-stats.html.
28 Digital Filipino Philippine Internet Demographics. See http://www.digitalfilipino.com/content.asp?FileName= per cent5Cstatistics per cent5Cdemographics.ini.
29 Data for Hungary from BBC in www.bbb.com; data for the US, Uganda and Bangladesh from Wired News (2001) in www.wired.com.
30 'Asian trends: internet usage across the Asia Pacific region', Netvalue press release, http://hk.netvalue.com.
31 See http://www.emarketer.com/estatnews/estats/easia/20010508_tns_asia.html?ref=asw.
32 Gartner Dataquest, as reported in eMarketer's *eAsia Weekly*, Issue 11, 2001.
33 See http://www.nua.ie/surveys/index.cgi?f=VS&art_id=905356940&rel=true.
34 See eMarketer's *Access Technology Weekly*, Issue 2, 2001.
35 From Japan's Ministry of Post and Telecommunications as reported in eMarketer's eStat Database.
36 See eMarketer's *eAsia Weekly*, Issue 8, 2001.
37 See eMarketer's *eAsia Weekly*, Issue 11, 2001.
38 See eMarketer's *Access Technology Weekly*, Issue 3, 2001.
39 See eMarketer's *eDemographics Weekly*, Issue 8, 2001.
40 The following information is from eMarketer's *eAsia Weekly*, Issue 19, 2001.
41 See www.cnnic.net.cn/dev/stl/rep200107-e.shtml.
42 See http://www.chinaonline.com/topstories/010207/1/B101020601.asp.
43 See eMarketer's *eAsia Weekly*, Issue 16, 2001.
44 See http://www.chinaonline.com/topstories/001221/1/C00121507.asp.
45 'White paper on the ASEAN information infrastructure'. Available at www.fit-ed.org/easean/index.html.
46 See eMarketer's *Global Weekly*, Issue 9, 2001.
47 See eMarketer's *eAsia Weekly*, Issue 15, 2001.
48 The following information is from eMarketer's *eAsia Weekly*, Issue 17, 2001.
49 See eMarketer's *eAsia Weekly*, Issue 17, 2001.
50 See eMarketer's *eAsia Weekly*, Issue 14, 2001.
51 See eMarketer's *eAsia Weekly*, Issue 16, 2001.
52 See e-Japan Priority Policy Program, Document No. 3.1, East Asia ICT Cooperation Conference, 17 September 2001, Okayama, Japan.
53 Korean National Computerization Agency IT e-Newsletter, 25 September 2001.
54 See '8 lessons from the telecom mess', *Businessweek* 13 August 2001, pp.42–49.
55 See the fuller discussion of the e-agreement in a later section of this paper.
56 Clearly, e-government is not only the use of ICT by a government for its internal operation. As the UN-ASPA 'Global survey of e-government' argues, it is important to add to the definition of e-government its object to improve 'citizen access to government information, services and expertise to ensure citizen participation in, and satisfaction with the governing process'. See http://www.aspanet.org/solutions/egovbrochure.PDF.
57 World Markets Research Center *Global e-Government Survey 2001*, p.1. See www.worldmarketsanalysis.com.
58 See http://www.schoolnet.ca/home/e/whatis.asp.
59 See http://www.school.net.th/about/milestones/index.html.
60 See http://www.world-links.org/english/html/overview.html.
61 Telemedicine Information Exchange lists 204 telemedicine projects worldwide. See http://tie.telemed.org/.

62  See www.healthnet.org. See also www.opt-init.org for a discussion.
63  See http://www.bridge.org/internet.html.
64  See www.un.org/esa/coordination/ecosoc/itforum/icttaskforce.htm.
65  See http://sdnhq.undp.org/it4dev.
66  See www.apdip.net.
67  See www.apdip.net.
68  See www.worldbank.org/ict/.
69  See www.infodev.org.
70  See www.developmentgateway.org.
71  The breakdown of the forty-three participants is as follows: seventeen government representatives (eight from developing countries), seven representatives from international and multilateral organisations (ECOSOC, ITU, OECD, UNDP, UNCTAD, UNESCO, World Bank), eleven representatives from the private sector (one representative per each G-8 country and three global networks (GIIC, GBDE, WEF), and eight representatives from the non-profit sector (one representative from each G-8 country).
72  See www.benton.org.
73  See www.digitaldividenetwork.org.
74  See www.bridges.org.
75  See http://www.microsoft.com/giving/iprog.htm.
76  'Spanning the digital divide: understanding and tackling the issues', p.4. Available at www.bridges.org/

## REFERENCES

Accenture, Markle Foundation and UNDP (United Nations Development Programme) (2001) 'Creating a development dynamic. Final report of the digital opportunity initiative', http://www.opt-init.org.

APEC (Asia Pacific Economic Cooperation) (2001) 'Meeting new challenges in the new century', APEC Economic Leaders' Declaration, Shanghai, China, 21 October 2001: http://www.apecsec.org.sg/apec_organization/sectriat/sectriat.html.

Castells, Manuel (1996) *The Information Age: Economy, Society and Culture. Volume 1: The Rise of Network Society*, Oxford: Blackwell: 92–93.

Chandler, Alfred (2001) *Inventing the Electronic Century: The Epic Story of the Consumer Electronics and Computer Science Industries*, New York: Free Press.

Dhawan, Rajat, Dorian, Chris, Gupta, Rajat and Sunkara, Sasi K (2001) 'Connecting the unconnected', in *'Emerging Markets'*, *The McKinsey Quarterly*, 2001, 4: www.mckinseyquarterly.com.

DDN (Digital Divide Network) (n.d.) 'What do we mean when we say "digital divide"?': http://www.digitaldividenetwork.org/tdd.adp

DOT Force (Digital Opportunity Task Force) (2001) 'Digital opportunities for all: meeting the challenge', report of the DOT Force, including a proposal for a 'Genoa Plan of Action'. Available at http://www.g7.utoronto.ca/g7/summit/2001genoa/dotforce1.html

Easterly, William (2001) *The Elusive Quest for Growth: Economists' Adventures and Misadventures in the Tropics*, Cambridge: MIT Press (cited in DeLong, Bradford (2001) 'Crisis of development', *WorldLink*, 2001 (September/October).

ITU (International Telecommunication Union) (1998) 'World telecommunication development report', http://www.itu.int/ITU-D/ict/publications/wtdr_98.

Levy, Brian and Spiller, Pablo T. (eds) (1996) *Regulations, Institutions and Commitments: Comparative Studies of Telecommunications*. Cambridge: Cambridge University Press.

Lipsey, Richard (2000) 'Some implications of endogenous technological change for technology policies in developing countries', unpublished typescript of 18 January 2000.

Macklin, Ben (2001) 'Global broadband dominated by five countries', www.emarketer.com/analysis/technologies/20010927_at.html?ref=ed

Minges, Michael (2000) 'Counting the net: internet access indicators', http://www.isoc.org/inet2000/cdproceedings/8e/8e_1.htm

Morrison, David (2001) 'An outlet to growth', *WorldLink* 2001 (September/October).

Naughton, John (1999) *A Brief History of the Future: The Origins of the Internet*, London: Weidenfeld & Nicolson.

Norris, Pippa (2000) 'The worldwide digital divide: information poverty, the internet and development', paper for the Annual Meeting of the Political Studies Association of the UK, London School of Economics and Political Science, 10–13 April 2000. Available at http://www.ksg.harvard.edu/iip/governance/psa2000dig.pdf

OECD (Organisation for Economic Co-operation and Development) (2001), 'Understanding the digital divide', available at http://www.oecd.org/dsti/sti/prod/Digital_divide.pdf

Savage, James (n.d.) 'International public programs to provide broadband access to the internet: a report prepared for Industry Canada', available at http://broadband.gc.ca/english/resources/inter_summ_jan05.pdf.

Stiglitz, Joseph E., Orszag, Peter R. and Orszag, Jonathan M. (2000) (October) 'The role of government in a digital age', report commissioned by the Computer and Communications Industry Association, available at http://www.ccianet.org/govt_comp.php3/

Werry, Chris and Mowbray, Miranda (eds) (2001) *Online Communities: Commerce, Community Action, and the Virtual University*, Upper Saddle NJ: Prentice Hall PTR.

# 13 E-commerce: the work program of the World Trade Organization

*Edsel T. Custodio*

## INTRODUCTION

This chapter reports on the scope and progress of work on electronic commerce (e-commerce) arising under agreements of the World Trade Organization (WTO). It also presents a broad review of the work done by other intergovernmental organisations and private sector groups in the area of e-commerce by way of positioning the WTO's role and future direction of work in the broad spectrum of e-commerce development. For the purpose of this discussion, I define e-commerce as the 'production, distribution, marketing, sale or delivery of goods and services by electronic means'.[1]

The Uruguay Round of trade negotiations was launched in Punta del Este in 1986 and finished eight years later, in 1994. E-commerce was in its infancy while negotiations were going on, but its subsequent development increased the urgency for a stronger WTO agenda on e-commerce. Since the conclusion of the Uruguay Round, the WTO has achieved a number of landmarks. In December 1994, the Agreement on Basic Telecommunications Services (Basic Telecom Agreement) was agreed to in Marrakesh. This agreement put competitive regulatory principles in place and opened markets to international competition. By 2003, it will apply to the provision of international services. In December 1996, in Singapore, ministers agreed to continue negotiations towards the elimination of tariffs and, subsequently, consultations on non-tariff barriers on all information technology (IT) products. In March 1998 in Geneva, the Ministerial Declaration on Trade in Information Technology of the WTO (the Information Technology Agreement or ITA) was signed after forty-six countries (representing 93 per cent of IT world trade) agreed on it. At the Geneva Ministerial Meeting in May 1998, the WTO brought out the Declaration on Global Electronic Commerce, calling for a WTO work program on e-commerce and a moratorium on customs duties on e-commerce transmission. This work is being carried forward into the Doha Round of trade negotiations.

The work program was designed to be comprehensive on 'all trade-related issues relating to global electronic commerce, taking into account the economic, financial, and development needs of developing countries' and those involved were 'to report on the progress of the work programme, with any recommendations for action, to the Third Session' (in Seattle). It was also intended to clarify how commitments in liberalisation and the existing WTO legal framework apply to e-commerce, identify any lacunae therein, and see if other trade-related issues relevant to e-commerce were not covered.

In September 1998, in Geneva, the WTO General Council provided relevant bodies with a mandate to analyse issues relating to e-commerce under their jurisdiction. In July 1999, subsidiary bodies provided their first reports. Various articles in the Uruguay Round agreements – the 1994 General Agreement on Tariffs and Trade (GATT), the General Agreement on Trade in Services (GATS), the Agreement on Trade-Related Aspects of Intellectual Property Rights (the TRIPs Agreement) and the Agreement on Government Procurement – are potentially affected by developments in e-commerce, but WTO members are not fully in accord on how these will be addressed. In October 2000, a proposal to create a task force on cross-cutting issues did not receive consensus, so the General Council had to deal with it in special session. The e-commerce work program was reinvigorated in May 2001, and there has been consultation on the treatment of cross-cutting issues by the General Council.

This chapter will focus on four main areas of interest. The first concerns the trade-related issues within the WTO legal system, including Annex 1A of GATT, GATS, the TRIPs Agreement and the work of the Committee on Trade and Development. The second concerns reports of the WTO subsidiary councils and cross-cutting issues recommended for General Council attention. The third concerns access and enablement issues in e-commerce. Finally, I will summarise the issues and make some recommendations.

## E-COMMERCE ISSUES IN THE WTO LEGAL FRAMEWORK

### GATT 1994 and other multilateral agreements on trade in goods

Until the WTO legal framework emerged in 1994, the multilateral framework (GATT 1947) covered only trade in goods. Seven successive rounds of trade negotiations were undertaken to reduce tariffs and liberalise non-tariff barriers. During the Tokyo Round (ending in 1979), various rules and disciplines governing the conduct of trade came within the purview of GATT. The Uruguay Round advanced the scope of market access and rules commitments (covering agriculture, services, intellectual property and investment measures) and consolidated these disciplines on a single-undertaking basis.

E-commerce stands to benefit from these advances by its impact on the traditional trade in goods and in the promising area of trade in services. In the goods trade, the application of e-commerce has tremendously improved production processes, reduced order times and transaction costs and led to greater efficiency in distribution and marketing. As a service provider and media carrier, e-commerce has provided global coverage at the click of a keyboard, reduced costs and allowed greater affordability for a wide range of traded products and services.

E-commerce transactions were still in their infancy at the time of the Uruguay Round negotiations, so negotiators did not have to wrestle with jurisdictional problems such as which rules in the WTO legal framework applied to e-commerce and how. The assumption was that the products determined which law applied, but this created difficulties in the case of new types of products (for example, printed matter, software, music and other media, films and video games) that could be transmitted and remain in cyberspace but could compete in use and therefore be substitutable with real-life versions offered for sale and distribution through the traditional trade in goods. This raised a gamut of trade policy questions to do with classification and rules of origin, most favoured nation (MFN) treatment, national treatment, valuation and other rules. Much of the present confusion in trade policy application to e-commerce could be traced to this situation (see Panagiriya 2000).

Thus the main thrust of the work program of subsidiary councils is to identify the provisions in their respective jurisdictions which have relevance to e-commerce, analyse their impact, make appropriate recommendations and report to the General Council. Let us look at the mandate of the Council for Trade in Goods. My comments encompass and overlap with subsequent discussions in other bodies.

First, the Council for Trade in Goods is required to examine and report on aspects of e-commerce relevant to the provisions of GATT 1994, Annex 1A of GATT 1994 and the approved work program. This will include examination of market access for and access to products related to e-commerce; valuation issues arising from the Agreement on Implementation of Article VII of GATT 1994; issues arising from the Agreement on Import Licensing Procedures; the Agreement on Customs Duties and Other Charges as defined under Article II of GATT 1994; standards in relation to e-commerce; rules of origin issues; and classification issues.

GATS and its annexes include many market access commitments for e-commerce products (e.g. financial services and telecommunications) as well as the ITA. The original two-year 'Moratorium on Customs Duties on Electronic Transmission' (which began at the end of 1999) fell into this category. In goods trade, however, the liberalisation commitments laid out in individual member schedules remain unaffected by the emergence of e-trade. Therefore, the market access work program in goods provides the opportunity for members to embark on new initiatives that could benefit e-commerce development.

The next five agenda items in the work program refer to specific rules in goods trade which should be examined for applicability on goods ordered and paid for on the internet, whether they are delivered physically or remain in cyberspace. It should be borne in mind that there is a difference in some of the disciplines used under GATT and GATS.[2] For example, GATT contains a general obligation in respect of MFN and national treatment. Under GATT, customs duties and other charges, internal taxes and the procedures surrounding the imposition of these items are the core elements addressed by the MFN and national treatment obligation. As to the manner of imposing customs duties, GATT is bound to Article VII (transaction value) and its interpretation.

The GATS obligations on MFN and national treatment, on the other hand, apply only on sectoral and other specific commitments as well as the modes of supply covered by individual member schedules of specific commitments and the conditionalities therein. GATT also embodies a general prohibition on quantitative restrictions, while GATS allows their use as a limitation (on the basis of an 'economic needs test') to market access. GATT has a well-defined safeguards mechanism, but GATS is still struggling with the issue.

One of the main issues concerning e-commerce is whether, or how, taxes or duties should be levied on digital transactions. This issue gained importance when the Moratorium on Customs Duties from Electronic Transmission[3] achieved support in 1998 irrespective of its impact on the structure of WTO rights and obligations and the financial situation of many developing countries which rely on customs revenues for the bulk of their governmental and development requirements.

Standards, be they technical or voluntary, are the key to the interoperability of hi-tech equipment and access to competitive infrastructure essential to e-commerce and the internet. GATT seeks to facilitate this through such measures as mutual recognition arrangements (under the Agreement on Technical Barriers to Trade, also known as the TBT Agreement) and harmonisation and/or equivalence (under the Agreement on the Application of Sanitary and Phytosanitary Measures). GATT has not considered standards on electronic transmission, although physical goods pertaining to e-commerce are assumed to be covered by the TBT Agreement and technical or voluntary standards. In the case of the TBT Agreement, GATS refers to the possibility of harmonisation and bilateral or plurilateral agreements and the principle of equivalence.

The harmonised system of tariff nomenclature and the rules of origin pave the way for clearer and definitive classification of goods and their origin for the appropriate application of GATT rules on the goods in question. However, electronic transmission is not classified in the harmonised system or in members' individual tariff schedules.

If e-commerce is classified as trade in goods, then the obligations and commitments in most of the rules illustrated above should apply. If not, then GATS or something else applies. Since some of the GATT rules are absent or applied under different circumstances in the GATS framework, such rules cannot be legally enforced if the transactions are classified as trade in services. Hence, for a large number of member governments, it is deemed essential to see priority resolution of the classification issue (discussed in the Council for Trade in Goods and the Council for Trade in Services) although a few contend that other issues, such as development and competition dimensions, can be discussed in parallel.

## The Agreement on Trade in Services and e-commerce

Of all the agreements in the WTO framework, GATS undoubtedly contains the provisions that have the greatest impact on e-commerce. E-commerce is essentially a revenue-producing service and a medium for intangible service transmission and transaction. To say, however, that an electronic transmission and/or its contents can be characterised as a service is fraught with what might have been unintended implications. If the transmission and its contents remained in cyberspace (e.g. books read on screen or music played from the computer), they might be conveniently treated as services. But complications arise when they cross borders, when they are downloaded and consumed in physical form, or when 'like' or 'similar' products are available for sale, distribution and consumption in the same market. If e-commerce trade were to be continually duty free and classified as trade in services, no other GATT 1994 disciplines on trade in goods would be applicable. Similar or like products, conveyed in the normal physical way, would have been treated less 'favourably', hence violating the fundamental principle of non-discrimination, with the additional risk of distortion through a possible shift in transaction (to benefit from the treatment) from the normal way to electronic transmission. This raises numerous economic, legal and practical complexities.

Member governments, therefore, attach the greatest importance to the GATS Council mandate on classification (or 'characterisation' of electronic transmission and its digitalised contents). The mandate includes the classification of electronic transmission or digitalised products sent electronically and the scope of GATS with respect to delivery of services and digitalised products, particularly in the areas of technological neutrality and the distinction between cross-border trade and commercial presence.

Intrinsically linked to the issue of classification is the principle of technological neutrality. In intellectual property protection, technological neutrality suggests equal protection for similar or like products: for example, member governments have the obligation to enforce copyright laws on unauthorised copies of copyrighted materials. This requires online enforcement without boundaries. In e-commerce, technological neutrality suggests there should be no discrimination in the application of tariffs and taxes to 'like' products with different media of transmission. 'Like' products are those that have essentially the same end use (although recent WTO dispute settlement decisions would not be consistent with this simplistic treatment of this subject).[4] The term 'technological neutrality' also suggests that the existing WTO legal framework should, to the fullest extent, be made applicable to e-commerce transactions: no exclusive set of rules would have to be devised for it.

GATS distinguishes four modes of delivery: cross-border trade (mode 1), consumption abroad (mode 2), commercial presence (mode 3) and movement of natural persons (mode 4). Cross-border trade and commercial presence are the modes of delivery that are most frequently involved in e-commerce. However, a decision on whether e-commerce would be treated under mode

1 or mode 2 is fraught with controversy because of the impact of liberalisation commitments made during and after the Uruguay Round negotiations. This is because the services negotiation yielded more liberalisation commitments under mode 2 than under mode 1. If e-commerce were treated under the cross-border trade mode, more service exports and providers would be robbed of liberalised treatment. In disputes, there would also be practical problems as to the origination of the product (e.g. multi-country processing), where the transaction was finalised and under whose legal jurisdiction it belonged. Under the present GATS rules for supply under mode 1, the transaction is deemed to have taken place in the country where the buyer resides. Under mode 2, the applicable regulatory regime would be that of the country in which the supplier resides (Panagiriya 2000).

The work program also requires the Council on Trade in Services to address the following issues: MFN (Article II) and national treatment (Article XVII) as applied to 'like products'; transparency (Article III); increasing participation of developing countries (Article IV); domestic regulation, standards and recognition (Articles VI and VII); competition (Articles VI, VII and IX); protection of privacy and public morals and the prevention of fraud (Article XIV); market-access commitments on the electronic supply of services, including commitments on basic and value-added telecommunications and on distribution services; access to and use of public telecommunications transport networks and services (Annex on Telecommunications); and Customs duties.

As discussed above, GATS does not consider MFN and national treatment as a general obligation, but rather applies them only on sectors, specific service measures and modes of supply covered in each member's schedule of specific commitments and the conditionalities stipulated therein. This came about as a result of the bottom-up approach taken during the negotiations. The obligation on transparency, which refers to the entirely different subject of marks and labelling in the case of GATT, applies to these specific commitments.

Under GATS, developed and developing countries are treated equally. No a priori special and differential treatment is made available to developing countries, except as agreed to and resulting from the negotiations for progressive liberalisation and certain technical assistance possible on a multilateral level to be decided by the Council for Trade in Services. Through these negotiated commitments, developing countries are expected to strengthen their capacity and market competitiveness, improve access to global distribution channels and information networks, and progressively liberalise market access in sectors and modes of supply of export interest to them. However, Article XIX provides that:

> there shall be appropriate flexibility for developing members for opening fewer sectors; liberalizing fewer types of transactions and progressively extending market access in line with their development situation ...

The GATS framework provides for clearer and greater scope for competition rules than does GATT: anti-competitive behaviours such as monopolies, exclusive service suppliers and restrictive business practices are targeted, albeit in the spirit of transparency and consultation rather than dispute settlement processes. Again, the competition provisions are not a general obligation but are only applicable to sectors, measures, transactions and modes of supply covered by members' schedules of specific commitments. These, together with the canopy of other liberalisation, deregulation and privatisation measures unilaterally or multilaterally adopted so far, are designed to provide a competitive environment in which services in general, and e-commerce in particular, will flourish. A highly competitive environment is the key to developed countries gaining access to the widening world markets for the sale of costly e-commerce and internet technology and infrastructure and for developing countries being able to access them. Open markets and competence in e-commerce and internet activities allow all stakeholders, especially developing countries and small and medium enterprises (SMEs), a fair share in the business.

Finally, both governments and the private sector identify as priorities issues to do with protection (intellectual property, privacy), fraud prevention and the protection of public morals. They also see trust as being essential to the wider business and other applications of e-commerce.

## The Agreement on Trade-related Aspects of Intellectual Property Rights

E-commerce embraces a complex amalgam of technologies, infrastructure, processes and products, and trade conducted electronically has a relatively high intellectual property content. The council is therefore required to examine and report on the following intellectual property issues: copyright and related rights; trademarks; and new technologies and access to them.

Those who classify e-commerce as trade in services contend that a transaction starts with buyer and seller agreement and ends when the service or digitalised products are conveyed by electronic transmission. The sellers' responsibilities expire at the latter point, and what transpires thereafter is the responsibility of the buyer. If the buyer consumes them in cyberspace (e.g. reads books or hears music without transferring them to a media carrier such as a diskette), the transaction is completed, without complication, as services trade. But if the product is transferred, reproduced and sold in competition with 'like' physical products, there is an issue of intellectual property rights. As discussed above in the section on 'technological neutrality', this is consistent with the obligation to enforce copyright laws on unauthorised copies of copyrighted materials in 'The agreement on trade in services and e-commerce', above.

## The Committee on Trade and Development

The committee is required to examine and report on the development implications of e-commerce, taking into account the economic, financial and development needs of developing countries. Five main issues are to be examined. The first is the effects of e-commerce on trade and the economic prospects of developing countries, especially on their SMEs, and the means of maximising possible benefits accruing to them. The second is access to e-commerce infrastructure. The third is challenges to and ways of enhancing the participation of developing countries in e-commerce, particularly as exporters of electronically delivered products, through improved access to infrastructure and transfer of technology, and in the movement of natural persons. The fourth is the use of information technology in the integration of developing countries in the multilateral trading system. The fifth is the implications for developing countries of the possible impact of e-commerce on the traditional means of distribution of physical goods, and the financial implications of e-commerce for developing countries.

## REPORTS OF THE WTO SUBSIDIARY COUNCILS AND CROSS-CUTTING ISSUES RECOMMENDED FOR GENERAL COUNCIL ATTENTION

The first reports of the subsidiary councils were submitted to the General Council in July 1999. In general, the reports affirmed that various provisions in the WTO legal framework are potentially affected by development in e-commerce, but members were not in accord on how to address the issues. In the run-up to Seattle, a few member governments (including the United States, the European Union and Singapore) submitted proposals on some issues, particularly the issues of classification, access and enablement. However, the confusion in Seattle did not allow the WTO to discuss them. Only in the middle of 2000 did the General Council revisit the e-commerce work program and require subsidiary council reports to be updated. Discussions highlighted the limitations of any further work, without dealing first with the cross-cutting issues, especially the issue of classification.

A proposal to create a task force under the General Council was rejected by the general membership in favour of a dedicated series of discussions in a special session of the General Council. Consultations have focused on settling on the list of cross-cutting issues and the appropriate procedure for dealing with them. In the lead-up to the fourth ministerial council meeting in Doha, the General Council also considered the possibility of a renewed and reinvigorated ministerial mandate for e-commerce work in the WTO, taking into account a tentative list of cross-cutting issues in five main categories. The first category, of general issues, included e-commerce definition; the classification of electronic transmissions and contents (digitalised products or physical downloads); the concept of technological neutrality (non-discriminatory treatment of real-world and cyber products); and issues of 'like products'. The second category concerned the fiscal implications of e-commerce, including possible substitution effects between e-commerce and traditional forms of commerce and the imposition of customs duties on e-transmission. The third category concerned issues related to development, including access to infrastructure and technology; enablement and participation of developing countries; capacity building; technical assistance; technology transfer versus protection; access for developing country producers and suppliers to developed markets; promoting the use of information technology; how to ensure that benefits of e-commerce accrue to developed and developing countries equally; and the movement of natural persons. The fourth category concerned legal issues, including jurisdiction and applicable laws. The fifth category concerned competition policy, including ensuring that monopolies, cartels and restrictive business practices do not act as constraints on the development of e-commerce due to concentration of market power; competition and domestic regulation; and competition and intellectual property protection.

It is worth commenting on a few of the above elements. On the e-commerce definition, members agreed that the September 1998 definition provided by the General Council ('the production, distribution, marketing, sale or delivery of goods and services by electronic means') was a useful working definition. Since e-commerce is in a state of rapid evolution, they felt that a more precise and legally binding definition could wait until e-commerce had matured and until it was understood by a wider spectrum of stakeholders. To the private sector, the issues of definition and classification are considered complex and the possibility of consensus is remote at the moment.[5]

On the fiscal implications of e-commerce, one area of concern has been the remote possibility of capturing transactions that occur in cyberspace and imposing, if need be, appropriate duties, other charges and internal taxes. If this is considered impractical, it might be wiser to extend the moratorium. However, this decision has economic, legal and practical implications, as set out earlier. Various members believe the post-Doha work program should consider the fiscal and financial implications of the moratorium, particularly to developing countries, as a priority matter.

WTO members agree that development is a fundamental element in the work program, because e-commerce promises for developing countries and SMEs the greatest and quickest avenue for economic growth and progress. However, it cannot be assumed that these benefits would flow automatically or evenly between developed and developing countries. The issue of the digital divide and how it is addressed is central to this debate. It involves the twin elements of access (to basic infrastructure, technology, hard goods, interconnection and so on) and enablement (education, training, internet exposure, software and so on). Through the services negotiations, the ITA and other goods liberalisation commitments, the WTO may have played out its role as a stimulating agent: the issues may now be continuing implementation and perhaps remaining work concerning access to the internet and web sites.[6]

Various governments, intergovernmental organisations, professional groups and other business organisations are now deeply involved on the enablement side, particularly in the areas of trust and protection, albeit on a voluntary and 'self-regulatory' basis. This might require the WTO to take a back seat, to take stock of developments in their entirety. Member governments will then be able to assess the forms and time frames in which WTO involvement can provide impetus and value added.

## ACCESS AND ENABLEMENT ISSUES IN E-COMMERCE

There are five major WTO agreements affecting infrastructure access. The first is the GATS Agreement, which requires competitive access to telecommunications networks with provisions on domestic regulation (Article VI); monopolies and exclusive suppliers (Article VIII); and business practices (Article IX). The second is the Agreement on Basic Telecommunications Services, under which seventy-nine governments have agreed to open markets to achieve competitive public telecommunications services. This has paved the way for building up physical infrastructure through heavy investments, for dramatically reducing the cost of the leased circuits that internet service providers require for their customers and for providing more easily available, more affordable and higher-quality (digital and broadband) networks. Such facilities are necessary for the expansion of e-commerce. The third is the 'Reference Paper on Regulatory Principles in the Agreement on Basic Telecommunications Services', which applies to sixty countries that have adopted the reference paper into their specific commitments.[7] The fourth is the ITA of March 1998, under which fifty-two members, covering 96 per cent of global trade, have pledged duty-free trade in information technology products such as computers, telecommunications equipment, semiconductors and software. The fifth is the 'Moratorium on Customs Duties on Electronic Transmission': the 1998 ministerial declaration in Geneva agreed for members not to impose customs duties on electronic transmissions until the third ministerial conference in Seattle, at which time the matter was to be reviewed. The moratorium was further extended at the fourth ministerial conference in Doha.

Various intergovernmental and private sector organisations are involved in the area of enablement. The International Telecommunication Union deals with access to infrastructure, telecommunications interconnectivity, the promotion of the commitment to the reference paper on regulatory principles, technical standards for the industry, and the 'Electronic Commerce for Developing Countries' project for transfer of e-commerce know-how. The United Nations Commission on International Trade Law's 'Model Law on Electronic Commerce 1996', with the 'Guide to Enactment Draft Rules for Electronic Signatures and Authentication', deals with enablement issues. The World Intellectual Property Organization (WIPO) has a ten-point agenda in e-commerce, including broadened access of developing countries to intellectual property information, global policy formulation and the use of intellectual property assets in e-commerce; a WIPO arbitration and mediation centre; and a domain name dispute resolution service in cooperation with the Internet Corporation for Assigned Names and Numbers. The United Nations Conference on Trade and Development deals with development aspects of e-commerce and the Global Trade Point Program.

The Organisation for Economic Co-operation and Development is primarily interested in taxation issues, and the Basle Committee on Banking Supervision in the electronic payments system. The International Trade Center promotes general awareness of e-commerce opportunities and challenges to business and deals with business applications of e-commerce through the internet, especially for SMEs. The United Nations Centre for Trade Facilitation and Electronic Business (UN/CEFACT) has developed a model for electronic transaction agreement, and deals with issues such as electronic data interchange and e-business standards development. The International Chamber of Commerce deals with uniform commercial codes and the Uniform Rules for Electronic Trade and Settlement (URETS) deal with uniform rules in e-transactions. The 'General Usage for Digitally Ensured Commerce' (GUIDEC) initiative deals with terms used in internet transactions capable of being directly incorporated into business contracts and internet advertising guidelines. Finally, the International Organization for Standardization (ISO) provides technical standards on infrastructure.

The legal framework is an important issue for enablement. The central objective is trust (in the documents used, the electronic signatures, consumer protection, privacy, intellectual property protection and dispute resolution). Private sector business associations, professional groups and intergovernmental organisations, in cooperation with some governments, are devising the tools and necessary standards and protocols for this. With respect to the government approach to e-commerce work with the private sector, the 'default' policy is believed to be least interventionist – that is, private initiative should be encouraged to the utmost, with the 'least' restriction from government, but each should be ready to help each other where specific needs arise.[8]

## SUMMARY AND RECOMMENDATIONS

E-commerce transactions expanded rapidly after the conclusion of the Uruguay Round of trade negotiations in 1994. This has made urgent the need for a stronger WTO work program on e-commerce, focusing on greater commitments to liberalisation and assessment of the implications of the existing WTO legal framework for this rapidly expanding area.

Since the Uruguay Round, the WTO has contributed to greater market access in goods and services trade (with consequent increases in e-commerce transactions) through negotiated commitments such as the ITA, the Moratorium on Duties on Electronic Transmissions and increased access to competitive infrastructure and technology resulting from the Basic Telecoms Agreement and other services protocols. There might be remaining work to be done in web site and internet development, especially on behalf of small economies and SMEs, but recent technological developments and market signals indicate a need for some caution in moving in this direction.

The unpredictable and imprecisely defined nature of e-commerce trade up to this juncture, however, has presented a major policy challenge in determining whether there is a need for more or fewer rules. Aside from privacy, security and protection, the needs indications of e-commerce practitioners have focused on infrastructure access and enablement. There has been wider and deeper private sector involvement in the latter, albeit on a voluntary and self-regulating basis. The WTO might, therefore, need to take a backseat for a period, take stock, monitor the situation on behalf of its members and assess in what time frame and under what conditions it can ensure that greater intervention in this area can provide the appropriate stimulus and value added. There is a danger that regulation – both overregulation and underregulation – will give the wrong trade-distorting signals.

After nearly three years, the WTO work program, designed to address some of the issues considered essential (but not necessarily validated as such by the private sector), has not moved beyond identifying certain cross-cutting issues. How necessary is it to intensify the work program? If the objective of the WTO is to contribute more positively to the rapid development of e-commerce, the relevant question is whether e-commerce is in fact impeded without appropriate WTO rules in place. Is it necessary to ensure the effective application of rules for e-commerce to flourish?

The answers to these questions are still unknown. Until there are known cases of difficulties encountered in the trade and there is sufficient information to build policy options appropriate to handling these difficulties, it is not possible to come to any valid conclusion. Cases of electronic fraud, market dominance and hardcore cartels, as well as IPR violations, certainly abound, and these are being dealt with under existing national laws and international protocols. Private sector arbitration processes are also slowly addressing these issues. This does not, of course, detract from the importance of ongoing WTO vigilance to ensure that national regulatory practice is consistent with WTO principles.

E-commerce is a business system that has expanded and flourished to date with 'least regulation'. Private sector initiatives have, so far, provided sufficient 'self-regulation' to build up trust and enable the relevant stakeholders to participate the business. At this juncture, there is probably little relevance and little value added in an elaborate new framework of rules and standards.

Earlier initiatives to launch a negotiating agenda on e-commerce rules have recently been muted, probably for the above reason. The Doha Ministerial Declaration has now called instead for reinvigoration of the work program. What should be the appropriate elements in a reinvigorated work program?

From a development perspective, the issue of the digital divide, reflected in the access and enablement process, is crucial. Technology transfer is an important element in access, as is capacity building and technical assistance.

The movement of natural persons is also important, and is embraced already under the negotiations on trade in services mandated by the Uruguay Round. E-commerce could have both positive and negative implications for this issue – as information technology workers might need to move to countries that lack information technology or service providers offer offshore services or move operations abroad.

It is also worth considering how e-commerce could benefit from increased liberalisation commitments in related IT sectors that will be covered in the new multilateral round of negotiations launched at Doha. It is worth asking which sectors are of interest to developing countries.

E-government is an important element in this equation. It plays the key role in creating a favourable domestic e-enabling environment through the elaboration of national strategies for its overall development and a sound regulatory and policy framework. There is also a wide range of government applications in e-commerce that will showcase the efficacy, efficiency, transparency and speed of the system. Government provides an important market also for software development and application, particularly for budding SME and e-commerce/internet entrepreneurs. Certainly, the implications of e-commerce for negotiation of a new multilateral agreement on transparency in government procurement deserve some attention. The same goes for competition policy and rules affecting the behaviour of multinational services firms and suppliers. The GATS competition framework might need to be vigorously implemented in this sector.

By itself, the WTO does not possess the sole, or even the most relevant, mandate for most of these development and other issues. It must network and encourage supportive initiatives from individual governments, other intergovernmental organisations and private sector business associations. The WTO might also have some responsibility in providing technical cooperation and institutional capacity building, particularly for the least developed countries.

The WTO, nevertheless, is the only intergovernmental organisation with authority in international rules making. It is essential, therefore, that the WTO monitor developments in e-commerce and internet trade in all areas. It must provide the networking mechanisms to keep its members up to date on these developments as they impact on global trade policy in general and market opportunities in particular. The workshops, seminars and discussions held under the auspices of the work program, particularly those relating to development issues, have been very useful.

Through such mechanisms, the WTO must provide the necessary practical and intellectual underpinnings for a continuing call for deeper and wider liberalisation. WTO vigilance can help to prevent a proliferation of new trade-distorting regulations in this vital new area. Through negotiations in relevant related sectors, the WTO can also bring about the adoption and implementation of relevant rules and disciplines to ensure ongoing rapid growth in e-commerce, in related goods and services trade and in global economic welfare more generally.

## NOTES

1. General Council meeting, 25 September 1998.
2. Table 13.1 compares the treatment of various disciplines under GATT 1994 and GATS. A more elaborate discussion is available in Telcher (2001).
3. *Declaration on Global Electronic Commerce*, 2nd Ministerial Council of the WTO, Geneva May 1998.
4. See Appellate Body decisions in the EC asbestos and US lamb, frozen and chilled lamb meat cases and others.
5. WTO Seminar on Electronic Commerce and Development: Report. World Trade Organization Committee on Trade and Development, Geneva, WT/COMTD/18, 23 March 1999.
6. WTO Seminar on Electronic Commerce and Development: Report. World Trade Organization Committee on Trade and Development, Geneva, WT/COMTD/18, 23 March 1999.
7. WTO Seminar on Electronic Commerce and Development: Report. World Trade Organization Committee on Trade and Development, Geneva, WT/COMTD/18, 23 March 1999.
8. WTO Seminar on Electronic Commerce and Development: Report. World Trade Organization Committee on Trade and Development, Geneva, WT/COMTD/18, 23 March 1999.

## REFERENCES

Panagariya, Arvind (2000). *E-commerce, WTO and Developing Countries*. United Nations Conference on Trade and Development, Geneva.

Telcher, Susan (2001). *Tariffs, taxes and electronic commerce: revenue implication for developing countries*. United Nations Conference on Trade and Development, Geneva.

Table 13.1  Comparative treatment of various disciplines under GATT 1994 and GATS

| Discipline | GATT 1994 | | | GATS | |
|---|---|---|---|---|---|
| Most favoured nation | Art. I | General obligation applied on customs duties/taxes | | Art. II | Only on measures in the schedule of commitments |
| National treatment | Art. III | General obligation applicable on internal taxes/trade regulations | | Art. XVII | Only on measures in the schedule of commitments |
| Transparency | Art. IX | Marks of origin and labelling | | Art. III | Only on measures in the schedule of commitments |
| Quantitative restrictions | Art. XI | General prohibition | | Art. XVI | Other regulations and 'economic needs' test allowed through the four modes of supply as per schedule of commitments |
| | Art. XIII | May be imposed on imports, but non-discriminatory | | | |
| Increasing participation of developing countries | Art. XXVIII | Bis Tariff Negotiations on reciprocal and mutually advantageous basis with due to regard to the varying needs of individual contracting parties | | Art. IV | Facilitated through negotiated liberalisation and specific commitments in sectors and modes of supply of export interest to them, strengthening domestic services capacity and access to distribution channels and information networks |
| Import licensing | Per the Agreement on Import Licensing Procedures | Transparent, fair, equitable and not restricting trade | | None | |
| Domestic regulations | None | | | Art. VI | Reasonable, objective and impartial |
| Standards | TBT Agreement, through MRAs | | | Art. VII | Recognition and licensing to consider qualification and technical standards |
| | Art. XI:2(b) | On agricultural commodities, SPS equivalence | | | |
| Valuation | Art. VII | | | None | |
| Rules of origin | Agreement on Rules of Origin | Harmonisation of non-preferential rules of origin | | None | |
| Emergency safeguards | Art. XIX and Agreement on Safeguards | | | Art. X | Emergency safeguards still subject to negotiations |
| Competition | None | | | Art. VI | Domestic regulation |
| | | | | Art. VIII | Exclusive business suppliers |
| | | | | Art. IX | Business practices |
| Subsidies | Art. XVI Agreement on Subsidies and Countervailing Duties | Income or price support | | Art. XV | Consultations but no multilateral trade discipline in place |

*Notes*: SPS – 'Sanitary and Phytosanitary Measures'; TBT Agreement = Agreement on Technical Barriers to Trade; MRAs = mutual recognition agreements.

# 14 Implications for APEC

*Mari Pangestu and Sung-Hoon Park*

## INTRODUCTION

APEC was buffeted by the outbreak of the Asian financial crisis in 1998. Under the immediate influence of the financial crisis, APEC members were unable to launch the 'early voluntary sectoral liberalisation' (EVSL) initiative, for which they had negotiated throughout that year. This was counterproductive (because continued trade and investment liberalisation was necessary to underpin regional economic reform and recovery), and also dealt a blow to the global momentum then building towards new World Trade Organization (WTO) negotiations. The failure of the EVSL diminished APEC's credibility in constructing a relationship with the rest of the world with open regionalism as a conduit to the WTO. The Asian crisis also triggered rethinking in many Asian developing APEC members about their international strategies to promote development.

In achieving its goals, APEC has to find an appropriate balance between its two mutually reinforcing cooperation pillars: the 'Trade and Investment Liberalisation Framework' (TILF) and 'Economic and Technical Cooperation' (Ecotech).

Developing and/or transforming countries assume the rotating presidency of APEC until 2005 – China in 2001, Mexico in 2002, Thailand in 2003 and Chile in 2004. Accordingly, there is some expectation that Ecotech activities may receive a stronger focus than previously. There is also a risk that excessive focus on Ecotech alone could diminish the momentum within APEC for trade and investment liberalisation, and undermine APEC's ability to make a constructive contribution to success of the new WTO round recently launched at Doha.

It is against this general background that we look at the implications of the new economy for APEC's future agenda. We examine the regional opportunities and challenges presented by the new economy, we discuss the importance of liberalisation and deregulation in the information, communication and technology (ICT) sector and we set out some priorities for regional capacity building.

## THE NEW ECONOMY: OPPORTUNITIES AND CHALLENGES FOR APEC

The 'new economy' has been changing the landscape of the world economy over the last decade. A recent OECD study (OECD 2001a) suggested that the new economy now plays a critical role in the growth performance of both member and non-member countries and points, in particular, to the importance of new investment in the ICT sector for achieving and sustaining long-term growth.

The APEC region is a diverse one, its members ranging from economies with leading-edge research and development (R&D) capacities to economies with the ability to adopt and adapt technologies, and economies with underdeveloped capacities to benefit from the new economy. The new economy provides both opportunities and challenges for APEC and its member economies.

On the one hand, and of critical importance, continued rapid advancement of the ICT sector could assist in securing full recovery from the effects of the Asian financial crisis and a return to long-term sustainable growth paths. There is a danger, on the other hand, that the 'digital divide' could widen, increasing the inequality between the technological haves and have-nots within APEC. The challenge is to minimise the digital divide by enhancing access to new technologies. APEC must orient its capacity-building activities to ensuring that all its members can enjoy the new digital opportunities, to facilitating the transfer of sustainable knowledge from advantaged to disadvantaged regional economies, and to supporting the development of appropriate domestic strategies.

APEC has risen to this dual policy challenge, and extensive work is under way on the new economy. Indeed since 2000, the new economy has become one of the central themes of APEC's work program and '*the*' focus of APEC capacity-building activities.

Table 14.1 presents selected APEC activities, categorised into five main areas of cooperation: TILF, Ecotech, strengthening markets, the new economy and education.

## THE IMPORTANCE OF LIBERALISATION AND DEREGULATION IN THE ICT SECTOR

The activities in Table 14.1 can be grouped roughly in terms of the dual policy challenges outlined above. The first category includes measures oriented to expanding macroeconomic growth potential – fostering knowledge-based economies, promoting entrepreneurship and start-up companies, developing the Asia Pacific information infrastructure (APII) and building a digital society more generally through the recent e-APEC initiative). The second category covers measures designed to bridge the digital gap (strengthening human capacity building, strengthening APEC social safety nets, establishing networks of skills development centres and fostering the use of information technology

Table 14.1 Selected APEC capacity building initiatives and activities

| Cooperation areas | Projects and initiatives | Initiating forum |
|---|---|---|
| TILF[a] | APEC strategic plan for capacity building for WTO agreements | Committee on Trade and Investment |
| Ecotech | Human capacity building | Senior Officials Meeting |
| | Knowledge-based economies | Leaders Summit |
| | Strengthening APEC social safety nets | Leaders Summit |
| Strengthening markets | Entrepreneurship and start-up companies | Committee on Trade and Investment |
| | Network of skills development centres | Human Resource Development Workgroup |
| | Asia Pacific Information Infrastructure | Telecom Ministerial Meeting |
| | Use of IT in a learning society | Human Resource Development Workgroup |
| New economy | Digital divide into digital opportunities | Senior Officials Meeting |
| | Readiness assessment evaluation partnership | Senior Officials Meeting/ Telecommunications and Information Working Group |
| | e-APEC: 'Building Digital Society' | e-APEC Task Force |
| Education | APEC Education Foundation | Leaders Summit |
| | APEC Cyber Education Cooperation | Leaders Summit |

Source: Park (2001) Information selected and rearranged by Sung-Hoon Park from a Korean government source.

Note
a Trade and Investment Liberalisation Framework.

in a learning society). APEC must find a balance between these two groups of activities.

A recent OECD study (OECD 2001b) provides some pointers on how this balance might best be struck. The study suggested that the *use* of ICT may be more important than the *production* of ICT in generating new economic growth potential. Promoting and enhancing the use of ICT, rather than merely the production of it, should therefore be at the core of any strategic approach taken by APEC, and its individual member economies should take this into account by promoting the use of ICT rather than merely its production.

Policies designed to make ICT products and services more available and cheaper for all member economies can be most effectively pursued through a more liberal flow of trade and investment. The TILF agenda, therefore, is an essential and conditional supporting mechanism for any capacity-building initiatives and activities associated with the new economy. The new economy is itself driving a powerful policy reform/liberalisation agenda, especially in the services sectors. There is, therefore, a tight connection between the two pillars of TILF and Ecotech. Traditionally, capacity building is seen as essential back-up support for liberalisation, but liberalisation, especially in the ICT-related sectors, is itself delivering both resources and capacity.

There have also been a number of explicit trade-related initiatives in APEC, focused on liberalisation of the new economy. One example was the consensus achieved within APEC to lower impediments to the trade of ICT-related goods and services. This meant that in 1996 APEC was able to assemble in the WTO the critical mass to support negotiations on the Information Technology Agreement (ITA) to reduce tariffs on ICT products. Similarly in 1999, APEC committed to a moratorium on the imposition of customs duties on electronic transmissions, providing the critical mass for wider ongoing WTO action.

Now that a new round of multilateral trade negotiations has been launched in the WTO, it is important that APEC continue to set an example by preserving – indeed enhancing – a least trade restrictive approach to the ICT sector.

Under the TILF agenda, and in the various APEC working groups, it has been recognised that much can also be done to set up the facilitating infrastructure for the new economy – for example, through trade and competition policy dialogue, investment facilitation and the sharing of international best regulatory practice. Such activities are already under way in APEC with regard to key backbone services such as telecommunications. An e-APEC task force formed in 2001 will bring many of these sorts of activities into one pool under one umbrella. It will be very important to ensure that the private sector is involved fundamentally in the task force's work.

## BUILDING CAPACITY FOR THE NEW ECONOMY

The concept of capacity building in APEC is evolving and the implementation and effectiveness of capacity-building activities are under review. By 2000, a number of shortcomings were apparent in the APEC approach. The actual identification and implementation of Ecotech projects had been left to individual members and to the many existing APEC working groups and other fora, resulting in a proliferation of projects without sufficient focus or concentration. The diversity of initiating fora, illustrated in Table 14.1, combined with insufficient coordination among them, led to overlapping of activities and jeopardised the effectiveness of their implementation. The ongoing activities also stood in an unfavourable relation to the budget of $2 million per year, and the scope of individual Ecotech projects was consequently limited.

On the more positive side, the variety of sources and the menu of capacity-building options available, from a sharing of policy experience to a sharing of resources, has meant that member countries have plenty to choose from.

Moreover, capacity building is not only a government-to-government activity: the private sector has been active in this area, and this must continue to be encouraged.

The relative ineffectiveness and lack of focus of Ecotech activities has been recognised, and in 2000 more emphasis was given to the role of capacity building, with APEC leaders in Brunei agreeing to focus capacity building on the new economy and on human capacity building in the new economy in particular. The follow-up work on this issue throughout 2001 in Beijing has again drawn attention, however, to the importance of open international markets to ensure cross-border flows of services and information. This is particularly vital in the provision of education and education services. If human capacity in the new economy is to be built effectively throughout the APEC region, all APEC economies will need access to the highest quality of education, training and skills; exchange of information; international best practices; and policy options.

After much debate, a new approach to Ecotech was devised in Brunei in 2000, and the notion of individual Ecotech action plans (EAPs) was introduced. At the October 2001 APEC Ministerial Meeting in Shanghai, ministers welcomed the fact that as many as sixteen EAPs have now been prepared and submitted on a voluntary basis, focusing on human resource development needs. Once the EAPs are drawn up, APEC, other APEC members or international donor agencies can help to implement them. Obviously, there is also a role for the private sector. In essence, the EAPs provide a means for economies to identify their own capacity-building needs and priorities. The underlying rationale is that there is a clearer idea at the receiving end of what kind of capacity building is most required.

## THE WAY AHEAD: A ROLE FOR APEC IN THE NEW ECONOMY

APEC's chief role in the new economy is to act as a catalyst for global trade and investment liberalisation and for capacity building and infrastructure development.

APEC itself cannot fund the region's huge needs for capacity building, which would require massive investment in infrastructure to ensure access as well as education and training. Some consolidation and improved clustering of capacity-building activities is required to enhance efficiency, and there is ongoing room for improvement in working structures and modes of delivery, especially for human capacity building in the new economy. APEC is making good progress on all these fronts.

Most importantly, APEC must, over the next three years, contribute positively to the success of the new WTO round of trade negotiations. The new round offers the single best chance for all APEC governments to reap the economic rewards associated with market opening and trade reform in the ICT-related sectors. There will be many opportunities for APEC economies to initiate contributions, including as a group, in many negotiating areas, especially services and industrial tariffs.

Finally, there is an emerging need for APEC to enter new areas of policy discussion and debate. Australia's experience, as a first mover in ICT usage, provides some insight. Studies undertaken by the Australian National Office for the Information Economy (Kennedy 2001) indicate that the application of e-commerce has very substantial impacts on individual firms, as well as the wider economy, and that there is much more to come. New thinking is required to deal with the fundamental discontinuities associated with the diffusion of ICT-based innovation (Rimmer 2001). Governments are in global competition for the capital and skills required to successfully execute the transition to the new economy.

There is an emerging need, therefore, to shift away from national approaches to industry policy, to focus more on international linkages and metropolitan capabilities in both a regional and global context. APEC governments will need to work more closely together to support internationally oriented entrepreneurial activity and new firm creation throughout the region. A role for APEC is emerging, perhaps, in industry policy.

One thing is clear. The use of ICT is only just beginning, even in the developed member economies of APEC. Despite the enormity of the change experienced to date, the scope for change is far from fully realised. There is much work to be done.

## REFERENCES

APEC (2001) 'APEC High Level Meeting on Human Capacity Building', *Beijing Initiative in APEC Human Capacity Building*, Beijing, 15–16 May 2001.

Kennedy, David (2001) 'Findings from 34 case studies on the use of e-commerce', paper presented at Pacific Trade and Development Conference 27, 'The New Economy: Challenges for East Asia and the Pacific', The Australian National University, Canberra, 20–22 August 2001.

OECD (2001a) 'The New Economy: Beyond the Hype: Final Report on the OECD Growth Project', DSTI/IND/STP/ICCP(2001)2/FINAL, 11 May 2001, Paris.

OECD (2001b) 'Firms, workers and the changing workplace: considerations for the old and the new economy', PAC/AFF/LMP(2001)3, 201, Paris.

Pangestu, Mari, 'The Role of APEC and the New Economy, paper presented at Pacific Trade and Development Conference 27, 'The New Economy: Challenges for East Asia and the Pacific', The Australian National University, Canberra, 20–22 August 2001.

Park, Sung-Hoon (2001) 'Implications for APEC and Regional Capacity Building', paper presented at Pacific Trade and Development Conference 27, 'The New Economy: Challenges for East Asia and the Pacific', The Australian National University, Canberra, 20–22 August 2001.

Rimmer, John (2001) 'Australian government policy for the new economy', paper presented at Pacific Trade and Development Conference 27, 'The New Economy: Challenges for East Asia and the Pacific', The Australian National University, Canberra, 20–22 August 2001.

# Index

Agreement on Trade-Related Aspects of Intellectual Property Rights (TRIPs) 212–5, 219n, 299, 304 *see also* intellectual property
agriculture 273, 277
aircraft industry 55
Amazon.com 231
America Online (AOL) 181
anti-trust 210, 235n, 253–4, 304 *see also* competition
APEC *see* Asia Pacific Economic Cooperation
application service provider (ASP) 192–4
Apple 222–3
ASEAN *see* Association of South East Asian Nations
Arpanet 75
Asia Pacific Development Information Program 288–9
Asia Pacific Economic Cooperation 1, 10, 267, 271n, 290, 313–18: e-APEC Task Force 2, 290, 314; Ecotech 315–18; ICT liberalisation and deregulation 314–16; policy role 314; presidency 313; projects 315; Trade and Investment Liberalisation Framework 315; *see also* new economy; World Trade Organization
Asia Pacific Regional Operations Centre Plan 169
Association of Southeast Asian Nations 257, 266–7, 269, 271n, 278, 283–90; e-ASEAN Task Force 290; Framework Agreement on Information and Communications Technology Products, Services and Investment 290
AT&T 23, 75, 291: break-up 79, 86n
auctions 232, 236n: online automobile auctions 257
Australia xiv, 24, 266, 280, 318
Automotive Network eXchange (ANX) 264
automotive industry *see* motor industry

B2B *see* business-to-business
B2C *see* business-to-consumer
Bangladesh 280
banking 7, 47, 225, 308: bank-cards (China) 132; Basle Committee on Banking Supervision 308; e-banking 123; i-flex (India) 190–3; Microbanker 190–3; retail 191

Bell Laboratories 75,135
Benton Foundation 291
biotechnology 55,112, 206: biogenetics 206; patents 206–8, 216
Boskin Advisory Committee (on consumer price index) 20
Bridges.org 291–2
broadband 4, 65–9, 78–80, 83, 84n, 162, 285: bandwidth 123, 245; cable television 79; connectivity 70, 74; consumers 72–4, 85n, 173; e-commerce 70; entertainment 71; home shopping 71; internet 247, 281; investment 83, 307; Japan 282; market-led strategy 286; music 282; penetration 67–8, 281; PlayStation 179–80; policy 285; price elasticities 73, 76; rural areas 286; South Korea 282; voice telephony 71
Brunei 267, 317
business-to-business (B2B) 228–33, 238: e-commerce 278, 294n; gofish.com 229; market research on exchanges 231; prices 231–2; software 231
business-to-consumer (B2C) 231–3, 278: online shopping 232

cable television services 67, 71, 75, 79: regulation 80, 86n
Canada: cable television 79; Canadian Radio-television and Telecommunications Commission 86n; e-learning 287; information highway 287; regulator 86n; SchoolNet 287; unbundled networks 78
capital investment: cross-country studies 217; economic life of capital equipment 225; information and communication technology 148; intellectual property protection 203–19; manufacturing companies 116, 120; *see also* innovation; investment; venture capital
capitalisation: falling investment (India) 114; R&D 93–4, 99, 101; venture capital 24, 102
cars *see* motor industry
cement production 120
Children's Partnership 276–7, 294n
China: China Internet Network Information Centre 282; computers 131, 208; Council for Economic Planning and Development

162; e-commerce 133,138; electronics industry 172, 218n; exports 132–7; geographical investment 136; Haier 135; imports 134; information technology 130, 172; information technology policy 133, 169n, 208; internet use 282; investment structure 136; IT companies 132; Konka Company 135; Legend 135; motor industry 257; 'one-China policy' 142; piracy 212; software 136, 208, 212, 218; Taiwan 141–5; telecommunications carriers 250; telemedicine 288; US economic slowdown, impact of 137; World Trade Organization 130, 137–9n; Zhongguancun 208, 218
Cisco 137, 181, 291–3
Citibank 189: Citicorp 173, 189, 193; Citicorp Overseas Software Limited 189
Citil *see* i-flex
competition: anti-competitive behaviour 210, 253–4, 304; competitive advantage 198, 221; drivers 249–55; first-mover advantage 207; five forces model of competition (Porter) 244–9; GATS framework 304; India 109, 114, 120, 196; Indonesia 246–68; information and communication technology policy 284; internet service providers 82, 246–8, 253–4; internet service provider industries 244–9; local-loop 78; local telephony 87n; markets 220–36, 244–50; monopolies 217, 250; motor industry 269; network tournaments 220–3; regulation 77, 223, 233, 250–6; telecommunications 77–8; vertical integration 79; world trade 306; *see also* markets
computer-aided design 118: motor industry 260, 271n
computers 23–5, 44, 64, 75, 124: Apple Macintosh PCs 222–3; architecture 222; China 131; computer-aided design 206; Dell 261; Galaxy (China) 132; games 174; industry convergence 243; penetration 274; *see also* electronics
Confederation of Indian Industries 115
contract law 206, 226: business-to-business agreements 228–33; intellectual property protection 206; *see also* intellectual property
contracting out 224, 227
copyright 204, 209, 302–4: China 212; copying 214; fair use exception 213; *see also* intellectual property
corporate strategy: chief executive officers 227; i-flex (India) 190, 197; information technology firms 172–202; marketing strategy 193–5; Sony PlayStation 176–83; uncertainty 224–5; value chain 193
corporate structure 224, 227, 269: vertical integration (telecommunications) 254
cost function (economics) 93–4, 103n: aggregate cost 97; cost elasticity 97–8, 100–1, 104n; Leontief cost function 96; production-cost approach 93; Sheppard's lemma 96

Council of Economic Advisers 22, 24, 30n, Covisint 264, 270n
critical success factors: manufacturing companies (India) 117; venture capital (India) 126
customer relationship management 192, 200, 214, 232; *see also* business-to-consumer
customs duties 300, 303: electronic transmission 300
cyberlaw 125: India 125–7; Information Technology Act 127; Malaysia 284
cyberspace 300–6

deregulation 23, 211, 304: ICT sector in APEC region 314–16; telecommunications 60; *see also* regulation
developing countries 56: Committee on Trade and Development (WTO) 304–5; e-commerce 298, 306, information and communication technology policy 273; information infrastructure 241, 304–6; literacy 276–8; market orientation 56; telecommunications policy 81, 240
digital divide 2, 64, 84n, 163–5, 170n, 240, 270–2, 294n–297n: access to information and communication technology 274, 278–84; APEC region 314; children 276–7, 294n; computer penetration 274; defined 274–6; Digital Divide Network 274, 291, 296n; disabled users 285; East Asia 273–97, 314; English-language speakers 276–8; gender 280–1; global and regional initiatives 288–90; health care 290; international development policy 293; internet 274, 279; Japan 274, 280; Korea 284–5; lifelong learning 292; literacy 276–7, 281; Microsoft 291; Organisation for Economic Co-operation and Development 274; Project Hope 291; private sector initiatives 290–2; state policy 284–8; teledensity (telephone density) 274–85; universal service 294
digital economy 171n, 206, 236n, 301
digital markets 236n, 301
Digital Opportunity Task Force 273
Digital Powerline 253
digital subscriber line 66–9
Doha Round 298, 306–13
dot.coms 22, 192, 211, 240

e-commerce 9, 61, 83, 213, 218n, 298–312: access and enablement issues 307–8; Asia Pacific 238, 278; authentication 307; broadband 70–1; business-to-business 125, 192, 227–8; China 133; commerce 305; Committee on Trade and Development 304–5; corporate strategy (India) 125; customs duties 300; Declaration on Global Electronic Commerce 298; digital transactions 301, 308; electronic data interchange 308; electronic signatures 307; English-language dominance 277; fraud 304; India 122–3, 129n; intellectual

property protection 206, 213; Latin America 278; Model Law on Electronic Commerce 307; motor industry 260; Singapore 283; supply chain management 125; Taiwan 141, 146, 158; World Intellectual Property Organization 307–8; World Trade Organization legal framework 298–312; World Trade Organization work program 298–312
economics: aggregate production function 39; economic value added 182–3; economy of scale 205, 209, 224, 243–4; economy of scope 175, 224–5, 243; externalities 103, 177; growth driven by technological change 33–59; information 220–36, 285; invisible hand 220, 236n; neoclassical theory 36, 51, 89, 234n; network models 222–3; Phillips curve 17; policy 41–3, 49, 120, 285; policy (US) 30n; productivity growth 21, 49; structuralist-evolutionary model 39, 54, 57n
economy: aggregate economy 93, 103; economic needs test 301; productivity growth 98, 120; *see also* new economy
e-government 283, 286
e-health *see* telemedicine
e-learning 287, 291
electricity 23, 38, 49–50, 96, 110: generation 42; Indonesia 253; markets 223, 226, 234n
electronic data interchange 259, 308; *see also* e-commerce
electronic data processing 192
electronics industry 34, 131, 141, 175: databases 209, 215; PlayStation (Sony) 176–83; Taiwan 142–3, 147, 149, 152–8; *see also* computers
employment 13–7, 101: chief executive officers 227; decentralisation 181; decline in manufacturing 26; employee monitoring 227; Indian manufacturing companies 116; labour costs 238; organisational structure 195–6; project management 183–7; training 114, 314; *see also* human capital
entrepreneurs 109, 145, 213, 217, 286, 318
e-trade *see* e-commerce
European Union (EU) 78, 214: Directive on the Legal Protection of Databases 215; internet use 280; telecommunications pricing 62, 84; telecommunications regulation 86n
exports: 27, 34–5; China 132, 137; software (India) 121

FDI *see* foreign direct investment
Federal Communications Commission 75
Flexcel International 192, 195
Flexcube 190–8
Ford Foundation 291
foreign direct investment 10, 29, 88–9, 112–13, 120, 125, 211, 216: China 134; Hong Kong 145; Indonesia 245, 255; Taiwan 142–5
Fortune 500 companies 122–4
Fortune 1000 companies 187

G-8 289–90: Charter on Global Information Society 289; Digital Opportunity Task Force 273, 289; Japan 2000 summit 289
games 174: Final Fantasy XI 199, 200n; Parappa the Rapper (Sony) 184–7, 198; PlayStation (Sony) 176–83; project management 183–7; Sega 177–8; Sony 176–87, 197, Ultima online 199; X-box 199; *see also* PlayStation
GATT *see* General Agreement on Tariffs and Trade
GATS *see* General Agreement on Trade in Services
gender 280–1
General Agreement on Tariffs and Trade 10, 299–304
General Agreement on Trade in Services 9, 10, 299–304; *see also* Uruguay Round
government: central government funding of S&T 110; digital divide policy 284–8, 297n; e-commerce 125, 308; information technology policy (China) 133; investment in R&D 93, 102; laboratories 102; procurement 310; World Trade Organization 302
gross domestic product 13, 18–20, 33, 88, 214, 240: China 130–1, 137; India 114; Taiwan 140–1, 149
Guangdong (China) 145

Harvard University, Centre for International Development 238
health care 290; *see also* telemedicine
high technology 12, 32, 106–7; parks (China) 134; Silicon Valley 106–7; software firms (India) 187; Taiwan 145
Hong Kong 59n, 142, 267: entertainment industry 213; investment in Taiwan 145; mobile telephony 274, 279
human capital 35, 41–2, 101, 111, 121–4: development 270; engineering specialists 227; English-language 121; i-flex (India) 190, 195; India competitive in 121, 188, 196; leveraging 173; network development 223; Taiwan 148

IBM 113, 135, 194, 222
ICT *see* information and communication technology
i-flex (India) 173, 187–98: consulting 192; Flexcube 190–8; global role 194; market incumbents 199; Microbanker 190; organisational structure 195
imports 27, 34, 110: information technology (China) 134; lowering duties 286
incubators (company) 107, 126, 192
India 5, 108–29: Bureau of Industrial Finance and Reconstruction 112; competitiveness 120; Confederation of Indian Industries 115; economic liberalisation 109; GE Capital 124; government support for e-commerce 125; i-flex 173, 187–98; Information Technology Act 127; informa-

## Index

tion technology software and services 121; internet 124, 249; IT-enabled services 123; low-cost labour 196; manufacturing 114–16; market entry 188; National Association of Software and Service Companies 121–4; National Renewal Fund 112; new economy 114, 128; offshore software development 123; quality ratings for industry 115; replicates Silicon Valley 108, 121, 128; software industry 187–96, 201n; Technology Development Board 113; Technology Development Fund 113; United States of America, exports to 122; venture capital 125–6

Indonesia 8, 237–56: cable 252–3; Director General of Post and Telecommunications (MOCT) 252; drivers of competition 249–55; foreign direct investment 245, 255; Indosat 248–51; information and communication technology 237–56; information service provider 244–5, 252–4; intellectual property 217; internet use 283; internet service providers 246–8, 284; Linknet 245–8; new technology 237–56; oligopoly 247; Pact Indonesia 291; Perusahaan Listrik Negara 253; price sensitivity 246, 254; RadNet 244–5; regulation 251; regulation of the internet 249–55; satellites 248–50, 254; Sekolah 2000 292; telecommunications law 251, 255; telecommunications reform 242–4; telephone services 253–5; Telkomnet 244–5, 254; Telkom 248, 250–4; warnets (internet kiosks) 246, 252, 255n

information: consumption 222; costs 7; economic value in the new economy 220–36; externalities 221; gatekeepers 229; government information 287; public good 222; utility 221

information and communication technology 1, 3, 9–10, 32, 37, 44, 88: adoption 146, 230; agriculture 273; APEC region 314–16; Asia Pacific Development Information Program 288–9; developing countries 273–97; digital divide 273–97; government policy 284–8, 318; impact on productivity and growth 1, 49, 230, 316; infrastructure 162, 284–5, 306–8; investment (Taiwan) 140, 148, 158–65, 205, 225, 315–8; IT-enabled services (India) 123; Japan 284; motor industry 8, 257–72; policy 10, 46, 204, 284, 289, 318; revolution 44–50, 205, 237, 242, 284; services for multinationals 188; World Bank programs 289; World Trade Organization 298–312

information technology: commercialisation 211; computer integration in manufacturing 117; consulting (India) 123; corporate strategy of IT firms 172–202; e-Japan Strategy 284; exports (China) 133n; hardware patents and trademarks 215; impact on costs 220, 230; integration in trade 305; intellectual property protection 203–19; investment 21, 116, 225, 230;

modularity 175; product barriers 298, 315–16; product commodification 196; product cycle 174; prototypes 175; quality standards 121; revolution (India) 120; Sony 176–83; standards 192–3; strategy 196, 230; supply chain management 232; World Trade Organization 298, 305

Infosys 188

innovation 88, 91, 104n, 106, 174, 318: diffusion 109, 121, 158, 165; financial returns 223; India 109; innovator's dilemma 198; intellectual property protection 203–19; patents 135; project management 183–7; total factor productivity 148; user-developer interaction 186; see also information and communication technology; technological change; Silicon Valley

Intel 122, 135, 194, 243, 291

intellectual property 7, 134, 138, 203–19, 228, 304, 309: Agreement on Trade-Related Aspects of Intellectual Property Rights 212–15, 219n, 299, 304; Asia Pacific region 204; biotechnology 208; capital markets 203–19; circumvention 206, 214–17; competition 210, 228; contract law 206, 231; copyright 209, 214, 302; databases 215; digital rights 206; distribution rights 205, 214; films 213; games (computer) 177; infringement 208, 309; innovation 203; intellectual property rights 120, 203; investment capital 207, 216–18, 231; joint ventures 231; music 213; new economy 203–19, 228, 304, 309; patents 28, 134–5, 206; piracy 212; reverse engineering 205–12; software piracy 123, 205, 208; technological neutrality 302; technology adoption 204; telecommunications equipment 215–16; trademarks 213–14; TRIPs 212; World Intellectual Property Organization 307–8; see also licensing

Interactive Audience Measurement Asia Ltd (IAMAsia)(Hong Kong) 282

International Chamber of Commerce 308

International Data Corporation 231, 282

International Organization for Standardization (ISO) 114, 118, 121, 308: 9000 compliance 118

International Telecommunication Union 250, 256n, 276, 279, 296n, 307

internet 60, 80, 192, 209, 213–15: access costs 241–2, 248, 280; Arpanet 75; barriers to diffusion 241–2; cafes 275, 279; connectivity 288; customers 245–6, 276; cyber-squatting 214; digitally ensured commerce 308; domain names 214, 275–6; dot.coms 192; economic exclusion 240, 276–7; economic indicators 240; India 124, 129n; Indonesia 237–56; intellectual property 209, 213–15; internet protocol 254; Japan 281; MP-3 music files 70; motor industry 262; penetration 163–4, 241, 252, 274–9; portals 194; providers 237–56; Singapore 283; subscribers 275–7; surfing 280–3; Taiwan 143, 162–4; transaction costs 239;

universal service 64, 274; usage 66; welfare effects 229; wireless application protocol 123; World Wide Web 229
Internet Corporation for Assigned Names and Numbers 308
internet protocol telephony 85n
internet service providers 80, 124, 209, 283, 307: Asia 283; carriers 251; China 133; consortia 249; economies of scale 244; fees 242, 248; Indonesia 237–56; oligopoly 247; PalapaNet 249; telephone lines 249, 307
investment 10, 26, 315: Asian miracle 36; capital in Indian manufacturing companies 116; capital markets 210; China 134; divestment 255; electronics industry 159; film 213; foreign 10, 112, 134; ICT sector in APEC region 314; India, falling in manufacturing 114; music 213; process technology (India) 116; R&D 94–5, 112, 221; return on investment from intellectual property 216; Sony 182; telecommunications regulation 76, 307; telephone networks 79; uncertainty 225–6, 235n; *see also* foreign direct investment; R&D
ISP *see* internet service provider
IT *see* information technology

Japan xiv, 9, 29–30n, 85n, 122–3, 161, 260–3: broadband 282; Computer Entertainment Software Association 174, 201; computer games 174, 199; e-Japan Strategy 284; intellectual property rights 212; internet use 281; mobile telephony 274, 281; motor industry 260–1; Nintendo 176; NTT DoCoMo Inc 258; infomediaries 259; information and communication technology 257–72; intelligent cars 258, 270n; internet 262, 269; Japanese Automobile Manufacturers Association 271n; Japan automotive Network eXchange (JNX) 265, 271n; joint ventures 264; National Association of Software and Service Companies (India) 121–4, National Renewal Fund 112, modular design 258–9; network exchanges 263; Nissan 258; online auctions 267; order cycles 260; procurement 257–61, 267; standardisation 261; supply chain management 258–67; Toyota 258; United States 263; web-based tools and services 261–8
joint ventures 112–13, 181, 194–5, 220–1, 234n; brokers 229; business-to-business agreements 228–3; information-enabled 220–1, 226, 230–3; motor industry 264; networks 225, 234n; price behaviour 230; software 231; Sony and Toshiba 181; technical collaboration 113; uncertainty 221, 226, 233

knowledge: capital 93–4; cooperative agreements 228; economy 101, 113, 206; engineering 227; enterprise 109; intensivity 89, 120; management 27, 230; market makers as information gatekeepers 229; search costs 233n; specialists in the firm 227; tacit knowledge 52, 230; technological 33–4, 230
Korea *see* South Korea

labour: capacity utilisation 111; costs of (India) 240; productivity 25, 43, 111; *see also* employment; *see also* human capital
Latin America 56, 213, 276: e-commerce 278; internet use 277, 280; *see also* South America
law:cyberlaw 125, 284; internet service providers 80, 124; *see also* contract; intellectual property
Lebanon 213, 216
Legend (China) 135–6
liberalisation 10–11n, 56, 299, 303: APEC region 314–6; e-commerce 299, 306–10; India 109; innovation driver 109; manufacturing 109, 112; trade 303, 306; Trade and Investment Liberalization Framework (APEC) 315; *see also* telecommunications; World Trade Organization
licensing: royalties 110, 177; technology 110, 112, 143, 208, 216, 254; *see also* intellectual property

macro endogenous growth theory 51
Malaysia 217, 241, 250, 267, 279, 283: cyberlaw 284; information technology development policy 284; internet use 283
manufacturing 109: automation 120; competitiveness (India) 116; computer integration 117–19; critical success factors 117; India 114–16; liberalisation 112; new technology impact 119; software systems 111; strategy (India) 111; Taiwan 141
marketing strategy 187, 193, 200, 217
markets: business-to-business market makers 228–33; changes in 176; collusion over prices 230; competition 220–36, 250, 309; digital 236n; incumbents 227, 244, 250; metals 229; monopsony 230; new economy 224–8; new entrants 225, 244–5; oligopoly 228–33, 247, 309; time to market 223; world markets and risk 237; *see also* competition
mergers 228, 235n: *see also* joint ventures
Mexico 216,
Microsoft 135–7, 189, 194, 199, 201n, 210: digital divide 291; licences 210; piracy 212; Project Hope 291; X-box 199
mobile telephony 5, 9, 130, 141, 280: Japan 281; networks 286; NTT DoCoMo (Japan) 179; penetration in Asia 280; *see also* telephony
monopolies 217, 220–9, 243,250–5, 304
motor industry 8, 33, 232, 257–72: Automotive Network eXchange (ANX) 264; build-to-order 258–261; China 257; competition policy 269; components 262; computer-aided design 260,271n; consumers 258–9;

Covisint 264, 270n; dealer networks 259; distribution 259; e-commerce 260–2; electronic data interchange 259; e-marketplaces 267–8; exports to Japan 257; General Motors 258; infomediaries 259; information and communication technology 257–72; intelligent cars 258, 270n; internet 262, 269; Japan 260–3; Japanese Automobile Manufacturers Association 271n; Japanese Automobile Research Institute 271n; Japanese automotive Network eXchange (JNX) 265, 271n; joint ventures 264; modular design 258–9; network exchanges 263; Nissan 258; online auctions 267; order cycles 260; procurement 257–61, 267; standardisation 261; supply chain management 258–67; Toyota 258; United States 263; web-based tools and services 261–8
MP-3 music files 70
multifactor productivity growth 89–90, 95, 100–3n: Solow residual 91; weakness 90
multinational companies 89, 108, 135, 187, 216: China 135; competition with 189, 196; India 200n; intellectual property rights 216; R&D joint ventures (India) 113; Silicon Valley 108; software services for 188; Taiwan 143–7; technological development 92
Music 213: copyright 213; internet 282

National Telecommunications and Information Agency (US) 84n
neoclassical economics 39–40, 51, 89–91, 103n
NetAid.org 293
networks: bandwidth 123; costs 224–5; critical size 210; economies 101, 209–10; externalities 62, 67, 73, 209, 222–3, 236n; Indonesia 237–56; interoperability 209; inter-plant 107; local area networks 72, 118; new economy 222–3, 285; obsolescence 64, 77; packet switching 75–9; private virtual network (motor industry) 264; software 209; standards 123; telecommunications 62, 285; unbundling services 77; wide area networks 118, 251–4
new economy 12–32, 88, 205–7, 217: Asia Pacific Economic Cooperation 313–8; intellectual property protection 203–19; intellectual property rights 212–6; India 114; markets 221–36; networks 222–3; overview 1–11; prices 232; Taiwan 142; transaction costs 229–31
newly industrialised economies 35–6, 142, 161, 
new technology 22, 43, 101: diffusion 270; economic development 237; Indonesia 251–3; intra-modal competition 251–3; unpredictability 225–6; see also technological change; technological development
New Zealand 226, 234n, 235n, 267, 286
Nintendo 176–8
Nissan 258, 270n

OECD see Organisation for Economic Co-operation and development
offshore services 187–9
oligopoly 228–33, 234n, 247
online games see games
online shopping 232: see also business-to-consumer; customer relationship management
Organisation for Economic Co-operation and Development 60, 88, 104n, 143, 147, 153, 163, 169n, 170n, 202n, 219n, 308, 314–18; digital divide 274–5
outsourcing 8, 28, 113, 122, 239: India 122, 190

Pacific Trade and Development (PAFTAD) xiv
patents 28, 134–5, 206; Mexico 216; registration 210; telecommunications 215–6; see also intellectual property
pharmaceutical industry 208, 211
Philippines 241, 249–50, 274–7, 285: Cebu 279; ed.venture 292; Foundation for IT Education and Development 292
Phillips curve 17
PlayStation 176–83: broadband 179–80; market forecasts 199; marketing 187; market performance 183; market trends 178; PlayStation 2 129, 179–83, 199
pornography 282–3
Porter's five-forces model of competition 244–9
product cycles: i-flex (India) 191; information technology firms 172; life cycle 192, 205
production planning 112, 195: sourcing 227; standardisation 226
productivity 90: growth 98; multifactor productivity growth 90; United States 239
project management 183–7
protectionism 203: intellectual property protection 203–19; see also imports
prototypes (IT) 175, 184–6

quality: benchmarking 112; case studies (India) 114; India, ratings for industry 115; International Organization for Standardization 114; manufacturing 111; products 118, 121; software 121; standards 121; supply chain response 230; systems implementation 115; total quality management 114

R&D 5, 34, 52–4, 89, 314: APEC region 314; capitalisation 93–101; centres (India) 113; cost elasticity 97; employment 101; endogenous growth theory 148; expenditure (India) 110; gross domestic product 91; intellectual property protection 203–19; investment 103, 112, 132, 135, 216, 314; Taiwan 147; tax relief 55; see also innovation
regulation 4, 211, 290, 309: broadband 285; competition 223, 233, 244–55; consumer protection 80; intellectual property protection 218; internet 249–55; self-regulation 307–9; see also deregulation;

telecommunications
research parks *see* technology parks
risk: cooperative agreements 221, 225; corporate strategy 172, 192, 198; investment in new technology 225; joint ventures 221, 225–6, 231; venture capital (India) 126; world markets 237
rural areas 286, 288

satellites 248–9
science and technology 110: agriculture 110; defence 110; central government funding 110; space 100
science parks *see* technology parks,
Silicon Valley (California) 28, 106: China 208, 218n: clones 108, 121; India 5, 108, 121, 128; innovation 106; model 106–7
Singapore 59n, 89–92, 105n, 193, 266–7, 298: e-government 284; electricity 96; information technology development policy 284; internet development 283; R&D expenditure 97–8; teledensity 279, 283
small and medium-sized enterprises 111, 143, 304–9
software 111, 121, 174: business-to-business market exchanges 231; China 135, 212; circumvention 206, 214; client-server model approach 189; communication protocols 209; costs of labour 240; costs of production 223; design 202n; exports (India) 122; games 183–7; i-flex (India)173, 187–96; India 238; intellectual property rights 208–10; Java 192, 291; middleware 261; motor industry 261; offshore software development (India) 123; piracy 123, 212; project management 183–7; proprietary 210; reverse engineering 205–12; standards 121, 192–3; Taiwan 146; *see also* networks
Solow productivity puzzle 140, 168: residual 91
Sony 6, 173–5, 196–7, 202n: Idei, Nobuyuki (President) 179–80; investment 215; PlayStation 176–83; strategic alliances 182
South America 56 213, 276: *see also* Latin America
South Korea 34, 89–92, 217, 241: broadband 282; digital divide 284–5; electricity 96; internet use 282; patents 211; R&D expenditure 97–9; teledensity 279–80
sponsorship 229–30,
standards 301, 308: Agreement on Technical Barriers to Trade 301; International Organization for Standardization 308; international software 192–3; networking 123, 301; World Trade Organization 301
Stanford University 108
stock exchange: India 192; Taiwan 152
strategic alliances 194, 225; *see also* joint ventures
supply chain management 229, 232: bargaining power 230; motor industry 258–9; new technology 232, 238
sustainable development 288
synergy 106, 173, 180, 198

Taiwan 6, 34, 54, 89–93, 140–71: capital market 150–2; China 141–5; computer industry 143–6, 163–5; digital divide 163–5, 279; e-commerce 141, 279; economic growth 140–1, 156–7, 165; e-government 287; electrical industry 96, 149, 156–7; electronics industry 142–52, 156–61, 166–8; employment 151–3; exports 142–5, 151; foreign direct investment 142–3, 151–3; gross domestic product 140, 149, 153–60; Hong Kong 145; ICT companies 147–58; ICT investment 140–7, 158–165; ICT policy 141, 147–58, 168–9; ICT spill-over effects 165–8; imports 153, 166–7; industrial policy 147–8; infrastructure 141; Institute for Information Industry 147; Institute for Technological and Industrial Research 143, 147, 169; international trade 145; manufacturing 99, 141, 154–8, 168; mobile telephony 141, 279; Nangkang Software Science-based Park 148; network industries 140; R&D 97–9, 147, 153; SchoolNet 287; small to medium-sized enterprises 143; software industry 146–9; Taiwan Semiconductor Manufacturing Corporation 143; Taiwan Stock Exchange 152; telecommunications 141–2, 163; teledensity 279; United Micro Electronics Corporation 143; United States 143; value-added 143–9, 157; World Trade Organization 142
tariffs 300–2
technological change 33–59, 94, 221–3, 228, 233–8, 269: catch-up 52, 60; corporate strategy of IT firms 172; disruption 197; dynamism 52, 112, 228, 233; economic growth 34, 112, 244, 310; government policy 111; microeconomic approach 30, 228; neoclassical interpretation 234n; networks 223–4; policy 59n, 111, 244; powertrain 270n, 272n; structuralist-evolutionary model 39, 54–9n; *see also* new economy
technological development 91–2, 196, 210: innovation 91, 113, 135, 214; local residents 92; new markets 196; technology transfer 310
technology: adoption 204, 218; alternative 254; cycles 89, 102; diffusion 2, 27–9, 110, 238–41, 257, 284; indicators 93; indigenous (India) 109; institutions 111, 284–8; motor industry 257–72; neutrality in trade 302; strategy 196, 284–8; transfer 310; *see also* information technology
technology parks: India 114; Taiwan 145, 148; *see also* Silicon Valley
telecommunications: Agreement on Basic Telecommunications Services 298, 307; broadband 66–8, 74–8; cable (Indonesia) 252–3; carriers 69, 76; charges 62–8, 82; China 132; churn 69; competitive local exchange carriers 69; costs 8, 62, 76 ,81; death of distance 61; deregulation 60–1;

digital divide 293; e-commerce 61; economies of scale and scope 243; e-Japan Strategy 284; entry barriers 82; equipment 215–16, 224; fibre-optic cables 61, 242, 252; frequencies 78; Indonesia 242–4, 250; infrastructure 224; inter-connection 87n, 224, 251; joint ventures (India) 113; law (Indonesia) 251, 255n; liberalisation 77, 251, 285; local-loop 78, 243, 256n; networks 225, 307; network software 223; patents 215; policy 60–87, 206, 285; price elasticities 73–4; price regulation 76; prices 63, 68, 81; Reference Paper on Regulatory Principles in the Agreement on Basic Telecommunications Services 307; regulation 61–9, 76–9, 285, 307; subsidy 67–9; switching 63, 87n; Taiwan 141, 163; taxation 68, 87n; Telecommunications Act 69; transmission 249; unbundling services 77–8, 86n; universal broadband connectivity 74; universal service 61–9, 76, 243, 274; vertical integration 79; value-added 303; World Trade Organization Agreement 298
telecommuting 72
teledensity (telephone density) 274–6: Asian economies 278–84; see also digital divide
telemedicine 72, 287–8, 295n–6n; China 288; Safelife HealthNet 288
telephony 29: charges 242, 249; China 130–1; digital subscriber line 66–7, 78–9; Indonesia 253–4; internet protocol 85n; local competition 87n; networks 61–5; Taiwan 141, 162; teledensity 274–80; unbundling services 77; voice telephony 71, 75, 84n; see also mobile telephony
Thailand 283: internet use 283
Tokyo Round 299
total factor productivity 36, 89, 148, 151, 164–5
total quality management 114, 118, 129n
Toyota 8, 37, 258, 270n, 272n: Gazoo 270; value chain management 258
trade 298–312: cross-border trade 302, 317; see also World Trade Organization
transaction costs 8, 220–4, 231–3: internet 239
Tuvalu 275

United Nations 46, 288: Centre for Trade Facilitation and Electronic Business (UN/CEFACT) 308; Commission on International Trade Law 307; Conference on Trade and Development 308; Development Programme 288, 293, 296n; ICT Task Force 288; Information Technology Service 288; Sustainable Development Networking Program 288
United Nations Conference on Trade and Development 214, 218n
United States of America 110: broadband networks 285; Congress 16, 122; databases 215; Department of Commerce 158, 171n; deregulation of telecommunications networks 87n; dot.coms 240; economy 15, 30n, 137, 214; e-government 287; government and aircraft industry 55; monetary policy 18; motor industry 263; productivity 2, 25, 239; Telecommunications Act 69, 77, 85n
universities 7, 135, 208, 211
Uruguay Round 298–303, 308: see also General Agreement on Tariffs and Trade (GATT); see also General Agreement on Trade in Services (GATS)

venture capital 24, 102, 125, 210; Citicorp 189; critical success factors 126; India 125–6; Indian Venture Capital Association 126; Silicon Valley 108; Yozma Group (Israel) 126–7n; see also capital investment
Vietnam 26, 217
virtual networks see networks

wireless application protocol 123
wireless services 66, 282; broadband 282
World Bank 111, 213, 219n, 238, 256n, 288: digital divide initiatives 289; Global Development Gateway 289; Information for Development Program 289; telecommunications 289
World Intellectual Property Organization 214, 307–8; e-commerce 307–8
World Links SchoolNet 287
World Markets Research Centre 272n, 287, 295n
World Trade Organization 5, 9, 46, 57n, 77, 112, 120, 142, 219n, 286, 311n: Agreement on Basic Telecommunications Services 298, 307–8; Agreement on Technical Barriers to Trade 301; Agreement on Trade-Related Aspects of Intellectual Property Rights 299, 304; APEC contribution to trade negotiations 317; China 130, 137–8; Committee on Trade and Development 304–5; Council for Trade in Goods 300–4; Council on Trade in Services 303; cross-border trade 301–2; Declaration on Global Electronic Commerce 298; dispute settlement 302–3; e-commerce work program 298–312; General Agreement on Tariffs and Trade 299–304, 312; General Agreement on Trade in Services 299–304, 311n, 312; General Council 299, 305–7; Information Technology Agreement 10, 30n, 286, 298, 306–7; infrastructure agreements 307; IT products 298; trade and exchange rate policies 112, 301
WTO see World Trade Organization

Yozma Group (Israel) 127, 128n
Y2K 189

Zhongguancun (China) 208, 218n